Finite Strip Analysis of Bridges

OTHER BOOKS ON STRUCTURES FROM E & FN SPON

Aluminium Alloy Structures
F. M. Mazzolani

Alternative Materials for the Reinforcement and Prestressing of Concrete
Edited by J.L. Clarke

Bridge Bearings and Expansion Joints
D.J. Lee

Bridge Deck Behaviour
E.C. Hambly

Bridge Management 3
Edited by J. Harding, G.A.R. Parke and M.J. Ryall

Building the Future
Innovation in design, materials and construction
Edited by F. K. Garas, G. S. T. Armer and J. L. Clarke

Computer Methods in Structural Analysis
J. L. Meek

Concrete Structures
Stresses and deformations
A. Ghali and R. Favre

Concrete under Severe Conditions
Environment and loading
Edited by K. Sakai, N. Banthia and O.E. Gjørv

Constructional Steel Design
An international guide
Edited by P. J. Dowling, J. E. Harding and R. Bjorhovde

Continuous and Integral Bridges
Edited by B. Pritchard

Design of Prestressed Concrete
R. I. Gilbert and N. C. Mickleborough

Flexural-Torsional Buckling of Structures
N. S. Trahair

Strip Method Design Handbook
A. Hillerborg

Structural Design of Polymer Composites
EUROCOMP Design Code and Background Document
Edited by J. L. Clarke

Structural Dynamics for the Practising Engineer
H. M. Irvine

Finite Strip Analysis of Bridges

M.S. Cheung

Director, Ottawa–Carleton Bridge Research Institute
and
Adjunct Professor,
Department of Civil Engineering,
University of Ottawa,
Ottawa, Canada

W. Li

Post-Doctoral Fellow,
Department of Civil Engineering,
University of Ottawa,
Ottawa, Canada,

and

S.E. Chidiac

Research Officer,
Structures Laboratory,
Institute for Research in Construction,
National Research Council Canada,
Ottawa, Canada

CRC Press
Taylor & Francis Group
Boca Raton London New York

CRC Press is an imprint of the
Taylor & Francis Group, an **informa** business
A CHAPMAN & HALL BOOK

CRC Press
Taylor & Francis Group
6000 Broken Sound Parkway NW, Suite 300
Boca Raton, FL 33487-2742

First issued in paperback 2019

ISBN-13: 978-0-419-19150-6 (hbk)
ISBN-13: 978-0-367-86457-6 (pbk)

A catalogue record for this book is available from the British Library

Visit the Taylor & Francis Web site at
http://www.taylorandfrancis.com

and the CRC Press Web site at
http://www.crcpress.com

To our families

Contents

Preface

The finite strip method is widely recognized as a powerful and versatile analysis tool, and is very effective in the analysis of structures which have complex geometry, material properties and loading conditions but with relatively simple support conditions, such as bridges. This method has been extensively applied in the static and dynamic analysis of bridge structures for many years. Undoubtedly one of the reasons for its great appeal to bridge engineers and researchers is the fact that this method provides efficient and accurate analysis with minimum modelling effort and input requirements. One objective of this book is to attempt to bring together many of the specialized applications for different types of bridges and to present this material, along with the fundamentals of the method, in a unified and consistent manner. This book summarizes the current developments and advancements of the finite strip method and their applications to bridge engineering. Among the special applications considered are the following: linear and nonlinear analysis of slab, slab on girder, composite box girder, cable-stayed and other forms of medium and long span bridges.

The book comprises of three parts: I) basic concepts of numerical methods; II) fundamentals of the finite strip method; and III) applications of the finite strip method to bridge analysis. It is written for senior engineering students, practising and research engineers, and others who have acquired the knowledge of bridge analysis and mechanics of materials. The reader does not need an extensive background in more advanced techniques, such as the theory of elasticity, energy methods, plasticity and numerical analysis. The mathematical knowledge required is likewise not very great, since the presentation is based upon a physical structural point of view rather than a mathematical one. Differential and integral equations are required to a limited extent only in the first two parts, whereas applications of the theory in the derivations of solutions appear in the later chapters. Throughout, an attempt has been made to present the material in its entirety, that is, the developments begin with fundamental principles, followed by applications, numerical examples and the final results are translated into tables or graphs for convenient reference and use. Many solved examples for different types of bridges are given.

Part I consists of three chapters which are intended as an overview of the fundamentals of numerical methods and structural mechanics, including such areas as basic concepts of numerical methods, variational principles, convergence requirements and error estimations. Part II concentrates on the theory of the finite strip method and its development as well as potential applications in various areas of analysis. Specific details on conventional and spline finite strip methods are given in Chapters 4 and 6, while sophisticated finite strips such as higher order and compound finite strips are discussed in Chapters 5, 7 and 11. Three-dimensional finite strips in the form of finite layers or finite prisms are also derived and discussed in detail in Chapter 8. These strips can be applied not only to the analysis of thick concrete bridge decks but also to laminated composites, nuclear structures and concrete dams. Chapter 9 is devoted entirely to the dynamic and stability analyses of bridges using various types of finite strip models. Since dynamic and stability analyses are essential factors in bridge design, especially for steel bridges, this chapter provides important guidance and the necessary tools for practising engineers. Material and geometry nonlinearity considerations in the finite strip formulation are presented in Chapter 10. The remaining text presents applications of the finite strip method to specific types of bridge. These include slab and slab on girder bridges, curved and skewed bridges, box girder bridges, continuous haunched bridges and cable-stayed bridges.

An extensive list of reference documents and a subject index are given in the appendices. These appendices have been compiled to provide a quick reference to any subject areas discussed in the book.

The senior author of this book has devoted over 25 years of research to the finite strip method. A number of major finite-strip computer programs have been developed by him and his colleagues during this time. In order to promote the widespread use of this method by bridge-design engineers, especially in small design offices, the authors have consolidated all these programs into the comprehensive and easy use package BAFSM which will be available soon.*

The authors wish to express their gratitude to the Natural Science and Engineering Research Council of Canada for financial support of their research in this area during the last 12 years. The authors would also like to thank Mr G. Franche of the National Research Council of Canada for his help in preparing all the drawings and tables in this book.

M.S. Cheung
W. Li
S. E. Chidiac

* For information about BAFSM software, contact Dr. M.S. Cheung, Department of Civil Engineering, University of Ottowa, 161 Louis Pasteur, Ottowa, Ontario, Canada K1N 6NS.

Notation[1]

A	area of cross-section
	area of domain
$\{a\}$	vector of joint displacements
b	width of strip
	subscript for bending analysis
	superscript or subscript for beam
$[B]$	strain matrix
$[\bar{B}]$	strain matrix for large deformation
D	plate rigidity
$[D]$	elastic matrix
$[D_T]$	tangential elastoplastic matrix
$[D_c]$	tangential elastic matrix of concrete
e	prescribed tolerance of iteration
	eccentricity of beam
	subscript for elastic state
E	modulus of elasticity
E_o	tangential modulus at zero stress
E_c	secant modulus at maximum compressive stress
EA	axial stiffness of beam
EI	flexural rigidity of beam
f'_c	uniaxial compressive strength of concrete
f'_t	uniaxial tensile strength of concrete
$\{f\}$	vector of displacement components
$\{F\}$	vector of unbalanced forces
$[f]$, $[F]$	flexibility matrix
G	shear modulus
GJ	torsional rigidity of beam
$[G]$	slope matrix
h	length of longitudinal section in spline finite strip
	thickness of strip in vibration analysis
H'	strain hardening parameter
i	subscript for nodal line No. i
	subscript for principal stress σ_i
I	superscript or subscript for finite strip No. I

[1] This list may not include those symbols which are used only in an individual section.

j	subscript for node, knot or nodal line No. j
k_m	$m\pi/l$
$[k]$, $[K]$	stiffness matrix
$[K_G]$	geometrical stiffness matrix, initial stress stiffness matrix
$[K_T]$	tangential stiffness matrix
l	length of strip
$L_j(y)$	Lagrange interpolation expression of node j
m	subscript for the m-th term of series
	subscript for the m-th knot in spline finite strip
M	bending moment or twisting moment
$[M]$	mass matrix
$[N]$	matrix of shape functions
p	in-plane force per unit area
	subscript for plastic state
P	point force
	subscript for plane stress analysis
$\{p\}$, $\{P\}$	load vector
q	transverse load per unit area
	weight per unit length
r	number of series terms used in an analysis
	number of longitudinal sections in a spline strip
	redundant force
	radius in r–ϕ coordinate system
	curvature radius of curvilinear coordinate line
$\{r\}$	vector of redundant forces
R	curvature radius of bottom flange
$\{R\}$	vector of resistant forces to deformation
s	boundary coordinate
S	total number of strips in a structure
t	thickness of strip
	time in vibration analysis
	subscript representing the whole structure
$[t]$, $[T]$	coordinate transformation matrix for displacements
$[T_\epsilon]$	coordinate transformation matrix for strains
u, v, w	displacements in x, y and z directions, respectively
U	strain energy
W	potential energy of external loading
x, y, z	local cartesian coordinate system (for individual strip)
$\bar{x}$, $\bar{y}$, $\bar{z}$	global cartesian coordinate system (for structure)
	skew coordinates
X	x/b
$Y_m(y)$	the m-th eigenfunction of beam vibration

α	spanwise subtended angle in r–ϕ plane
	ratio σ_1/σ_2
	amplification factor of flexural stiffness
	skew angle
γ	shear strain
δ	displacement
$\{\delta\}$	vector of displacement parameters
	eigenvector in vibration or stability analysis
	vector of joint displacements due to external loads
Δ	increment
	settlement of support
	relative displacement of joint
ϵ	normal strain
$\{\epsilon\}$	vector of strains and curvatures
$\bar{\epsilon}$	equivalent strain
ϵ_c	uniaxial compressive strain at f'_c
ϵ_t	uniaxial strain at concrete cracking
ϵ_u	ultimate strain of steel
ϵ_{iu}	equivalent uniaxial strain
ϵ_{ic}	equivalent uniaxial strain at σ_{ic}
θ	transverse slope $\partial w/\partial x$
$\{\kappa\}$	vector of curvatures in bending and twisting
λ	eigenvalue in stability analysis
	relaxation coefficient in nonlinear solution
μ_m	the m-th eigenvalue of beam vibration
ν	Poisson's ratio
ξ, η	coordinates for subparametric or isoparametric transformation
	curvilinear coördinates for web strip
	dimensionless coordinates for combined analysis
Π	total potential energy
ρ	mass per unit volume
σ	normal stress
$\bar{\sigma}$	equivalent stress
$\{\sigma\}$	vector of stresses and moments
σ_1, σ_2	principal stresses
σ_s	yield strength
σ_{ic}	maximum compressive stress
τ	shear stress
ϕ	angle in r–ϕ coordinate system
$\Phi_m(y)$	B_3 spline function centered at y_m
χ	curvature in bending or twisting
ω	natural frequency

PART I

Mathematical Approach

CHAPTER 1

Introduction

1.1 Bridge analysis by refined methods

The art of designing and building bridges has evolved gradually. In early bridges, a non-composite slab and beam system was frequently used where the slab provided the riding surface, and the beams supported dead and live loads transmitted from the slab. For this non-composite system, force responses in the slab and the beams are very simple and easy to calculate. Composite design followed once the interactions between various bridge components was sufficiently understood. The main advantage of the composite system is the increased load carrying capacity due to the load sharing mechanism developed from its composite action. However, in this case, the stress responses and their distributions and redistributions at ultimate limit state become much more complicated. With increasing complexity of geometry and combination of different materials in bridge construction, the analysis of such structures becomes a very complex undertaking. In this case simplified analytical methods, such as equivalent beam theory, are not adequate. A refined analysis method which is capable of representing all structural responses and deformations must be used.

Bridges can have many different structural types and within each type they can have many variations. However, based on its geometry and main structural responses, they can be classified into five basic categories:

1. slab and slab on girder bridges;
2. box girder bridges;
3. arch and truss bridges;
4. rigid frame bridges;
5. cable-stayed and suspension bridges.

Within these basic categories, many variations are possible. To illustrate this, 11 commonly used bridge types in North America are shown in Fig. 1.1: bridges A, B, C, D and H belongs to category 1 (slab and slab on girder bridges); bridges I, J and K to category 2 (box girder bridges); bridges E and F to category 3 (arch and truss bridges); and bridge G belongs to category 4 (rigid frame bridges). Most arch, truss and rigid frame bridges were built in the 1940s and 1950s, and recent constructions are almost exclusively slab, slab on girder, box girder or cable-stayed bridges. Cable-stayed and suspension bridges are long span bridges and their designs are usually not covered by the design codes.

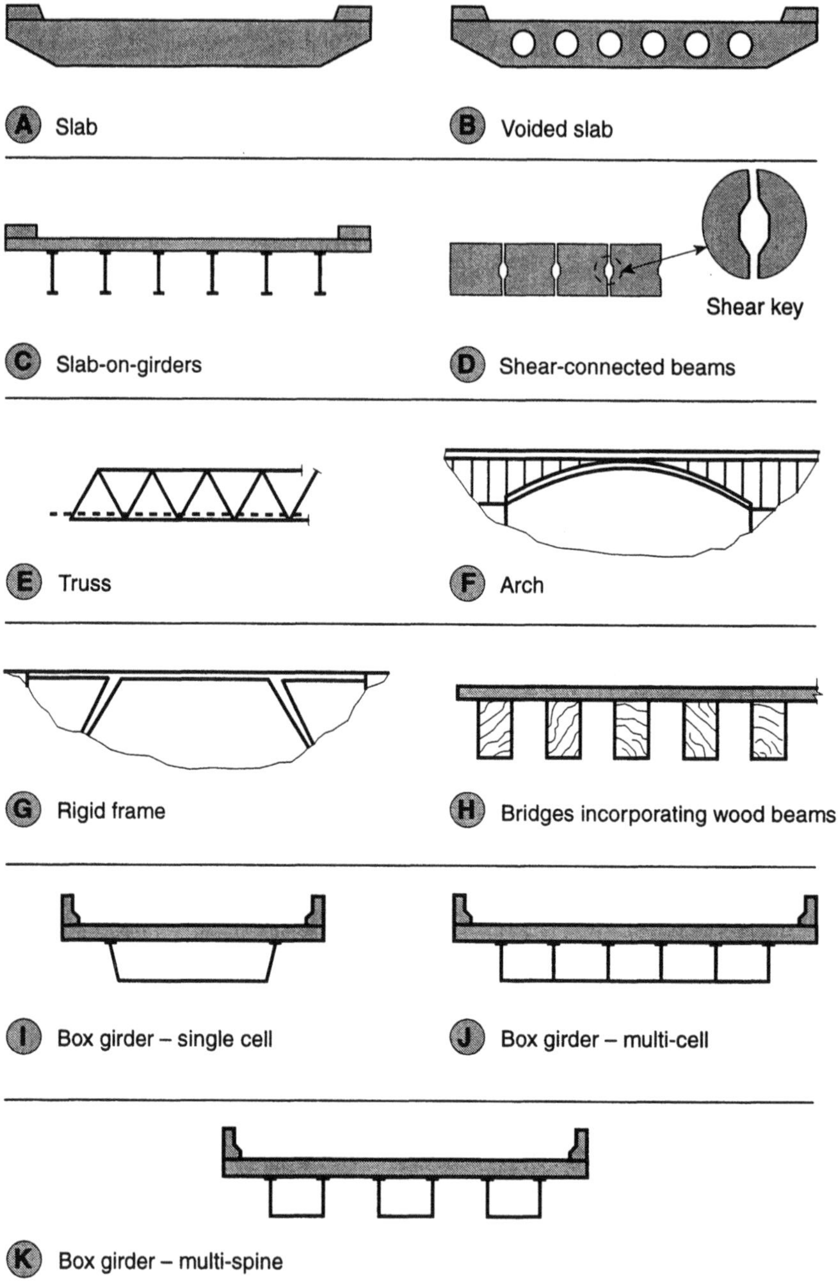

Figure 1.1 *Representative cross-sections and elevations of bridge types (Courtesy of OHBDC).*

In selecting a suitable refined method of analysis, all essential force responses must be carefully considered. For example, when analyzing a steel–concrete composite box girder bridge, the chosen method of analysis must be able to encompass longitudinal, transversal and punching shears, longitudinal and transversal moments, longitudinal torsion, distortion, torsional distortion, warping and local buckling. At the ultimate limit state, cracking of concrete, losing of composite action, material and geometry nonlinearity must also be considered in the analysis.

There are many refined methods available for bridge analysis. However, only the following five are recognized by the North American bridge design codes (CSA-S6 1988, OHBDC 1991 and AASHTO 1989):

1. grillage analogy methods;
2. orthotropic plate theory methods;
3. finite element methods;
4. finite strip methods;
5. folded plate methods.

Finite strip methods in conjunction with the other methods described in this book, can effectively analyze most of the bridge types shown in Fig. 1.1. However, in order to obtain accurate results, the correct type of strips, structural models and material properties must be used in the analysis. One of the objectives of this book is to provide this information to the user in a unified and consistent manner.

Part III of this book focuses its entire discussion on application of the finite strip method to the analysis of three modern types of constructions in North America, namely slab and slab-on-girder bridges, box girder bridges and cable-stayed bridges. A number of examples are provided and the results are compared with other analysis or testing results whenever available.

1.2 Development of the finite strip method

The finite strip method was first published by Y. K. Cheung (1968A) for analysis of simply supported plates. The finite strip method for rectangular slab bridge was also developed independently by Powell and Ogden (1969). Since then, considerable research and development on that method have been carried out in many countries. In the late 1960s and early 1970s, the field of research extended to many types of bridge and loading conditions, such as:

- rectangular slabs with general end boundary conditions (Y. K. Cheung, 1968B);
- box girder bridges (Y. K. Cheung, 1969A);
- finite layer analysis of thick slabs (Y. K. Cheung and Chakrabarti, 1971);
- finite prism analysis of thick bridge boxes (Zienkiewicz and Too, 1972);

- circularly curved slab and box girder bridges, (Y. K. Cheung, 1969B; Y. K. Cheung and M. S. Cheung, 1971);
- skewed slab and box girder bridges, (Brown and Ghali, 1972 and 1975);
- continuous slab bridges, flexibility approach, (M. S. Cheung, Y. K. Cheung and Ghali, 1970);
- continuous box girder bridges with transverse diaphragms, flexibility approach (Loo, 1975);
- continuous slab and box girder bridges, stiffness approach (Wu and Cheung, 1974; Delcourt and Cheung, 1978);
- free vibration analysis (M.S. Cheung and Cheung, 1971);
- dynamic response of slab bridges to moving loads (Smith, 1973);
- stability analysis (Przemieniecki, 1973; Wittrick and Plank, 1974);

The books by Y.K. Cheung (1976) and Loo and Cusens (1978) summarize the basic theory of the finite strip method and its applications in bridge engineering and the results of research work during that period.

Since the mid 1970s, enormous effort has been devoted to more complicated topics, the main subjects being:

- analysis of general plates (Bucco, Mazumdar and Sved, 1979);
- analysis of curved composite girders with incomplete interaction (Arizumi et al., 1982);
- behavior of composite box girder bridges during construction (Branco and Green, 1985);
- macrostrip method for plates with transversely variable thickness (Arbabi and Li, 1991);
- compound strip method for analysis of slab and box girder bridges with intermediate supports and stiffenings (Puckett, 1983; Maleki, 1987; Wiseman and Puckett, 1991);
- analysis of haunched continuous slab girder and box girder bridges (M.S. Cheung and Li, 1988B and 1989A);
- finite strip method combined with finite element method and boundary element method for the analysis of irregular plates and slab girder bridges (M. S. Cheung and Li, 1991 and 1992; M. S. Cheung, Akhras and Li, 1994);
- post buckling behavior and geometrically nonlinear analysis of plate structures, (Graves–Smith and Sridharan, 1978; Hancock, 1981; Langyel and Cusens, 1983; Gierlinski and Graves–Smith, 1984; Azizian and Dawe, 1985);
- material nonlinear analysis of steel structures (Olawale and Plank, 1988; M. S. Cheung, Ng and Zhong, 1989; Ng et al., 1991);

- material nonlinear analysis of reinforced concrete slabs (Guo et al., 1988; M. S. Cheung and Li, 1990A);
- large deflection elastoplastic analysis of plate structures (Abayakoon et al., 1989);
- nonlinear analysis of cable-stayed bridges (M. S. Cheung, Li and Jaeger, 1988 and 1990).

In 1982, the B_3 spline function was introduced as the longitudinal displacement function by Y.K. Cheung et al. to form the spline finite strips for analysis of rectangular plates.

The spline finite strip method can include any types of boundary condition and more easily simulate peak values of bending moments at loaded sections and intermediate supports. By means of coordinate transformation, this method can also be used to model arbitrarily curved slab bridges. Therefore, this method is more universal than the semi analytical finite strip method.

In later years, the spline finite strip method has been extended to right box girder bridges by Y. K. Cheung and Fan (1983), to skew plates by Chen, Tham and Cheung (1984), to arbitrarily curved slab bridges by Y. K. Cheung et al. (1986), to circular or noncircular curved box girder bridges by Li, Tham and Cheung (1988), to haunched continuous bridges by M.S. Cheung and Li (1990B), to compound strip analysis by Chen, Gutkowski and Puckett (1990 and 1991), to vibration and stability analysis by P. Cheung et al. (1987) and Mizusawa (1988), to postbuckling analysis by Zhu and Cheung (1989), to inelastic postbuckling analysis by Lau and Hancock (1989), to material nonlinear analysis of concrete slab bridges by M. S. Cheung, Ng and Zhao (1993).

Indeed, since 1968, a very large number of researchers and engineers have made so many important contributions to the development of the finite strip method and to its application in engineering that it is not possible to list all of their accomplishments here.

1.3 Reading suggestions

This book is divided into three parts to provide a complete understanding of the finite strip analysis of bridges. However, each chapter is written relatively independently of the others so that the reader can quickly acquire the necessary information without having to read all the material that precedes it. Nevertheless, all the later chapters assume a knowledge of Chapter 4, which gives the basic theory essential to those who are new to this method. Also, the reader is strongly encouraged to read the book from the beginning, especially Chapters 2 and 3, which are written for research engineers and senior engineering students who require a reasonable background on the basic concepts of numerical methods and error estimation.

Having finished Chapter 4, the reader may consult selected chapters only, depending on the application. For instance, those who intend to analyze a circularly curved slab on girder bridge may turn to Chapter 13 directly. Then, if interested in the spline finite strip method and the compound strip method, turn to Chapters 6 and 7. After that, Chapter 17 may be consulted for advice on preparing an input file for a finite strip software. The reader who is calculating the free vibration frequencies of a skewed box girder bridge, need consult Chapters 4, 9 and 13 only. For those working on the static analysis of a continuous right box girder bridge, Chapters 4, 14 and 17 will provide sufficient information.

CHAPTER 2

Basic concepts of numerical methods

2.1 Introduction

In reality, most continuum mechanics problems have infinite degrees of freedom and frequently have complicated boundary conditions along with non-uniform material properties. By applying acceptable numerical procedures based on engineering principles, engineers are able to provide solutions to some of the most complex engineering systems. How to formulate the governing equilibrium equations of engineering systems and how to establish a solution are the objects of this chapter. The discussion will also provide a valuable basis for the presentation of the finite strip method in Part II.

Engineering systems are typically categorized as discrete or continuous. The term discrete refers to systems whose behavior can be exactly described using a finite number of state variables. On the other hand, an infinite number of elements in the form of differential equations govern the continuous systems responses. Obtaining exact solutions to the differential equations is only possible for simple systems with simple boundary conditions and, for most cases, the implementation of a numerical method is required to idealize a continuous system as finite discrete systems. This discretization procedure transforms a continuous engineering system to a discrete one. Numerous classical mathematical procedures of approximations have been developed for their analysis.

The methods of approximations used to solve the system governing differential equations can be grouped into three main approaches: the direct approach, the variational principle and the weighted residual methods. In the direct approach, direct physical reasoning is used to establish the system governing equilibrium equations including the boundary conditions in terms of finite state variables. On the other hand, the variational principle yields the governing differential equations and the non-essential boundary conditions of a problem by just following the calculus of variations. The attractiveness of using such a method for formulating equilibrium equations is the ability to interpret the functional, based on the physics of the problem. However, not all differential equations have a corresponding variational principle; for such equations, the weighted residual method may be adopted. The weighted residual method offers the most general procedure for solving the governing differential equations and can be applied to the majority of practical problems of science and engineering. The basic concepts of all three methods will be presented in this chapter. We shall also focus on the finite element method since it is the most commonly applied displacement-based formulation used to solve continuum problems and which also encompasses the finite strip method.

2.2 Problem formulation

The solution to a generalized engineering problem begins by first establishing the system of differential equations that governs its behavior, and then by finding an unknown function u that satisfies the following:

$$A(u) = 0 \quad in \quad \Omega \tag{2.1}$$

$$S(u) = 0 \quad on \quad \Gamma_q \tag{2.2}$$

Here Ω is the domain (volume, area, etc.) and Γ are boundaries of the domain, Fig. 2.1. In general, u can be a scalar quantity or a vector and the differential equation representing the system can be a single or a set of simultaneous equations.

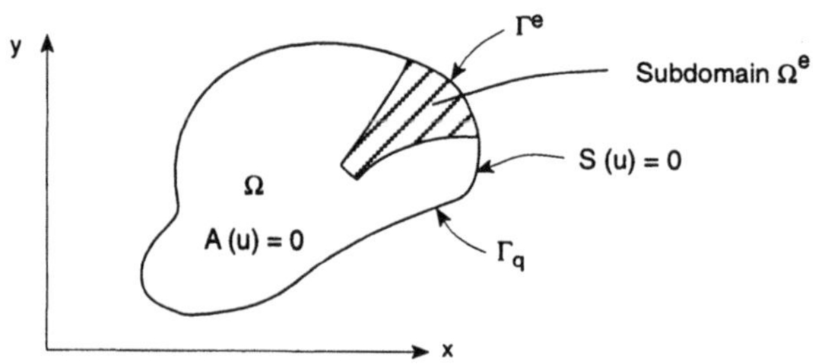

Figure 2.1 *Problem domain Ω and Γ boundary.*

Suppose that the differential equations are linear, then Equations 2.1 and 2.2 can be represented by

$$A(u) = Lu + p = 0 \; in \; \Omega \tag{2.3}$$

$$S(u) = Mu + q = 0 \; on \; \Gamma_q \tag{2.4}$$

in which L and M are the differential operators, u the independent variable–displacement of a point, and q the prescribed value on boundary Γ. Then we shall seek a numerical procedure by which a continuous problem in mechanics is reduced to a discrete problem, whose solution leads to a set of linear algebraic equations of the form

$$[K]\{u\} + \{f\} = 0 \tag{2.5}$$

where $[K]$ is the stiffness matrix, and $\{u\}$ and $\{f\}$ are the generalized displacement and load vector, respectively.

To acquaint the reader with definite problems, examples are included to illustrate the basis of Equations 2.3 and 2.4.

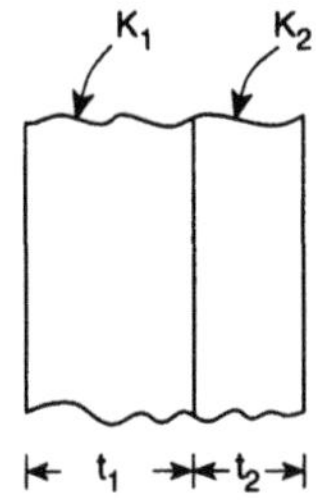

Figure 2.2 *Heat flow through a composite layered wall.*

Example 2.1 For steady state heat conduction problems such as a wall constructed of two bonded homogeneous materials (Fig. 2.2), the differential equations that govern the heat flow are given by

$$A(\Phi) = \frac{\partial}{\partial x_i}(k\frac{\partial \Phi}{\partial x_i}) + Q = 0 \quad in\ \Omega \tag{2.6}$$

$$S(\Phi) = k\frac{\partial \Phi}{\partial n} - \bar{q} = 0 \quad on\ \Gamma_q \tag{2.7}$$

$$G(\Phi) = \Phi - \Phi_o = 0 \quad on\ \Gamma_\Phi \tag{2.8}$$

where k represent the heat conductivity of the body, and Φ_o and $\bar{q}$ are the prescribed values of temperature and heat flow on the boundaries. The reader should note that for this example the unknown function u is a scalar and is represented by Φ.

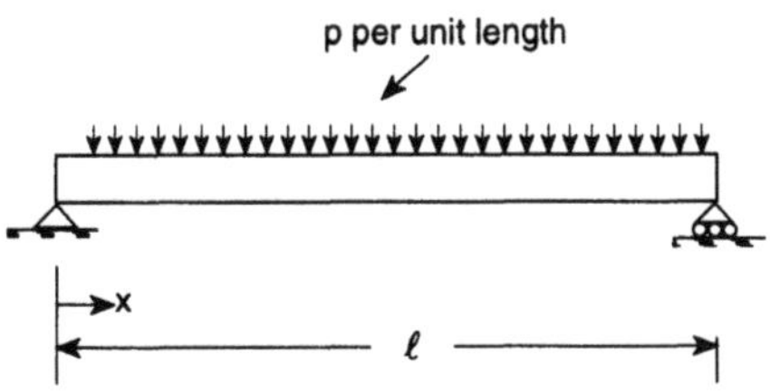

Figure 2.3 *A simply supported beam under a uniformly distributed load.*

Example 2.2 For a simply supported beam subjected to a uniformly distributed load p (Fig. 2.3), the deformational behavior is represented by the differential equations

$$A(w) = EI\frac{d^4w}{dx^4} - p = 0 \quad 0 \le x \le l \tag{2.9}$$

$$S(w) = EI\frac{d^2w}{dx^2} = 0 \quad at\ x = 0\ and\ x = l \tag{2.10}$$

$$G(w) = w = 0 \qquad at\ x = 0\ and\ x = l \tag{2.11}$$

where E and I are the modulus of elasticity and the second moment of inertia of the beam, respectively. The unknown function w, the out-of-plane deformation, is in this case a vector. The boundary conditions of the simply supported beam are zero bending moment and deflection at both ends and are represented by Equations 2.10 and 2.11, respectively.

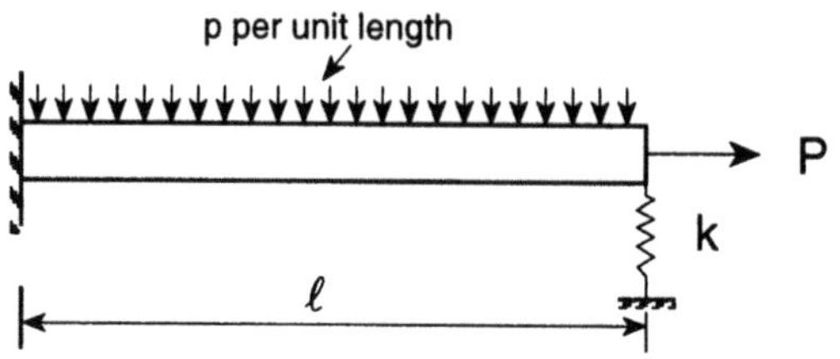

Figure 2.4 *A beam fixed at one end and supported by a spring on the other end subjected to an axial load P and a uniformly distributed load p.*

Example 2.3 By adding a spring of stiffness k to one end and constraining the other end of the beam, and then subjecting it to an axial force P and a uniformly distributed load p (Fig. 2.4), the deformational behavior is now represented by the differential equations

$$A(w) = EI\frac{d^4w}{dx^4} + P\frac{d^2w}{dx^2} - p = 0 \qquad 0 \le x \le l \tag{2.12}$$

$$S(w) = EI\frac{d^2w}{dx^2} = 0 \qquad at\ x = l \tag{2.13}$$

$$\bar{S}(w) = EI\frac{d^3w}{dx^3} + P\frac{dw}{dx} - kw = 0 \quad at\ x = l \tag{2.14}$$

$$G(w) = w = 0 \qquad at\ x = 0 \tag{2.15}$$

$$\bar{G}(w) = \frac{dw}{dx} = 0 \qquad at\ x = 0 \tag{2.16}$$

The boundary conditions of the beam shown in Fig. 2.4 are now represented by Equations 2.13–2.16. Equations 2.13 and 2.14, referred to as non-essential or natural boundary conditions, correspond to the equilibrium condition for the bending moment and shear at free end of the beam. Equations 2.15 and 2.16, referred to as the essential or forced boundary conditions, reflect a zero displacement and rotation at the constrained end of the beam. In brief, the essential boundary condition implies the prescriptions of primary variables usually the displacements in solid mechanics, and the non-essential boundary conditions becomes the prescription of secondary variables, i.e. stresses.

2.3 Direct approach

Physical procedures derived from direct reasoning, such as structural framework and network analysis, form one of the numerical approaches that engineers have used to solve mechanics problems. This approach, as will be illustrated, is only suited for simple problems with uncomplicated boundary conditions. The next two examples are used to demonstrate the advantages and disadvantages of this approach.

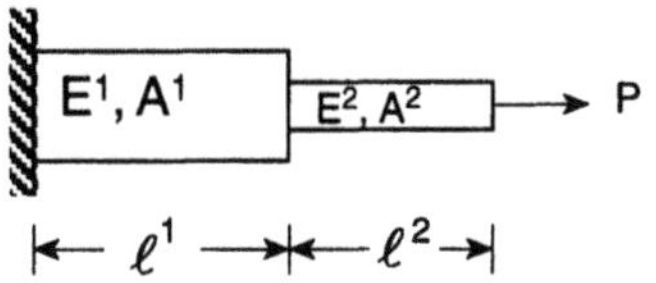

Figure 2.5 *A stepped bar under axial load.*

Example 2.4 Consider the stepped bar shown in Fig. 2.5 which is loaded by an axial force P at its free end. The solution of this problem begins by first dividing the system into segments referred to as elements not necessarily of equal length but whose material and geometrical properties are equal for a particular element, Fig. 2.6. The stepped bar is now divided into two elements by three nodes.

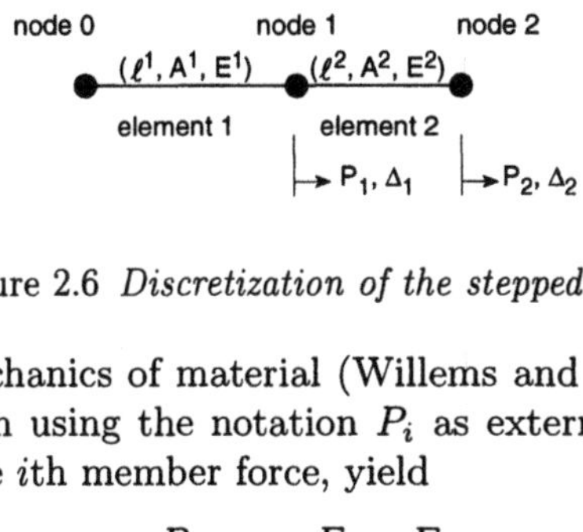

Figure 2.6 *Discretization of the stepped bar.*

Following the mechanics of material (Willems and Lucas, 1978), the x-direction equilibrium using the notation P_i as externally applied load at node i and F_i as the ith member force, yield

$$P_1 = F_1 - F_2$$

and

$$P_2 = F_2 - F_3 \tag{2.17}$$

Applying Castigliano's first theorem yields

$$\begin{aligned} F_1 &= K_{11}\Delta_1 + K_{12}\Delta_2 \\ F_2 &= K_{21}\Delta_1 + K_{22}\Delta_2 \end{aligned} \tag{2.18}$$

where K_{ij} is the member stiffness and Δ_i the displacement. The substitution of Equation 2.18 into 2.17 results in

$$P_1 = K_{11}^1\Delta_1^1 + K_{11}^2\Delta_1^2 - K_{12}\Delta_2^2$$

$$P_2 \quad = \quad -K_{21}^2\Delta_1^2 + K_{22}\Delta_2 \tag{2.19}$$

The compatibility condition implies

$$\Delta_i^1 = \Delta_i^2 \tag{2.20}$$

in which the superscripts represent the element number, and this leads to

$$\begin{aligned} P_1 &= (K_{11}^1 + K_{11}^2)\Delta_1 - K_{12}\Delta_2 \\ P_2 &= -K_{21}^2\Delta_1 + K_{22}\Delta_2 \end{aligned} \tag{2.21}$$

where $K_{ij}^k = \frac{E^kA^k}{L^k}$, and represents the force developed due to unit extension or shortening of the kth member. Thus, from the compatibility and equilibrium conditions imposed at each node, the system of equilibrium equations was derived and represented by the following matrix equation:

$$\begin{bmatrix} \frac{E^1A^1}{L^1} + \frac{E^2A^2}{L^2} & -\frac{E^2A^2}{L^2} \\ -\frac{E^2A^2}{L^2} & \frac{E^2A^2}{L^2} \end{bmatrix} \begin{Bmatrix} \Delta_1 \\ \Delta_2 \end{Bmatrix} = \begin{Bmatrix} 0 \\ P \end{Bmatrix} \tag{2.22}$$

Since the boundary condition has already been incorporated ($\Delta_0 = 0$), the deformation of the bar can be obtained by solving Equation 2.22.

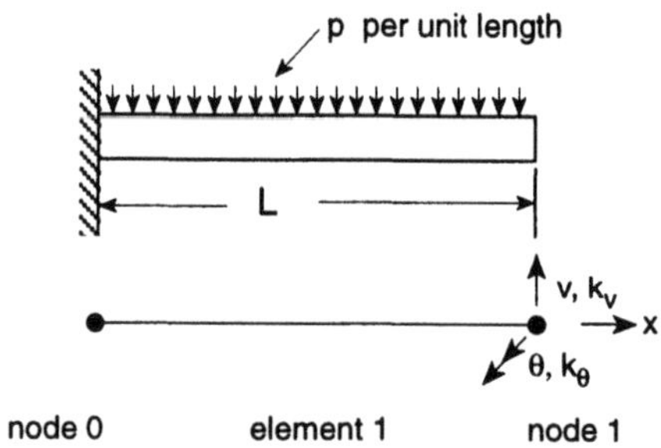

Figure 2.7 *A beam fixed at one end under a uniformly distributed load p.*

Example 2.5 Consider the simply supported beam bending problem presented in Example 2.2 but now assume one end of the beam fixed. In this case, we will divide the beam into one element and two nodes as shown in Fig. 2.7. Now, similarly to Example 2.4, we need to derive first the stiffness matrix. According to beam theory (Higdon et al., 1976)

$$M = EI\frac{d^2v}{dx^2} \tag{2.23}$$

where I is the second moment of inertia of the cross section about the z-axis. However, a general expression for M can be obtained as

$$M = V^1(L - x) + M^1 \tag{2.24}$$

Substituting Equation 2.24 into Equation 2.23 and integrating twice yields an expression for the elastic curve, i.e.

$$v = \frac{1}{EI}[V^1(L\frac{x^2}{2} - \frac{x^3}{6}) + \frac{M^1x^2}{2}] + C_1x + C_2 \tag{2.25}$$

Applying the boundary conditions

$$\frac{dv}{dx}\,|_{x=0} \quad = 0 \quad => C_1 = 0$$

and

$$v\,|_{x=0} \quad = 0 \quad => C_2 = 0 \tag{2.26}$$

leads to the elastic curve, where

$$v = \frac{V^1x^2}{2EI}(L - \frac{x}{3}) + \frac{M^1x^2}{2EI} \tag{2.27}$$

and its derivative is

$$\frac{dv}{dx} = \frac{V^1x}{2EI}(L - \frac{x}{2}) + \frac{M^1x}{EI} \tag{2.28}$$

Now substituting for $x = L$, the length of the member, and inverting Equations 2.27 and 2.28 yields

$$\begin{bmatrix} \frac{12EI}{L^3} & -\frac{6EI}{L^2} \\ -\frac{6EI}{L^2} & \frac{4EI}{L} \end{bmatrix} \begin{Bmatrix} v_1 \\ \theta_1 \end{Bmatrix} = \begin{Bmatrix} V^1 \\ M^1 \end{Bmatrix} = \begin{Bmatrix} \frac{PL}{2} \\ 0 \end{Bmatrix} \tag{2.29}$$

The solution of Equation 2.29 leads to the deformation of the free end of the beam. The development of the stiffness matrix for Example 2.5 is obviously more complex than the previous one. The order of complexity increases drastically as we proceed to solve two- and three-dimensional continuum mechanics problems. Further, we have not yet discussed the implications of complex boundary conditions, loading conditions and material properties. Thus to overcome these anticipated difficulties, we shall now proceed to describe a more elegant and versatile procedure, namely the variational method.

2.4 Variational method

The variational method is the most commonly applied tool to establish the governing equilibrium equations in structural analysis. In brief, the variational principle defines a scalar quantity J, also called functional, which takes the following integral form

$$J = \int_\Omega L(u, \frac{\partial u}{\partial x}, \frac{\partial u}{\partial y}, ...)d\Omega + \int_\Gamma M(u, \frac{\partial u}{\partial x}, \frac{\partial u}{\partial y}, ...)d\Gamma \tag{2.30}$$

Knowing the variational principle, one can apply the calculus of variation to derive the governing equilibrium differential equation and the accompanying boundary condition at the extremum. Therefore, the solution to a continuum problem is obtained by setting the first variation of the functional equal to zero, i.e.

$$\delta J = 0 \tag{2.31}$$

One of the attractions of using such a method is the ability to interpret the functional based on the physics of the problem, such as potential energy in solid mechanics, the least energy dissipation principle in viscous flow, etc. The advantages of the variational formulation of a continuum problem over the differential equation formulation have also been recognized as listed below (Rao, 1992).

1. *The functional possesses a clear physical meaning in most of the practical problems.*
2. *The functional contains lower order derivatives of the field variable compared to the governing differential equation and hence an approximation solution can be obtained using a larger class of functions.*
3. *For the problems that possess dual variational formulation, the solution can be obtained by minimizing the functional or its dual functional.*
4. *Using the variational formulation, it is possible to prove the existence of a solution in some cases.*
5. *The variational formulation permits us to treat complicated boundary conditions as natural boundary conditions.*

These advantages are illustrated in the subsequent sections starting with an overview of the calculus of variation as it relates to the variational principle.

2.4.1 Calculus of variation

The calculus of variation, when applied to a functional, provides a means with which to establish an approximate solution in the standard integral form, also referred to as extremum principle (Forray, 1968). In general, the functional contains both dependent and independent variables and the basic problem is to find the function u_i which makes the functional

$$J = \int_t \int_\Omega L(x_i, u_i, \frac{\partial u_i}{\partial x_j}, \frac{\partial^2 u_i}{\partial x_j \partial x_k}, \frac{\partial u_i}{\partial t}) d\Omega dt \tag{2.32}$$

stationary. Here, x_i and t are the independent variables, and the function u_i and the derivatives with respect to x_i and t are the dependent ones. The first variational due to an arbitrary changes in the solution u_i is best explained by first defining the variational operator δ. δu_i yields the change or variation in u_i at fixed values of x_i and t as expressed by

$$\delta u_i(x_i, i) = \bar{u}_i(x_i, t) - u(x_i, t) \tag{2.33}$$

where $u_i(x_i,t)$ is the exact solution and $\bar{u}_i(x_i,t)$ the tentative solution, which satisfy the essential boundary conditions defined by

$$\bar{u}_i(x_i,t) = u_i(x_i,t) + \epsilon\Phi_i(x_i,t) \tag{2.34}$$

where ϵ is a small parameter and $\Phi_i(x_i,t)$, is a differentiable function whose values must equal zero on the boundaries with essential boundary conditions. For a practical application, consider J with only three independent and one dependent variable, i.e.

$$J = \int_\Omega L(x,y,z,u,\frac{\partial u}{\partial x},\frac{\partial u}{\partial y},\frac{\partial u}{\partial z})d\Omega \tag{2.35}$$

Since the operation of variation is cumulative with respect to both integration and differentiation, the first variation of J due to small change in the solution u can be expressed as

$$\begin{aligned} \delta J &= \int_\Omega (\frac{\partial L}{\partial u}\delta u + \frac{\partial L}{\partial u_x}\delta u_x + \frac{\partial L}{\partial u_y}\delta u_y + \frac{\partial L}{\partial u_z}\delta u_z)d\Omega \\ &= \int_\Omega (\frac{\partial L}{\partial u}\delta u + \frac{\partial L}{\partial u_x}\frac{\partial}{\partial x}\delta u + \frac{\partial L}{\partial u_y}\frac{\partial}{\partial y}\delta u + \frac{\partial L}{\partial u_z}\frac{\partial}{\partial z}\delta u)d\Omega \end{aligned} \tag{2.36}$$

in which $u_x = \frac{\partial u}{\partial x}$ and $\delta u_x = \delta\frac{\partial u}{\partial x} = \frac{\partial}{\partial x}\delta u$. By integrating the second term by parts and applying Green–Gauss theorem yields

$$\begin{aligned} \int_\Omega \frac{\partial L}{\partial u_x}\frac{\partial}{\partial x}(\delta u)d\Omega &= \int_\Omega \frac{\partial}{\partial x}(\frac{\partial L}{\partial u_x}\delta u)d\Omega - \int_\Omega \frac{\partial}{\partial x}(\frac{\partial L}{\partial u_x})\delta u d\Omega \\ &= \int_\Gamma l_x\frac{\partial L}{\partial u_x}\delta u d\Omega - \int_\Omega \frac{\partial}{\partial x}(\frac{\partial L}{\partial u_x})\delta u d\Omega \end{aligned} \tag{2.37}$$

in which l_x is the direction cosine of the normal to the outer surface with respect to the x axis. By following the similar procedure on the third and fourth terms of Equation 2.36, δJ can be expressed as

$$\begin{aligned} \delta J &= \int_\Omega \left[\frac{\partial L}{\partial u} - \frac{\partial}{\partial x}\frac{\partial L}{\partial u_x} - \frac{\partial}{\partial y}\frac{\partial L}{\partial u_y} - \frac{\partial}{\partial z}\frac{\partial L}{\partial u_z}\right]\delta u d\Omega \\ &+ \int_\Gamma \left[\frac{\partial L}{\partial u_x}l_x + \frac{\partial L}{\partial u_y}l_y + \frac{\partial L}{\partial u_z}l_z\right]\delta u d\Gamma \end{aligned} \tag{2.38}$$

For arbitrary variations δu and for $\delta J = 0$ as a requirement for the stationary value, then the bracket terms must equal zero. This immediately produces the equations

$$A(u) = \frac{\partial L}{\partial u} - \frac{\partial}{\partial x}(\frac{\partial L}{\partial u_x}) - \frac{\partial}{\partial y}(\frac{\partial L}{\partial u_y}) - \frac{\partial}{\partial z}(\frac{\partial L}{\partial u_z}) = 0 \tag{2.39}$$

$$S(u) = \frac{\partial L}{\partial u_x}l_x + \frac{\partial L}{\partial u_y}l_y + \frac{\partial L}{\partial u_z}l_z = 0 \tag{2.40}$$

which are the Euler–Lagrange equations needed to solve the problem. Equation 2.40 is known as a natural boundary condition while the kinematic boundary conditions should be satisfied by function u a priori. Langhaar (1962) and Washizu (1982) provide additional information.

2.4.2 Euler equations from the natural variational principle

Using the calculus of variations, we have derived $A(u)$, the differential equations governing the problem and $S(u)$ the natural boundary conditions. If it can be shown that the first variation of the functional takes the form

$$\delta J = \int_\Omega \delta u^T A(u) d\Omega + \int_\Gamma \delta u^T S(u) d\Gamma = 0 \tag{2.41}$$

then the variational principle is a natural one (Zienkiewics, 1982; Bathe, 1982). We shall illustrate this using a clamped beam subjected to a uniformly distributed load $p(x)$ where the functional is given by

$$J = \int_0^l \left[\frac{1}{2} EI \left(\frac{d^2 w(x)}{dx^2} \right)^2 - p(x) w(x) \right] dx \tag{2.42}$$

By following rules of differentiation, the first variation can be written as

$$\delta J = \int_0^l \left[EI \left(\frac{d^2 w(x)}{dx^2} \right) \delta \left(\frac{d^2 w(x)}{dx^2} \right) - p(x) \delta w(x) \right] dx \tag{2.43}$$

Since $\delta \left(\frac{d^2 w(x)}{dx^2} \right) = \frac{d^2}{dx^2} (\delta w)$, by performing partial integration on the first term twice, we obtain

$$\begin{aligned} & \int_0^l \left[EI \left(\frac{d^2 w(x)}{dx^2} \right) \frac{d^2}{dx^2} (\delta w(x)) \right] dx \\ = \; & EI \frac{d^2 w(x)}{dx^2} \delta \left(\frac{dw(x)}{dx} \right) \Big|_{x=0}^{l} - EI \frac{d^3 w(x)}{dx^3} \delta w(x) \Big|_{x=0}^{l} \\ + \; & \int_0^l \left[EI \frac{d^4 w(x)}{dx^4} \delta w(x) \right] dx \end{aligned} \tag{2.44}$$

Thus, we can rearrange Equation 2.43 to give

$$\begin{aligned} \delta J = \; & \int_0^l \left(EI \frac{d^4 w(x)}{dx^4} - p(x) \right) \delta w(x) dx + EI \frac{d^2 w(x)}{dx^2} \delta \left(\frac{dw(x)}{dx} \right) \Big|_{x=0}^{l} \\ - \; & EI \frac{d^3 w(x)}{dx^3} \delta w(x) \Big|_{x=0}^{l} \end{aligned} \tag{2.45}$$

and immediately observe that the Euler equations are

$$A(w) = EI \frac{d^4 w(x)}{dx^4} - p(x) \qquad 0 \leq x \leq l \tag{2.46}$$

$$S_1(w) = EI \left[\frac{d^2 w(x)}{dx^2} \delta \frac{dw(x)}{dx} \right]_{x=0}^{l} = 0 \;\; => \;\; EI \frac{d^2 w(x)}{dx^2} = 0 \;\; or \;\; \frac{dw(x)}{dx} = 0$$

$$S_2(w) = EI\left[\frac{d^3w(x)}{dx^3}\delta w(x)\right]_{x=0}^{l} = 0 => EI\frac{d^3w(x)}{dx^3} = 0 \; or \; w(x)0$$

where $A(w)$ are the differential equations, and $S_1(w)$ and $S_2(w)$ provide the necessary boundary conditions. We have therefore demonstrated that a set of Euler equations can be established from the variational principle. Next we shall show which forms of differential equations are Euler equations of the functional.

2.4.3 Linear self-adjoint differential equations

The natural variational principle for linear differential equations can be established provided the operator is symmetric or self adjoint. Revisiting Equation 2.3 and for simplicity ignoring the natural boundary conditions given in Equation 2.4, i.e.

$$A(u) = Lu + p = 0 \; ; \qquad in \;\; \Omega \tag{2.47}$$

the symmetry and therefore the self-adjointness of the operator L can be demonstrated by showing that

$$\int_\Omega v^T Lu d\Omega = \int_\Omega u^T Lv d\Omega + boundary \tag{2.48}$$

for any two functions u and v. This proof automatically allows the establishment of the variational principle, i.e.

$$J = \int_\Omega \left(\frac{1}{2}u^T Lu + u^T p\right) d\Omega + boundary \tag{2.49}$$

Taking the first variation Equation 2.49 leads to

$$\delta J = \int_\Omega \left(\frac{1}{2}\delta u^T Lu + \frac{1}{2}u^T \delta(Lu) + \delta u^T p\right) d\Omega + boundary \tag{2.50}$$

and recalling that the operator is linear and self-adjoint, Equation 2.50 can be rewritten as

$$\delta J = \int_\Omega \delta u^T (Lu + p) d\Omega + boundary \tag{2.51}$$

The terms inside the bracket yield the Euler equation as required.

Consider again the simply supported beam subjected to a uniformly distributed load $p(x)$ presented in Example 2.2. the differential equation is given by

$$A(w) = EI\frac{d^4w}{dx^4} - p = 0 \quad 0 \le x \le l \tag{2.52}$$

and, according to Equation 2.3, $L = EI\frac{d^4}{dx^4}$. From Equation 2.49,

$$J = \int_0^l \left[\frac{1}{2}wEI\frac{d^4w(x)}{dx^4} - p(x)w(x)\right] dx \tag{2.53}$$

The first term will not lead to a symmetric operator; however, symmetry can be established by integrating the first term twice, i.e.

$$\begin{aligned} J &= \frac{1}{2}wEI\frac{d^3w(x)}{dx^3}\Big|_{x=0}^{x=l} - \frac{1}{2}\frac{dw}{dx}EI\frac{d^2w(x)}{dx^2}\Big|_{x=0}^{x=l} \\ &+ \int_0^l \left[\frac{1}{2}wEI\left(\frac{d^2w(x)}{dx^2}\right)^2 - p(x)w(x)\right] dx \end{aligned} \tag{2.54}$$

Now the functional is the same as the potential energy and the first two terms are defined by the end conditions of the beam. Taking the first variation shown in the previous section, the Euler equation can be obtained, thus

$$\begin{aligned} \delta J &= \int_0^l \left[\left(EI\frac{d^4w(x)}{dx^4} - p(x)\right)\delta w(x)\right] dx + EI\frac{d^2w(x)}{dx^2}\delta\frac{dw(x)}{dx}\Big|_{x=0}^{l} \\ &- EI\frac{d^3w(x)}{dx^3}\delta w(x)\Big|_{x=0}^{l} = 0 \end{aligned} \tag{2.55}$$

Additional information on the variational principle and self-adjointness of the differential equations can be found in Mikhlin (1964 and 1965) and Zienkiewicz (1982). Further illustrations of the variational principle, and a discussion of how to formulate the problem and solution for a vibration and an eigenvalue problem are presented in the next two examples.

2.4.4 Dynamic analysis

Before attempting to solve any dynamics problems, a quick overview of the principles that govern the dynamics behavior is given. The reader is assumed to have a reasonable knowledge in dynamics. Additional information is given in Clough and Penzien (1975).

The principle of conservation of energy requires that the total mechanical energy of a particle moving from location L_1 at time t_1 to location L_2 at time t_2 be conserved (Fig. 2.8), i.e.

$$\pi_1 + T_1 = \pi_2 + T_2 + \int_{L_1}^{L_2} F ds \tag{2.56}$$

Where π and T represent the potential and kinetic energy of the system, respectively.

The last term includes the non-conservative forces such as friction forces, and damping forces. These always act in the opposite direction of the motion of the particle. For simplicity, we will only deal with conservative forces.

Newton's second law provides the requirement for a dynamic equilibrium, i.e.

$$F = ma = m\frac{d^2x}{dt^2} \tag{2.57}$$

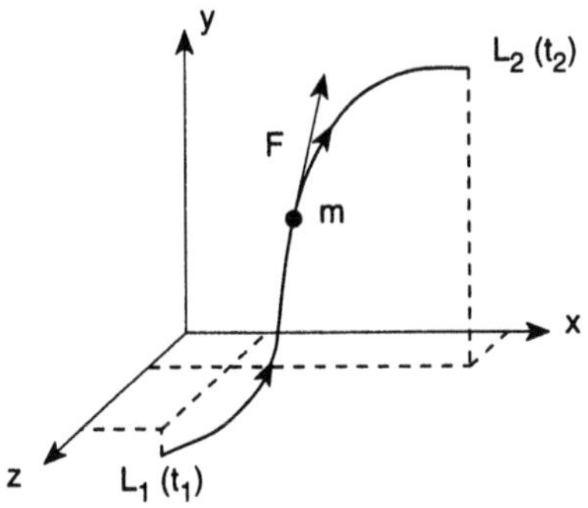

Figure 2.8 *Graphical representation of a particle of mass m, moving from location L_1 at time t_1 to location L_2 at time t_2.*

where m is the mass of the particle and a, the second derivative of x with respect to time, is the corresponding acceleration. If δx is assumed to be the change (or variation) in x along the path for the same time t, but not for δt, and satisfies the conditions

$$\delta x = 0 \quad at \quad t_1\, and\, t_2 \tag{2.58}$$

the work done during the variation of the path becomes

$$W = F\delta x = m\ddot{x}\delta x \tag{2.59}$$

Assuming that the mass m is constant and that $d\dot{x} = \ddot{x}dt$, the integration of Equation 2.59 by parts with respect to time, and taking into consideration Equation 2.58, yields

$$\begin{aligned}
\int_{t_1}^{t_2} F\delta x dt &= \int_{t_1}^{t_2} m\ddot{x}\delta x dx \\
&= \int_{t_1}^{t_2} m\frac{d\dot{x}}{dt}\delta x dt \\
&= \int_{t_1}^{t_2} m[\frac{d}{dt}(\dot{x}\delta x) - \dot{x}\frac{d}{dt}\delta x]dx \\
&= [m\dot{x}\delta x]_{t_1}^{t_2} - m\int_{t_1}^{t_2} \dot{x}\delta(\dot{x})dt \\
&= -m\int_{t_1}^{t_2} \dot{x}\delta(\dot{x})dt
\end{aligned} \tag{2.60}$$

Recalling

$$T = \frac{1}{2}m\dot{x}^2 \tag{2.61}$$

and

$$\delta T = m\dot{x}\delta\dot{x} \tag{2.62}$$

and

$$F = -\frac{\partial \pi}{\partial x} \tag{2.63}$$

this results in

$$F\delta x = -\delta \pi \tag{2.64}$$

By substituting Equations 2.62 and 2.64 into 2.60 we obtain

$$-\int_{t_1}^{t_2} \delta \pi dt = -\int_{t_1}^{t^2} \delta T dt \tag{2.65}$$

or

$$\int_{t_1}^{t_2} \delta(T - \pi)dt = \delta \int_{t_1}^{t_2} (T - \pi)dt = 0 \tag{2.66}$$

$L = T - \pi$ is also called the Lagrange potential of motion or Hamiltonian (Hartmann, 1985). The application of this principle leads directly to the equation of motion for any given system. This can be proved by simply recalling the expression for π in linear elasticity:

$$\pi = \frac{1}{2}\int_{\Omega} \sigma_{ij}\epsilon_{ij} d\Omega - \int_{\Gamma} T_i u_i d\Gamma - \int_{\Omega} f_i u_i d\Omega \tag{2.67}$$

where a double appearance of any index (i or j) in the same term implies summation over that index, σ_{ij} represents the stress field, ϵ_{ij} the strain field, u_i the displacement field, T_i the surface tractions and f_i the body forces. The kinetic energy of the system is

$$T = \frac{1}{2}\int_{\Omega} \rho \dot{u}_i \dot{u}_i d\Omega \tag{2.68}$$

in which ρ is the density and equal to mass per unit volume. Then by performing the first variation and applying Green's theorem, the Euler–Lagrange equations are obtained as

$$-\rho \frac{du_i}{dt} + \frac{d\sigma_{ij}}{dx_j} + f_i = 0 \quad in \quad \Omega \tag{2.69}$$

$$\sigma_{ij}\nu_j = T_i \quad on \quad \Gamma_t \tag{2.70}$$

and

$$\delta u_i = 0 \quad on \quad \Gamma_u \tag{2.71}$$

Here ν_j is the direction cosine of the normal to boundary Γ_t and, according to Hamilton's principle, Equation 2.69–2.71 must be satisfied at any instant of time during the interval (t_1, t_2).

Example 2.6 Let us determine the free vibration of the cantilever column shown in Fig. 2.9 assuming that both the density and stiffness are constant. Thus, kinetic and potential energies for this system are obtained

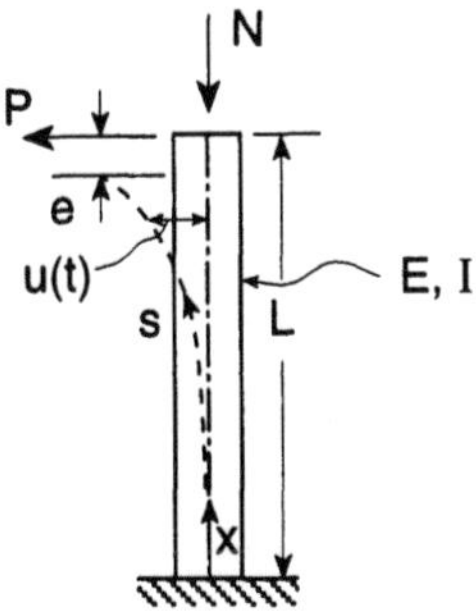

Figure 2.9 *A cantilever column subjected to an axial load N, and a horizontal force P.*

from Equations 2.67 and 2.68, respectively, i.e.

$$T = \frac{1}{2}\int_0^l \rho A\dot{u}_i\dot{u}_i dx = \frac{1}{2}\int_0^l \rho A\dot{u}^2 dx = \frac{1}{2}\int_0^l m\dot{u}^2 dx \tag{2.72}$$

and

$$\pi = U - W = \frac{1}{2}\int_0^l EI\left(\frac{d^2u(x)}{dx^2}\right)^2 dx - [Pu + Ne] \tag{2.73}$$

where e represents the shortening of the column due to curvature.

Note that

$$\begin{aligned} e &= l - \int_0^l ds \\ &= l - \int_0^l (1 - \frac{1}{2}\left(\frac{du}{dx}\right)^2)dx \\ &= \frac{1}{2}\int_0^l \left(\frac{du}{dx}\right)^2 dx \end{aligned} \tag{2.74}$$

By denoting $du/dx = u'$, we have

$$\begin{aligned} L &= T - \pi \\ &= \frac{1}{2}\int_0^l m\dot{u}^2 dx - \frac{1}{2}\int_0^l EI(u'')^2 dx + Pu + \frac{N}{2}\int_0^l (u')^2 dx \end{aligned} \tag{2.75}$$

Thus

$$\begin{aligned} \delta\int_{t_1}^{t_2} Ldt &= \int_{t_1}^{t_2} (\delta T - \delta\pi)dt \\ &= \int_{t_1}^{t_2} [\int_0^l m\dot{u}\delta\dot{u}dx - \int_0^l EIu''\delta u''dx + P\delta u + N\int_0^l u'\delta u'dx] \\ &= \int_{t_1}^{t_2} \{-\int_0^l [m\ddot{u} + (EIu'')'' + Nu'']\delta u dx - EIu''\delta u' \mid_0^l + \end{aligned}$$

$$+[(EIu'')' + Nu' + P]\delta u\,|_{x=l} - [(EIu'')' + Nu']\delta u\,|_{x=0}\}dt = 0 \tag{2.76}$$

and recalling that

$$\delta u = 0 \quad at \quad t = t_1 \quad and \quad t = t_2 \tag{2.77}$$

and using Lagrangre's lemma and the fact that δu is arbitrary (Hartman, 1985), the governing differential equation is obtained, i.e.

$$m\frac{d^2u}{dt^2} + \frac{d^2}{dx^2}(EI\frac{d^2u}{dx^2}) + N\frac{d^2u}{dx^2} = 0 \tag{2.78}$$

This is subject to the boundary conditions

$$either \quad EIu'' = 0 \quad or \quad \delta u' = 0 \quad at \quad x = 0 \quad and \quad x = l \tag{2.79}$$

$$\frac{d}{dx}(EIu'') + Nu' + P = 0 \quad at \quad x = l \tag{2.80}$$

$$\delta u = 0 \quad at \quad x = 0 \tag{2.81}$$

The solution can be obtained by separation of variables using $u(x,t) = \psi(x)Z(t)$, where $\psi(x)$ is a function representing the mode shape, and $Z(t)$ is a function of time alone.

Substituting $u(x,t)$ into Equation 2.76 and recalling that $\dot{u} = \psi\dot{Z}$, $\delta\dot{u} = \psi\delta\dot{Z}$, $\delta u'' = \psi''\delta Z$, $\delta u' = \psi'\delta Z$ and $\delta u = \psi\delta Z$ yields

$$\int_0^l m\psi^2 dx\ddot{Z} + [\int_0^l EI\psi''^2 dx - N\int_0^l \psi'^2 dx]Z = 0 \tag{2.82}$$

or

$$m^*\ddot{Z} + (K^* - K_G^*)Z = 0 \tag{2.83}$$

where m^* is refered as to the generalized mass, K^* the generalized stiffness and K_G the generalized geometric stiffness. Thus the frequency of the cantilever beam is

$$\omega = \sqrt{\frac{K^* - K_G^*}{m^*}} \tag{2.84}$$

which in the absence of axial load N yields the Rayleigh quotient.

$$\omega = \sqrt{\frac{K^*}{m^*}} = \sqrt{\frac{\int_0^l EI(\psi'')^2 dx}{\int_0^l m\psi^2 dx}} \tag{2.85}$$

Equation 2.85 can be used to obtain the approximate value of frequency by assuming a mode shape $\psi(x)$, which satisfies the kinematic boundary conditions. It can be proven that the true mode shape yields the exact solution for the frequency, which is the minimum value of the Rayleigh quotient obtained from all the possible mode shapes (Washizu, 1982).

2.4.5 Stability analysis

In general, there are at least four different approaches that can be used to determine the critical load of a system, namely as a bifurcation problem, the response problem, the dynamic method and the energy method. Here, we only discuss the energy method. Again, the reader is assumed to have a reasonable knowledge of stability theory (Dym and Shames, 1973; Timoshenko and Gere, 1961).

The solution begins by first establishing a functional for the potential energy π using a kinematically admissible function that is also a point function in the configuration space. Then we shall examine the extremum conditions of the functional, keeping in mind that the system is said to be at equilibrium if this point is stationary. A conservative holonomic system is in a configuration of stable equlibrium if, and only if, the value of the potential energy is a relative minimum, i.e.

$$\delta\pi = 0 \quad for \quad stationary$$

and

$$\delta^2\pi > 0 \quad for \quad minimum$$

The procedure is demonstrated next.

Example 2.7 Let us determine the critical load for the beam column shown in Fig. 2.10 with one end fixed and an axial load P applied to the free end. The properties E and I are assumed constant. The functional can be established from the potential energy of the system,

$$\pi = U - W$$

where the strain energy is defined as

$$U = \frac{1}{2}\int_0^l EI\left(\frac{d^2w(x)}{dx^2}\right)^2 dx$$

and the external work is given by

$$W = Pu$$

To solve this problem, we need to express the variable u in terms of w. Since the length along the beam does not change

$$l = \int ds \tag{2.86}$$

where

$$ds = \sqrt{(dx)^2 + (dw)^2} = \sqrt{1 + \left(\frac{dw}{dx}\right)^2}\, dx \tag{2.87}$$

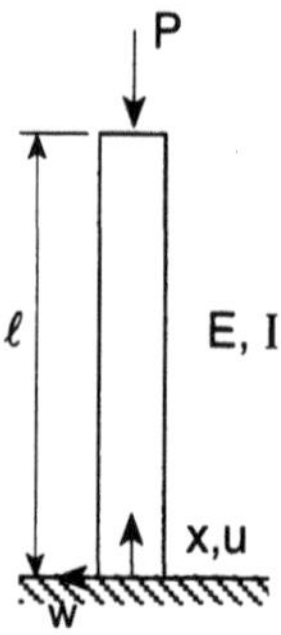

Figure 2.10 *Beam column subjected to an axial load P.*

Therefore

$$l = \int_0^{l'} \sqrt{1 + \left(\frac{dw}{dx}\right)^2}\, dx \approx \int_0^{l'} \left[1 + \frac{1}{2}\left(\frac{dw}{dx}\right)^2\right] dx \tag{2.88}$$

and, since u can be expressed as

$$u = l - l' = \frac{1}{2}\int_0^{l'} \left(\frac{dw}{dx}\right)^2 dx \tag{2.89}$$

and u is small, Equation 2.89 can be rewritten as

$$u = \frac{1}{2}\int_0^{l} \left(\frac{dw}{dx}\right)^2 dx \tag{2.90}$$

The functional becomes

$$\pi = \frac{1}{2}\int_0^{l} EI\left(\frac{d^2w(x)}{dx^2}\right)^2 dx - \frac{P}{2}\int_0^{l} \left(\frac{dw(x)}{dx}\right)^2 dx \tag{2.91}$$

To solve, we need to define a function $w(x)$ that satisfies the kinematic boundary condition, i.e. if $w = a + bx + cx^2$ and

$$w(x) = 0 \ \ and \ \ \frac{dw(x)}{dx} = 0 \ \ \ at\ x = 0 \tag{2.92}$$

then

$$w = cx^2 \tag{2.93}$$

Substituting Equation 2.93 into 2.91 and integrating along the length of the beam yields

$$\pi = 2c^2 EIl - \frac{2}{3}c^2 Pl^3 \tag{2.94}$$

Taking the first variation gives

$$\frac{d\pi}{dc} = 4l(EI - \frac{1}{3}Pl^2)c = 0 \tag{2.95}$$

This leads to

$$P_{cr} = \frac{3EI}{l^2} \tag{2.96}$$

and the exact solution is $\dfrac{\pi^2 EI}{4l^2} = 2.467\dfrac{EI}{l^2}$ (Timoshenko and Gere, 1961).

2.5 Weighted residual method

The weighted residual method is a technique that can be used to approximate the solution to linear and nonlinear differential equations. Its largest application remains for those engineering problems whose functional cannot be established and for those that contain nonlinear material responses, boundary and loading conditions (Finlayson, 1972) .

Suppose that the governing differential equations and non-essential boundary conditions of the problem as before are represented by

$$A(u) = \quad Lu + p = 0 \quad in\ \Omega \tag{2.97}$$

$$S(u) = \quad Mu + q = 0 \quad on\ \Gamma_q \tag{2.98}$$

$$G(u) = \quad u + u_o = 0 \quad on\ \Gamma_u \tag{2.99}$$

and, as an approximate solution, we assume a function for u such that

$$u \approx \bar{u} = \sum_i N_i \alpha_i \tag{2.100}$$

where N_i and α_i are the interpolation functions and undetermined coefficients that must be established as part of the solution procedure, respectively. Substituting Equation 2.100 into Equations 2.97–2.99 yields in general the residuals

$$R_L = A(\bar{u}) \neq 0 \quad in \quad \Omega \tag{2.101}$$

$$R_q = S(\bar{u}) \neq 0 \quad on \quad \Gamma_q \tag{2.102}$$

$$R_u = G(\bar{u}) \neq 0 \quad on \quad \Gamma_u \tag{2.103}$$

For $\bar{u} = N_i\alpha_i$ to be an acceptable solution, we must define the coefficients α_i such that R_L, R_q and R_u are all small at all points of the solution domain. The various weighted residual methods provide a solution by minimizing the residuals over the domain using a weighting function $\bar{\omega}$ (Finlayson and Scriven, 1966). These methods are best described as

$$E = \int_\Omega \bar{\omega} R_L d\Omega = 0 \tag{2.104}$$

and they differ in their definition of $\bar{\omega}$ (Crandall, 1956). Some popular schemes are discussed next and the corresponding procedure is illustrated through a simple example.

1. Collocation – $\bar{\omega}$ is a unit delta function that is not zero only at the collocation points, i.e.

$$\bar{\omega} = \delta(x - x_i) \quad = \infty \quad x = x_i$$
$$= 0 \quad x \neq x_i \tag{2.105}$$

The property of the Dirac delta δ is defined by

$$\int_{-\infty}^{\infty} f(x)\delta(x - x_i)dx = \int_{x_i - a}^{x_i + a} f(x)\delta(x - x_i)dx = f(x_i) \tag{2.106}$$

Thus, we have $E_i = R_{Li} = 0$.

2. Least squares – $\bar{\omega}$ is set equal to the residual, then the integral of the square of the residual is minimized with respect to the coefficients α_i, i.e.

$$E = \int_{\Omega} R_L^2 d\Omega \tag{2.107}$$

and that

$$\frac{\partial E}{\partial \alpha_i} = 0 \tag{2.108}$$

3. Least squares collocation – $\bar{\omega}$ is set equal to the residual, then the sum of the residual squares evaluated at the allocation points is minimized with respect to the coefficients α

$$E = \sum_{i=1}^{n} R_{Li}^2 \tag{2.109}$$

and that

$$\frac{\partial E}{\partial \alpha_i} = 0 \tag{2.110}$$

4. Galerkin – $\bar{\omega}$ is defined as the partial derivative of the function $\bar{u}$ with respect to the coefficients α_i, i.e.

$$E_i = \int_{\Omega} \bar{\omega}_i R_L d\Omega = 0 \tag{2.111}$$

with $\bar{\omega}_i = N_i$.

These methods of approximation lead to the generation of a set of simultaneous equations that form the solution of the problem. The selection of appropriate method is critical to ensure that the resultant matrices are positive definite and symmetric.

Example 2.8 Consider a bar with a linearly varying cross-sectional area $A = A_1 + (A_2 - A_1)x/l$, length l, elastic modulus E, fixed at one end $x = 0$ and subjected to a uniformly distributed axial load q and end load σ_0 at $x = l$ (Fig. 2.11). The governing differential equations and non-essential boundary conditions are

$$\frac{d}{dx}\left[EA(x)\frac{du(x)}{dx}\right] + q = 0 \tag{2.112}$$

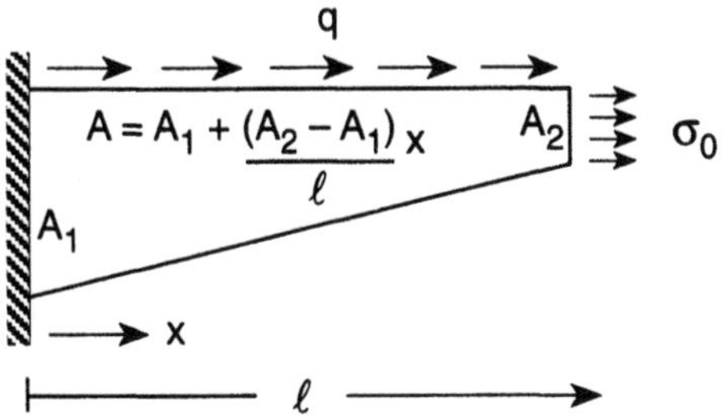

Figure 2.11 *A bar with linearly varying cross-sectional area under a uniform axial load q and an end axial stress σ_0.*

$$u\mid_{x=0}=0 \tag{2.113}$$

$$\left[EA(x)\frac{du(x)}{dx}\right]_{x=l}=\sigma_0 A_2 \tag{2.114}$$

To calculate the displacement using the above-mentioned methods with only two undetermined coefficients, the displacement field is approximated by the trial function

$$\bar{u}=a_1x+a_2x^2 \tag{2.115}$$

where $\bar{u}$ satisfies the essential boundary condition, i.e. $\bar{u}=0$ at $x=0$. Substituting Equation 2.115 back into Equations 2.101 and 2.103 we obtain

$$R_L=E\frac{d}{dx}\left[(A_1+\frac{A_2-A_1}{l}x)\frac{d\bar{u}}{dx}\right]+q \tag{2.116}$$

Since

$$\begin{aligned}\frac{d\bar{u}}{dx} &= a_1+2a_2x\\ \frac{d^2\bar{u}}{dx^2} &= 2a_2\end{aligned} \tag{2.117}$$

This leads to

$$\begin{aligned}R_L &= a_1\frac{A_2-A_1}{l}+2a_2(A_1+\frac{A_2-A_1}{l}2x)+\frac{q}{E}\\ R_q &= a_1+2a_2l-\frac{\sigma_0}{E}\end{aligned} \tag{2.118}$$

<u>Collocation solution</u>

Since R_q is evaluated at $x=l$, R_L is evaluated at $x=L/4$. Thus for $R_L=R_q=0$ this yields

$$\begin{aligned}a_1 &= \frac{1}{E(3A_1-A_2)}[-2ql+\sigma_0(5A_1-3A_2)]\\ a_2 &= \frac{1}{E(3A_1-A_2)}[q+\frac{\sigma_0}{l}(A_2-A_1)]\end{aligned} \tag{2.119}$$

Least squares solution

Equation 2.107 becomes

$$E = \int_{\Omega} R_L^2 d\Omega + \bar{\omega}^2 R_q^2 A_2 = 0 \tag{2.120}$$

$\bar{\omega}^2$ is chosen equal to $1/l$ so that both terms have the same units. This yields

$$\begin{aligned} E &= \int_0^l R_L^2 (A_1 + \frac{A_2 - A_1}{l} x) dx + \frac{A_2}{l} R_q^2 \\ &= \int_0^l [a_1 \frac{A_2 - A_1}{l} + 2a_2 (A_1 + \frac{A_2 - A_1}{l} 2x) + \frac{q}{E}]^2 (A_1 + \frac{A_2 - A_1}{l} x) \\ &\quad + \frac{A_2}{l} (a_1 + 2a_2 l - \frac{\sigma_0}{E})^2 \end{aligned} \tag{2.121}$$

From $\partial E/\partial a_i = 0$ for i=1 and i=2, one can obtain a_1 and a_2.

Galerkin Solution

The residual for the ith term is obtained from

$$E_i = \int_{\Omega} \bar{\omega}_i R_L d\Omega - \bar{\omega}_i R_q A_2 \tag{2.122}$$

To solve the above equation, it is standard practice to reduce the order of differentiation by integrating by parts (Strang and Fix, 1973). This reduces the requirements on the continuity of the trial function to unity, i.e. only the displacement field across the subdivision boundaries need to be continuous.

Here

$$\bar{\omega}_i = \frac{\partial \bar{u}}{\partial a_i}$$

i.e.

$$\bar{\omega}_1 = \frac{\partial \bar{u}}{\partial a_1} = x \quad and \quad \bar{\omega}_2 = \frac{\partial \bar{u}}{\partial a_2} = x^2 \tag{2.123}$$

This results in two equations, E_1 and E_2, from which a_1 and a_2 are obtained from the condition $E_1 = 0$ and $E_2 = 0$. The solution is left as an exercise to the reader.

We have so far presented three general procedures, the direct approach, the variational approach and the weighted residual approach, as methods that can be employed to solve continuum mechanics problems. Given the limitations of the direct approach, most readers will find that the variational principle is the easiest and most applied method to formulate a solution to a continuum problem provided that a functional already exists. This is further demonstrated using the finite element method.

2.6 Finite element method

The previous sections show that equations of equilibrium can be established from the principle of minimum total potential energy. The finite element method can also be defined as a Rayleigh–Ritz method (Cook et al., 1989), in which the continuum is divided into a finite number of subdomains called finite elements. These finite elements are used to approximate the response of the continuum and are interconnected at the nodal points on the element boundaries. The displacements $\{u\}$ at any point inside an element are related to the nodal displacements through the displacement field assumed over an element, and is given by

$$\{u\} = [N]\{\delta\} \tag{2.124}$$

where $[N]$ is the matrix of interpolation functions and $\{\delta\}$ the vector of nodal displacements. Further, there are additional requirements on the matrix $[N]$, namely that it is conformable and complete to yield monotonic convergence. This is discussed further in Section 2.7.

Next, the strain–displacement and stress–strain relationships need to be defined. In this case, the basic parameter is the strain which represents the rate of change of deformation. For axial deformation and assuming small deformation, the strain displacement relationship is given by

$$\epsilon_x = \frac{du}{dx} \tag{2.125}$$

or, in general form, by

$$\{\epsilon\} = [L]\{u\} = [L][N]\{\delta\} = [B]\{\delta\} \tag{2.126}$$

For the stress field, we rely on the material contitutive law. In matrix form, the stress field is expressed as

$$\{\sigma\} = [D]\{\epsilon\} = [D][B]\{\delta\} \tag{2.127}$$

where $[D]$ is the elastic matrix. Substitution of Equations 2.124, 2.126 and 2.127 into

$$\delta\pi = 0 \tag{2.128}$$

and recalling Equation 2.67 for π yields

$$\begin{aligned}\int_V \delta\{\delta\}^T[B]^T[D][B]\{\delta\}dV &= \\ \int_V \delta\{\delta\}^T[N]^T\{f\}dV &+ \int_S \delta\{\delta\}^T[N]^T\{T\}dS \\ &+ \sum_i \delta\{\delta\}^T[N]^T\{F\}_i\end{aligned} \tag{2.129}$$

where $\{f\}$ is the body force, $\{T\}$ is the surface traction and $\{F\}_i$ represents the point load in the element. Since $\delta\{\delta\}$ are arbitrary and summing over all the finite elements, the following discretized equations of equilibrium are obtained

$$\left[\sum_e \int_V [B]^T[D][B]dV\right]\{\delta\} = \sum_e \int_V [N]^T\{f\}dV + \sum_e \int_S [N]^T\{T\}dS + \sum_e \sum_i [N]^T\{F\}_i \qquad (2.130)$$

in which all the vectors and matrices have been expanded to the structure size before summation, i.e. $\{\delta\}$ includes the nodal displacements of the entire structure, both $[N]$ and $[B]$ have the corresponding number of columns. Equation 2.130 can also be written in the matrix form

$$[K]\{\delta\} = \{P\} \qquad (2.131)$$

where the stiffness matrix

$$[K] = \sum_e \int_V [B]^T[D][B]dV \qquad (2.132)$$

and the load vector $\{P\}$ is due to the following

$$\{P\} = \sum_e \int_V [N]^T\{f\}dV + \sum_e \int_S [N]^T\{T\}dS + \sum_e \sum_i [N]^T\{F\}_i \qquad (2.133)$$

We have so far described the procedure to formulate the element stiffness matrix and load vector without much consideration to the kinematics conditions. However, from Section 2.6, it is apparent that the accuracy of the solution depends on how well the interpolation function meets the kinematics conditions. Selection of the appropriate interpolation function in order to meet the convergence requirements is covered in the next section.

2.7 Convergence requirements for interpolation functions

In general, an interpolation function representing a displacement field is admissible provided it does not violate the internal compatibility nor the essential boundary conditions (Melosh, 1963; Arantes e Oliveira, 1968; Mebane and Stricklin, 1971; Taylor, 1972). The compatibility requirement implies that the displacement field within the element and across the element boundaries must be continuous, i.e. there must be no gaps or overlaps between elements. For beam, plate and shell analyses, not only the deflection but also its first derivatives must be continuous in order to ensure a continuous displacement field over the entire volume of the structure. Moreover, the element formulation must be complete and compatible in order to ensure a monotonic convergence, i.e. the accuracy of the solution increases consistently with the number of elements. The definition of 'complete' means that the interpolation function that describes the displacement field can represent both the rigid body displacements and the constant strain states. The rigid body displacements are those rigid translations and rotations that the element must be able to undergo without generating any strains and stresses in it. As an example, a two-dimensional plane-stress element must be able to translate uniformly in either direction of its plane and to rotate about its out-of-plane axis without developing any stresses. For a simple element, the rigid body modes can be identified by inspection and they are equal to the element degrees of freedom less the element constant straining modes. The rigid body modes and straining modes of a four node element are shown in Fig. 2.12.

The constant strain mode is required to ensure that in the limit as the size of the element decreases due to an increase in the number of elements, the strain does not keep on changing but rather approaches a constant value within each element. This requirement also provides assurances that the element is capable of representing complex deformation modes.

In some cases, it is also desirable to have the element geometrically invariable, i.e. spatially isotropic (Desai and Abel, 1972). This becomes a requirement when analyzing two- or three-dimensional continua with a spatially isotropic behavior. In general, this condition is satisfied by ensuring that the spatial distribution of the primary variables is equally distributed in the interpolation function.

2.8 Generalized procedure for a finite element solution

We have so far presented the approach along with the essential requirements to properly formulate a solution to a generalized continuum mechanics problem. In summary, a numerical representation of a continuum problem is made up of three main steps, the definition of the physical problem, the identification of its mechanical idealization and then the solution of the discretized problem. We shall now illustrate these steps for the finite element solution process.

2.8.1 Problem description

The physical problems must be first defined and this involves the geometry, the material composition, the loading conditions and interactions with the surroundings. This identification process will assist in formulating the problem and subsequently obtaining an approximate solution.

2.8.2 Problem formulation

Formulation of the problem is the key to establishing the proper solution. From the description of the geometry, the necessary kinematic conditions can be established, i.e. the anticipated deformational behavior and state of

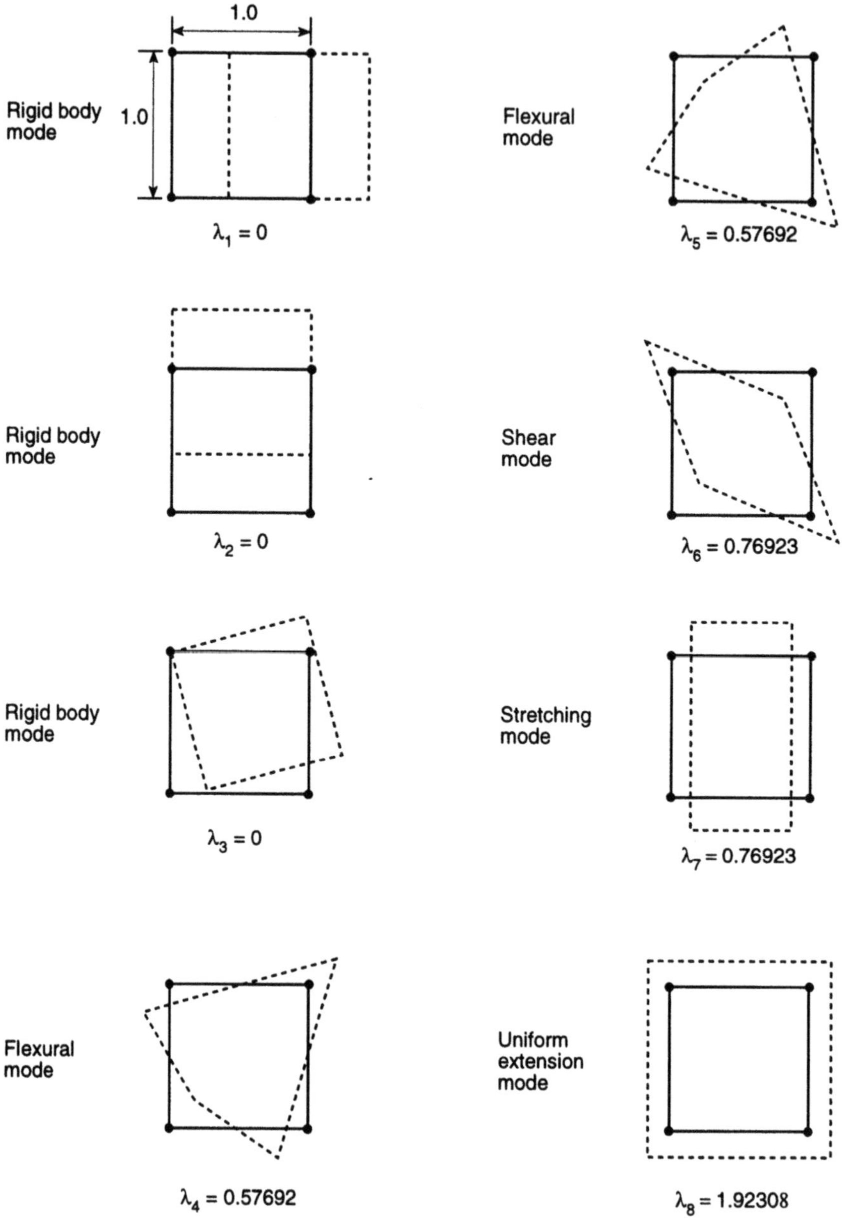

Figure 2.12 *Eigenvalues and eigenvectors of a four-node plane stress element.*

stress. For instance, a geometric domain with a thickness much less than the length of the other two sides implies that its deformational behavior can be described using either plate or shell theory. If the thickness is very small, so that shear deformations can be neglected, then Kirchhoff's plate theory can be used. Similarly, the material composition of the problem will dictate whether the response will be isotropic, linear, elastic, plastic etc. For example, if the continuum is made up of rubber material, then the Mooney–Rivlin or Ogden material models will be best suited to describe its response.

The other two important factors are the loading and boundary conditions. The former is deduced from the application and is represented by concentrated loads, pressure, centrifugal loads and gravity loads. However, specifying the boundary conditions in most cases is not a straight forward process especially if the continuum being studied is part of a larger system. A variety of the boundary conditions is available such as fixed boundary, prescribed boundary and contact boundary. The final results yield the governing equation of motion such as

$$EI\frac{d^4w}{dx^4} - p(x) = 0 \tag{2.134}$$

that supports the boundary conditions.

2.8.3 Finite element solution

The response of the continuum is then obtained by approximating the response of its mechanical idealization. Using the finite element that complies with the kinematic conditions, i.e. thin plate or thin shell element for the problem presented in Section 2.8.2, the set of equilibrium equations are then obtained according to Section 2.5. The solution will depend on the type of material response, boundary and loading conditions. For a linear behavior, the solution procedure is straightforward using the Gaussian elimination technique (Strang, 1976). However, if the response is nonlinear due to geometric, material, boundaries or loading factors, an iterative procedure is then required in order to ensure that the solution converges to the equilibrium conditions. The nonlinear solution procedure is discussed further in Chapter 10.

For an approximate solution, an assessment of the errors is always required. Chapter 3 presents the sources of numerical errors inherent in the finite element solution and discusses how to first approximate their magnitude and then how to minimize their effect on the solution of continuum problems.

2.9 Finite strip method

The discussion has so far been limited to the displacement-based finite element method since this is the most widely applied and recognized method of solution. The finite strip method can be regarded as a degenerate form of the finite element method which is used to primarily model the response of prism-like structures such as plates and solids. For example in plate bending problems, the finite element method requires that the element be capable of simultaneously modeling the deformation pattern in both directions, whereas the finite strip method models the problem as a one-dimensional strip in the longitudinal direction using an interpolation function for the displacement field that satisfies a priori the boundary conditions at both ends. The finite element requirements on the admissibility of the interpolation function also apply to the finite strip method with the exception of the rigid body modes. The exception arises from the finite strip requirements to satisfy the support boundary conditions a priori.

The advantages of the finite strip method over the finite element method include reduced computer resources and a significant reduction in the time taken to model the problem. However, as a degenerate form of the finite element method, there are restrictions on its application to problems with arbitrary geometry, boundary conditions and material variations. State of the art knowledge on the finite strip formulation and its application to bridges are the subject of Parts II and III of this book, respectively.

CHAPTER 3

Numerical errors

3.1 Introduction

In the preceding chapter, the basic concepts of numerical methods were presented as solutions to real physical problems. These methods are used to solve problems of practical significance which do not possess derivable or accurate closed-form solutions. Such numerical methods provide an approximation to the governing equations in terms of values at a finite number of discrete points within the computational domain. The current practice is reflected by the assertion of Irons and Shrive (1983): 'if there is an opportunity for improving the design, then somebody is attempting to do so using the finite element method'. However, to meet such expectations, it is also essential that numerical errors can be quantified so that solutions of engineering projects especially where life safety must be ensured (such as bridges), meet certain permissible accuracy standards.

As early as 1910 (Richardson, 1910), engineers have attempted to quantify the errors of approximation in the analysis of structures in order to prevent catastrophic failures. The sources of numerical errors may be attributed to operator error or to the idealization of the solution procedure. Operator error refers to the inability of the approximation equation to correctly represent the governing equation. Additionally, there are errors caused by differences between the physical boundary and the approximated function such as contact boundaries, fracture mechanics and discrete crack. For errors arising from the discretization and solution procedure, the terms traditionally used in error analysis are *priori* and *posteriori*. A *priori* error estimate is applied to predict the convergence rate prior to the execution of the solution, whereas *posteriori* error analysis estimates the accuracy of the solution by implementing a sequence of refinements of the mesh. This chapter deals only with the *priori* and *posteriori* errors. Computational errors are discussed first, followed by a discussion of the sources that lead to their introduction. This is followed by a brief review of test methods that can be used to detect and avoid errors. A review of the relationship between the element size and the solution error (from the energy sense) is discussed along with a presentation of *posteriori* error analysis.

3.2 Definition of computational errors

The numerical solution is an approximation of the exact response of an idealized engineering system, and thus the first source of computational errors occurs due to differences between the physical system and its mathematical model–*modeling errors*. Modeling errors also arise from solution errors in the constitutive model, temporal and spatial integration and iteration procedures. *Discretization errors* are caused by representing the response of a system of differential equations by a finite number of degrees of freedom. *Rounding errors* are dependent on the smallest floating point number of the computer used, where the least significant digit may be either rounded or truncated. The manifestation of these errors leads to contamination of the assembled global matrices, which can result in an ill-conditioned system of equations or an erroneous response of the idealized system.

3.2.1 Ill-conditioning

A set of algebraic equations is said to be ill-conditioned if their solutions are too greatly influenced by small changes in either the coefficient matrix or the load vector. Consider the simple two-degrees-of-freedom structure shown in Fig. 3.1. The system equations are given by

$$[k]\{\delta\} = \{F\} \tag{3.1}$$

or

$$\begin{bmatrix} k_1 & -k_1 \\ -k_1 & k_1 + k_2 \end{bmatrix} \begin{Bmatrix} u_1 \\ u_2 \end{Bmatrix} = \begin{Bmatrix} F_1 \\ 0 \end{Bmatrix} \tag{3.2}$$

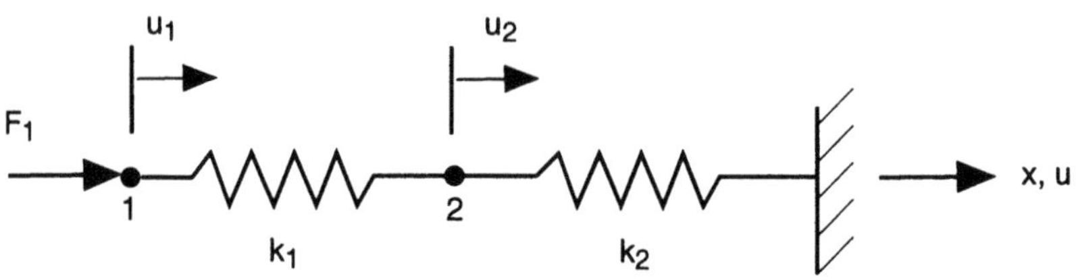

Figure 3.1 *A one-dimensional structure composed of two linear springs of stiffness k_1 and k_2.*

For $k_1 \gg k_2$, the system is then made up of two equations that are linearly dependent and the ratio k_2/k_1 will dictate the accuracy and sensitivity of the solution. However, if $k_2 \gg k_1$, this results in a well behaved system of equations. If one plots Equation 3.2, the resulting lines (Fig. 3.2) will be almost parallel when $k_1 \gg k_2$ and thus the solution, which is the intersection of the two lines, either does not exist if they are parallel or is very sensitive to small changes in the magnitude of k_2.

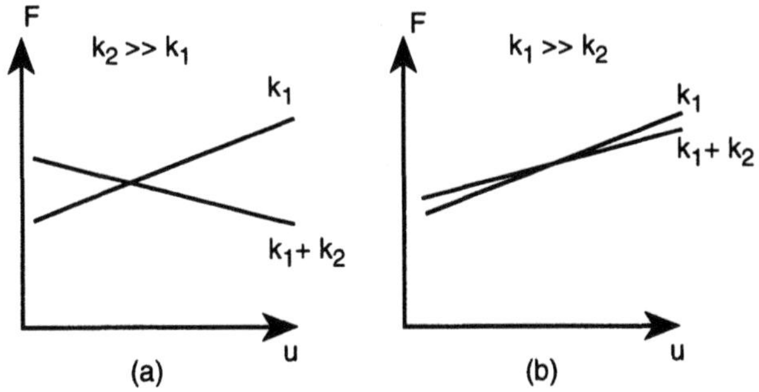

Figure 3.2 *(a) Well-conditioned and (b) ill-conditioned systems of two linear equations in two unknowns.*

The ill-conditioning in practical finite element models occurs due to large differences in stiffness of the material caused by flexible support, inclusion of a relatively soft or rigid material in comparison to the rest of the continuum, structural members with large differences between the shear, membrane or bending stiffness, or elements that have large aspect ratio or are distorted. One way to avoid such difficulties is to reduce the large discrepancies in the stiffness by minimizing the causes listed above and by using the longer computer words, i.e. double precision, to reduce rounding errors and solution sensitivity. The next section concentrates on test methods used to detect an ill-conditioned matrix.

3.3 Assessing computational errors

Three simple test methods are presented for the detection of potential solution errors. As will be demonstrated, the result of these tests will seldom agree due to the complexities of the assembled system of equations and the simplicity of the test methods. However, they provide an early warning for an ill-conditioned matrix.

3.3.1 Residuals

The equilibrium of a system yields

$$\{F\} - [K]\{\delta\} = \{\Delta R\} \tag{3.3}$$

where the residual $\{\Delta R\}$ is equal to zero if the equations can be solved exactly. Unfortunately, ill-conditioning caused by rounding and discretization errors tend to contaminate $\{\delta\}$ and yield $\{\Delta R\} \neq 0$. A measure of the error can be obtained for static analysis from

$$e = \frac{\{\delta\}^T\{\Delta R\}}{\{\delta\}^T\{F\}} \tag{3.4}$$

where e is small if $\{\Delta R\} \ll \{F\}$. Similarly for vibration analysis, the error can be estimated using

$$e = \frac{\{\delta\}^T([K]\{\delta\} - \omega^2[M]\{\delta\})}{\{\delta\}^T[K]\{\delta\}} \tag{3.5}$$

where the influence of ill-conditioning is measured by the combination of the frequency when compared to the system mass and stiffness. It should be noted that a smaller residual does not necessarily imply that the solution is more accurate (Nobble, 1969)

3.3.2 Diagonal decay

The inclusion of variable stiffness and/or rounding errors causes ill-conditioning in practical finite element models. However, a measure of decay of the diagonal term of the stiffness matrix can be used to test the accuracy of the assembled system of equations. The diagonal decay is obtained from

$$\phi_i = K_{ii}/P_{ii} \tag{3.6}$$

in which K_{ii} is the original diagonal term before pivoting commences and P_{ii} is the reduced value of K_{ii}. Equation 3.6 is used to measure the loss of accuracy in the number of digits, i.e. for a computer whose words contain p digits, the remaining accuracy in the number of digits for $\phi = 10^3$ is $p - 3$. However, this simple test will not detect loss of accuracy due to the mixing of element sizes, nor will it detect other ill-conditioning causes such as the structure of Fig. 3.1; only when $k_2 \gg k_1$ will there be a well behaved system of equations (Cook et al., 1989).

3.3.3 Condition number

The condition number (CN[K]) is another measure of the ill-conditioning in a coefficient matrix [K]; a large value suggests that the solution may be contaminated. One approach to computing the CN[K] is by adopting the spectral condition number of [K] where

$$CN[K] = \lambda_{max}/\lambda_{min} \tag{3.7}$$

in which λ_{max} and λ_{min} represent the largest and smallest eigenvalues of the matrix [K]. According to Rosanoff et al. (1969), the loss of accuracy can be estimated from:

$$\begin{aligned} No.\,of\;accurate\;digits &= p - loss\;of\;accurate\;digits \\ &\approx p - \log_{10}(\frac{\lambda_{max}}{\lambda_{min}}) \\ &= p - \log_{10} CN[K] \end{aligned} \tag{3.8}$$

Thus for a system with a $CN[K] = 10^7$, the remaining accurate digits for a computer word of 15 digit capacity is eight digits. Furthermore, the loss in accuracy due to rounding errors as a result of the solution algorithm is not compensated for in Equation 3.8 nor does it estimate the accuracy of the stress field since the load vector is completely ignored.

Fried (1972) has found that the errors introduced due to element sizes can be estimated from

$$CN[K] = b\left(\frac{\Delta_{max}}{\Delta_{min}}\right)^{2m-1} N^{2m/n} \tag{3.9}$$

in which b is found to be strongly dependent on the Poisson ratio

$$b = \frac{b_1}{1-2\nu} \tag{3.10}$$

where b_1 is a positive constant, independent of Δ_{max}, Δ_{min} and the modulus of elasticity. Δ_{max} and Δ_{min} are the largest and smallest node spacing in the discretized domain, N the number of elements, $2m$ the differential equation order and n the dimension of the problem.

We have so far presented simple tests that can be used to measure the potential ill-conditioning of the matrix [K]. As suggested by Anon (1986), it is preferable to measure the likelihood of an ill-conditioning by applying the CN[K], the decay and the residual tests, and that the solution will probably be unacceptable if two or more of these tests indicate trouble. Further, if the resulting set of equations of a particular problem are found ill-conditioned, it is advisable to rework the discretization of the model rather than to attempt improving its condition by trial and error (Cook et al., 1989).

3.4 Assessing discretization errors

It is easier to idealize a physical system using the same size and type of elements. However, in many engineering applications, to capture the physical phenomenon requires further refinement of the mesh in small localized regions, e.g. regions of abrupt geometrical or material properties changes and boundary conditions. Further, to encounter monotonic convergence, the elements must be complete and compatible (Melosh, 1963; Key, 1966; Arantes e Oliveira, 1968; Felippa and Clough, 1968; Strang and Fix, 1973; Oden and Reddy, 1976) and must be proportional to avoid directional bias solution. If these conditions are met, the increase in the refinements of the discretized domain results in an improvement of the accuracy of the solution. The requirements of completeness and compatibility were presented in Section 2.7 and the proof from the energy sense is presented in the next section.

To achieve optimal mesh in which the error is nearly uniformly distributed in the domain of analysis, adaptive mesh generation techniques have been developed and can be divided into two categories.

Category I covers mesh refinement strategies in which the number of degrees of freedom increases within the calculation domain:

1. *h refinement*_The size of element is adjusted (reduced) suitably for minimum error. Element subdivision and complete mesh regeneration are the two alternatives of h refinement. Element subdivision, however, possesses uneconomic and complex transitions. Complete mesh regeneration, which is governed by the mesh density, is more favorable.
2. *p refinement*_Changing of the order of the polynomial, of order p, defining the shape function, in each element. When the governing equations are elliptic, this method has many advantages, such as relative ease of error estimation and fast convergence.
3. *r refinement*_Relocation of nodes while the number of nodes in the mesh remains constant.
4. *h–p refinement*_h refinement is followed by a uniform increase of p (Zienkiewicz et al., 1989).

In category II the whole domain of computation is rediscretized, based on the existing mesh and node spacing function from the error estimates of the previous analysis:

1. Boundary nodes are generated first. Interior nodes are then added in selected sites of initial mesh to obtain a refined mesh (Blacker and Stephenson, 1991).
2. In the advancing front technique the element size in each direction is controlled by the node spacing function. The node spacing function, which specifies the element size of the next discretization, is based on the current geometry of the body. Elements with different size functions over two orthogonal axes are generated (Lo, 1991).

3.4.1 Element size versus solution error

The application of the potential energy theorem to a beam problem yields

$$\begin{aligned} \pi(\delta) &= \frac{EI}{2}\int_0^l (\delta_{xx})^2 dx - \int_0^l p(x)\delta(x)dx \\ &= U(\delta) - W(\delta) \end{aligned} \tag{3.11}$$

Now if we assume that δ is the exact solution of the beam problem and $\delta + \Delta\delta$ is an approximate field that satisfies the same kinemtic boundary condition as δ, i.e.

$$\Delta\delta = 0 \quad and \quad \Delta\frac{d\delta}{dx} = 0 \tag{3.12}$$

at the boundary for the approximate field to be admissible. Thus

$$\begin{aligned}\pi(\delta + \Delta\delta) &= \frac{EI}{2}\int_0^l (\delta_{xx} + \Delta\delta_{xx})^2 dx - \int_0^l p(x)(\delta(x) + \Delta\delta)dx \\ &= \pi(\delta) + U(\Delta\delta) + \frac{EI}{2}\int_0^l 2\delta_{xx}\Delta\delta_{xx} dx - \int_0^l p(x)\Delta\delta dx \quad (3.13)\end{aligned}$$

Integrating the third term of Equation 3.13 by parts and summing of the results with the last term of Equation 3.13 yields

$$\pi(\delta + \Delta\delta) = \pi(\delta) + U(\Delta\delta) \tag{3.14}$$

since δ is the exact solution and satisfies the equation

$$EI\frac{d^4\delta}{dx^4} - p = 0 \tag{3.15}$$

Recalling that the strain energy U is a positive definite quantity and $U(\Delta\delta)$ is greater than zero for any non-zero $\Delta\delta$, this gives

$$\pi(\delta + \Delta\delta) > \pi(\delta) \tag{3.16}$$

This shows that of all the fields that satisfy compatibility and kinematic boundary conditions, only the field which satisfies the equilibrium equation and stress boundary conditions provides a minimum $\pi(\delta)$.

Now if we let $\bar{\delta}(x)$ be an approximate solution to a given beam problem with N degrees of freedom, $\tilde{\delta}(x)$ is also an approximate solution but is forced to take the value of the exact solution for each of its degrees of freedom, and $\delta(x)$ is the exact solution. From Equation 3.14 we have

$$\pi(\bar{\delta}) = \pi(\delta) + U(\bar{\delta} - \delta) \tag{3.17}$$

$$\pi(\tilde{\delta}) = \pi(\delta) + U(\tilde{\delta} - \delta) \tag{3.18}$$

Since the approximate solution yields a minimum π, then

$$\pi(\bar{\delta}) \leq \pi(\tilde{\delta}) \tag{3.19}$$

where $\pi(\tilde{\delta})$ represents the constrained solution. Also, from Equation 3.16,

$$\pi(\delta) \leq \pi(\bar{\delta}) \tag{3.20}$$

which implies that

$$\pi(\delta) \leq \pi(\bar{\delta}) \leq \pi(\tilde{\delta}) \tag{3.21}$$

Substituting Equations 3.17 and 3.18 into Equation 3.21 yields

$$0 \leq U(\bar{\delta} - \delta) \leq U(\tilde{\delta} - \delta) \tag{3.22}$$

This implies that the approximate solution is bounded. The convergence of $\bar{\delta}$ to δ in the energy sense is satisfied if we can show that $U(\tilde{\delta} - \delta)$ approaches zero in the limit.

The difference in potential energy between approximate and exact solution is obtained from

$$\begin{aligned}\Delta\pi &= \pi(\bar{\delta}) - \pi(\delta) \\ &= \pi(\delta) + U(\bar{\delta} - \delta) - \pi(\delta) \\ &= U(\bar{\delta} - \delta)\end{aligned} \tag{3.23}$$

The error in potential energy is therefore linearly proportional to $U(\bar{\delta} - \delta)$. Also, since

$$\pi(\delta) = -U(\delta) \tag{3.24}$$

it can be shown that the error in strain energy is linearly proportional to $U(\bar{\delta} - \delta)$, i.e. that

$$\Delta\pi = -\Delta U = U(\bar{\delta} - \delta) \tag{3.25}$$

and, knowing that the strain energy is always positive, this yields

$$\pi(\bar{\delta}) > \pi(\delta) \quad \textit{and} \quad U(\bar{\delta}) < U(\delta) \tag{3.26}$$

which means that the approximate solution provides an upper bound on potential energy and a lower bound on strain energy.

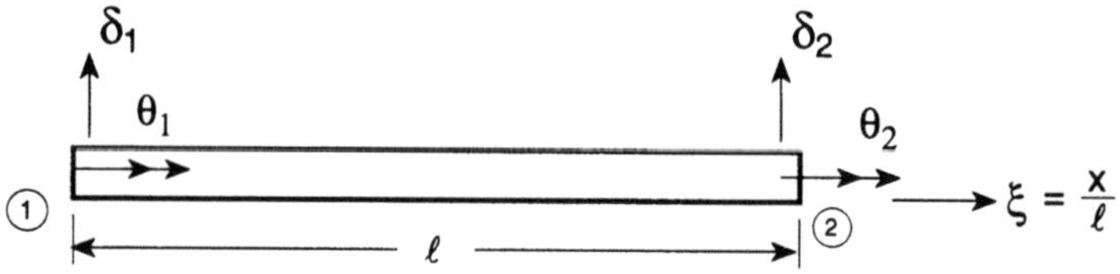

Figure 3.3 *Representation of the beam element's degrees of freedom.*

In the beam problem, we shall analyze the error of discretization by examing the approximate solution, where for the beam of Fig. 3.3 (Zienkiewics, 1982),

$$\bar{\delta}(\xi) = (1-3\xi^2+2\xi^3)\bar{\delta}_1+(\xi-2\xi^2+\xi^3)l\bar{\theta}_1+(3\xi^2-2\xi^3)\bar{\delta}_2+(\xi^3-\xi^2)l\bar{\theta}_2 \tag{3.27}$$

and

$$\tilde{\delta}(\xi) = (1-3\xi^2+2\xi^3)\tilde{\delta}_1+(\xi-2\xi^2+\xi^3)l\tilde{\theta}_1+(3\xi^2-2\xi^3)\tilde{\delta}_2+(\xi^3-\xi^2)l\tilde{\theta}_2 \tag{3.28}$$

where $\tilde{\delta}_1$, $\tilde{\delta}_2$, $\tilde{\theta}_1$ and $\tilde{\theta}_2$ are the exact solutions. Using Taylor's series, the expansion of δ yields

$$\delta(\xi) = \delta_0 + \delta_{x0} l\xi + \delta_{xx0}\frac{l^2\xi^2}{2!} + \delta_{xxx0}\frac{l^3\xi^3}{3!} + \ldots \tag{3.29}$$

Solving $\tilde{\delta}_1$, $\tilde{\delta}_2$, $\tilde{\theta}_1$ and $\tilde{\theta}_2$, using Equation 3.29, one obtains

$$\begin{aligned}\tilde{\delta}_1 &= \delta(\xi = 0) = \delta_0 + 0(l^4) \\ \tilde{\theta}_1 &= \frac{d}{dx}\delta(\xi = 0) = \delta_{x0} + 0(l^3)\end{aligned}$$

$$\tilde{\delta}_2 = \delta(\xi = 1) = \delta_0 + l\delta_{x0} + \frac{l^2}{2}\delta_{xx0} + \frac{l^3}{6}\delta_{xxx0} + 0(l^4)$$
$$\tilde{\theta}_2 = \frac{d}{dx}\delta(\xi = 1) = \delta_{x0} + l\delta_{xx0} + \frac{l^2}{2}\delta_{xxx0} + 0(l^3) \quad (3.30)$$

Substituting Equation 3.30 into Equation 3.27 yields

$$\bar{\delta}(\xi) = \delta_0 + l\xi\delta_{x0} + \frac{l^2\xi^2}{2}\delta_{xx0} + \frac{l^3\xi^3}{6}\delta_{xxx0} + 0(l^4) \quad (3.31)$$

A similar expression can also be obtained for Equation 3.28:

$$\tilde{\delta}(\xi) = \delta_0 + l\xi\delta_{x0} + \frac{l^2\xi^2}{2}\delta_{xx0} + \frac{l^3\xi^3}{6}\delta_{xxx0} + 0(l^4) \quad (3.32)$$

Subtracting Equation 3.31 from Equation 3.27 gives

$$\tilde{\delta} - \bar{\delta} \approx 0(l^4) \quad (3.33)$$

i.e. the error is of the order l^4. For $U(\bar{\delta} - \delta)$, we can assume that

$$\Delta\tilde{\delta} = \tilde{\delta} - \bar{\delta} \approx 0(l^4)$$
$$\frac{d}{dx}\Delta\tilde{\delta} \approx 0(l^3)$$
$$\frac{d^2}{dx^2}\Delta\tilde{\delta} \approx 0(l^2) \quad (3.34)$$

Therefore,

$$U(\Delta\tilde{\delta}) \approx \frac{EI}{2}\int_0^l \Delta\delta_{xx}^2 dx \approx 0(l^4) \quad (3.35)$$

If we further assume that the beam is approximated using N points that are equally spaced, then

$$l = \frac{L}{N} \quad (3.36)$$

and Equation 3.34 becomes

$$U(\Delta\tilde{\delta}) \approx 0\left(\frac{1}{N^4}\right) \quad (3.37)$$

Recalling Equation 3.25

$$\Delta\pi = -\Delta U \approx 0\left(\frac{1}{N^4}\right) \quad (3.38)$$

This implies that the strain energy and potential energy converge asymptotically as N^{-4} and that the convergence is monotonical.

The error can be generalized to other elements as given in Table 3.1 using the following definitions: l is the length of the largest segment connecting two points, $q-1$ is the degree of the highest complete polynomial in the element displacement field, $2m$ the order of the highest derivative in the governing equilibrium equation expressed in terms of displacements, and r

Table 3.1 *Order of error for displacement, stress and strain energy of various finite elements*

Parameter	Order of error		
	General	Plane & solid elements	Beam & plate elements
r		1	2
q-1		1	3
$2m$		2	4
Displacement	$0(l^q)$	$0(l^2)$	$0(l^4)$
Stress	$0(l^{q-r})$	$0(l^1)$	$0(l^2)$
Strain energy	$0(l^{2q-2m})$	$0(l^2)$	$0(l^4)$

is the order of the derivative of the displacements for obtaining the stress field (Cook et al., 1989). In the next section, a quantification of *posteriori* error will be presented.

3.4.2 Posteriori *error estimation*

In finite element analysis, two major types of *posteriori* error estimators have been developed, namely the residual type of error estimator by Babuska and Rheinboldt (1978) and the postprocessing type of error estimator proposed by Zienckiewicz and Zhu (1987), which have introduced new impetus into the analysis of discretization errors. The former yields a local error estimation, whereas the latter provides a global error estimation of the numerical solution. Let us first present the background information for a proper comparison of the two methods. The governing partial differential equation representing the equilibrium relation of the continuum body in domain Ω is given by

$$L(\delta) \equiv (S^T D S)\delta = b \quad in \quad \Omega \tag{3.39}$$

subject to the boundary conditions

$$\begin{aligned} M\delta &= q \quad on \quad \Gamma_q \\ \delta &= \delta_0 \quad on \quad \Gamma_q \end{aligned} \tag{3.40}$$

The compatibility and constitutive relationship of the continuum media are respectively defined by the following equations:

$$\begin{aligned} \epsilon &= S\delta \\ \sigma &= DS\delta = D\epsilon \end{aligned} \tag{3.41}$$

From Chapter 2, we have shown that if matrix $[D]$ is positive definite, operator $(S^T DS)$ will also be symmetrical and positive definite. This is a favorable property for error measures.

Introducing the finite element displacement approximation, $\tilde{\delta}$, for δ and using shape function $[N]$, we have:

$$\delta \approx N\tilde{\delta} = \sum_{i=1}^{m} N_i \delta_i \tag{3.42}$$

where i represents a nodal contribution and m is the total number of nodes per element. Stresses are calculated from

$$\tilde{\sigma} = (DSN)\tilde{\delta} \tag{3.43}$$

The resulting set of algebraic equation is represented by

$$L(\tilde{\delta}) - \tilde{b} = 0 \tag{3.44}$$

where L is the numerical operator representing the finite element equations. The residual is defined by

$$L(e) = L(\delta) - L(\tilde{\delta}) = R \tag{3.45}$$

The local operator error at any point in the element is defined by

$$e = \delta - \tilde{\delta} \tag{3.46}$$

However, direct pointwise estimation of error is often difficult and methods to measure the error using the energy norm were introduced. Nevertheless, we need to first define the energy norm production as

$$\begin{aligned} \| \delta \| &= \left(\int_\Omega (S\delta)^T D(S\delta) d\Omega\right)^{0.5} \\ &= \left(\int_\Omega \epsilon^T D\epsilon d\Omega\right)^{0.5} = \left(\int_\Omega \sigma^T D^{-1} \sigma d\Omega\right)^{0.5} \end{aligned} \tag{3.47}$$

and the error in the energy norm is written as:

$$\begin{aligned} \| e \| &= (\| \delta \|^2 - \| \tilde{\delta} \|^2)^{0.5} \\ &= \left(\int_\Omega (S\delta)^T D(S\delta) d\Omega - \int_\Omega (S\tilde{\delta})^T D(S\tilde{\delta}) d\Omega\right)^{0.5} \\ &= \left(\int_\Omega \delta^T b d\Omega + \int_{\Gamma_t} \delta^T t d\Gamma - \int_\Omega \tilde{\delta}^T b d\Omega - \int_{\Gamma_t} \tilde{\delta}^T t d\Gamma\right)^{0.5} \end{aligned} \tag{3.48}$$

The first type of error estimator was proposed by Babuska and co-workers (1978–1984). In this procedure, the error is measured by the response to a set of residual forces. Domain residuals and jumps across the boundaries are used to indicate the error in the energy norm. This method of analysis leads to local rather than global error estimation. As an example of this method, the error for bilinear elements $p = 1$ is obtained as

$$\epsilon^2 = \sum_{i=1}^{m} \epsilon_i^2 = \sum_{i=1}^{m} \left(\frac{h_i^2}{24K} \int_{\Omega_i} r^2 d\Omega + \frac{h_i}{24K} \int_{I_i} J^2 dI \right) \tag{3.49}$$

where ϵ_i is the element error estimator, I_i the element interface, r the residual of Equation 3.39 and J the inter-element traction jump. $K = E/(1-\nu)$ and E and ν are Young's modulus of elasticity and Poisson's ratio, respectively. Explicit determination of the interelement traction jump function and an integral along element interface is required for estimation of error using Equation 3.47.

In the second type of error estimator, the Zienkiewicz–Zhu post-processing method (Zhu et al., 1990; Zienkiewicz and co-workers, 1987–1993), the error is measured by using a recovered higher order finite element solution. In analogy to Equation 3.47, the error in the energy norm is defined as

$$\begin{aligned} \| e \| &= (\int_\Omega (Se)^T D (Se) d\Omega)^{0.5} = (\int_\Omega (\epsilon - \tilde{\epsilon})^T D (\epsilon - \tilde{\epsilon}) d\Omega)^{0.5} \\ &= (\int_\Omega (\sigma - \tilde{\sigma})^T D^{-1} (\sigma - \tilde{\sigma}) d\Omega)^{0.5} \end{aligned} \tag{3.50}$$

The exact solution of the stresses σ at each node is usually not available. Although the Galerkin approximation of displacement $\tilde{u}$ is C^0 continuous, the stresses approximation $\tilde{\sigma}$ does not possess interelement continuity and has a low accuracy at nodes and element boundaries. However, a continuous stress field may be obtained by a recovered higher order solution of stresses σ^* in place of the exact solution. The local norm error, therefore, is written as

$$\begin{aligned} \| \tilde{e} \| &= (\int_\Omega (\epsilon^* - \tilde{\epsilon})^T D (\epsilon^* - \tilde{\epsilon}) d\Omega)^{0.5} \\ &= (\int_\Omega (\sigma^* - \tilde{\sigma})^T D^{-1} (\sigma^* - \tilde{\sigma}) d\Omega)^{0.5} \end{aligned} \tag{3.51}$$

σ^* can be obtained by nodal averaging (Zienkiewicz and Zhu, 1987), least-squares smoothing (Ainsworth et al., 1989; Hinton and Campbell, 1974) or super convergence recovery methods (Zhu and Zienkiewicz, 1990). The recovered higher order solutions, i.e. C^0 continuous approximation solutions of stress, are summed to provide global values and can be expressed in terms of the smoothed nodal stress of an element $\tilde{\sigma}^*$:

$$\sigma^* = \sum_{i=1}^{m} \tilde{N}_i \tilde{\sigma}_i^* \tag{3.52}$$

σ^* is the current best estimate of the exact stress field. To minimize the energy norm error, the interpolation function $\tilde{N}$ is best chosen to be the same order as the shape function N, i.e. $\tilde{N} = N$. With n being the number of elements connected to node i, $\tilde{\sigma}^*$ can be established by the nodal averaging as

$$\tilde{\sigma}_i^* = \frac{1}{n}\sum_{j=1}^{n}(\tilde{\sigma}_i)_j \tag{3.53}$$

Alternatively, in superconvergent recovery method (Zhu and Zienkiewicz, 1990), $\tilde{\sigma}^*$ belongs to a polynomial expansion σ_p^* of the same order of the shape function N. $\tilde{\sigma}^*$ is valid over an element patch surrounding the particular assembly node considered:

$$\sigma_p^* = Pa \tag{3.54}$$

The above expression is valid over an element patch surrounding assembly node considered for error analysis: P contains the polynomial terms (linear or quadratic expansion). a parameters are obtained by a least-squares fit to the sampling points in the element patch, which results in a system of algebraic equations. The patch is assembled for each point inside the region. The weighted 'root mean square (r.m.s.)' percentage error, with reference to energy norm, is represented as

$$\eta\,\% = \frac{\| e \|}{\| \delta \|} \times 100 \tag{3.55}$$

The estimated r.m.s. percentage error (permissible relative energy norm error) is calculated as

$$\tilde{\eta}\,\% = \frac{\| \tilde{e} \|}{(\| \tilde{\delta} \|^2 + \| \tilde{e} \|^2)^{0.5}} \times 100 \tag{3.56}$$

The error estimator is the ratio of estimated error to true error. This effectivity index of the estimator is defined as

$$\theta = \frac{\| \tilde{e} \|}{\| e \|} \tag{3.57}$$

Zienkiewich–Zhu's (1990) post-processing finite element method provides a global error estimation of the relative energy error.

It should be noted that the error estimation method, based on the stress recovery procedure, cannot be easily extended to large deformation formulations. Employing strain differences rather than stresses, Bonet (1994) has suggested a *posteriori* error estimation method for large deformation problems.

3.5 Concluding remarks

Numerical errors originate at the beginning while deriving the problem formulation (operator error, incompatible boundary functions etc.), increase their magnitude when the governing system of equations is assembled for the discretized domain (discretization errors, rounding errors) and become even larger as the response of the approximated model is calculated (rounding errors, solution procedure errors). The operator error and incompatible boundary functions were not discussed here since these are more continuum mechanics issues than computational errors. Test methods were discussed that can be used to detect errors in the solution. However, we have also shown that these test are too simple to capture all the errors and it should be noted that passing these tests does not guarantee an accurate model. To reduce rounding errors, it is strongly advised that double-precision be used for both the data for all constants and for all the arithmetic steps of computing the element matrices, assembling the global matrices and solving them (Cook et al., 1989).

The convergence and accuracy of the solution depends, respectively, on the completeness and compatibility of the elements and on the size of the largest segment connecting two nodes. *Posteriori* error analysis, a promising method with which to quantify solution errors, is only presented briefly as it is still evolving.

PART II

Finite Strip Method

CHAPTER 4

Finite strip method

4.1 Introduction

In Chapter 2, our attention was focused on the basic concepts of the numerical methods and in particular the finite element method. The mathematical approach was introduced to gain an understanding of the finite strip method since the latter can be considered as a special form of finite element method.

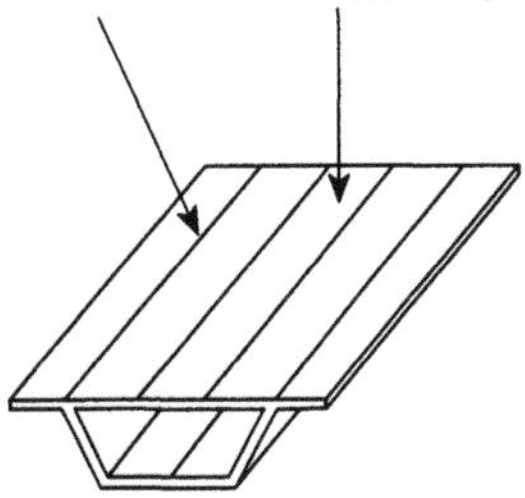

Figure 4.1 *Box structure and finite strips.*

For a structure with constant cross-section and end boundary conditions that do not change transversely, stress analysis can be performed using finite strips (Fig. 4.1) instead of finite elements. In each strip, the displacement components at any point are expressed in terms of the displacement parameters of nodal lines by means of simple polynomials in the transverse direction and continuously differentiable smooth series in the longitudinal direction, with the stipulation that such series should satisfy the boundary conditions at the ends of the strip *a priori*. Using strain–displacement relationships, the strain energy of the structure and the potential energy of external loads can be expressed in terms of the displacement parameters. At equilibrium, the values of the displacement parameters should make the total potential energy of the structure become minimum. This yields a set of linear algebraic equations with the displacement parameters as unknowns. Then, by solving the equations for the displacement parameters, the displacement and stress components at any point in the structure can be obtained.

In this chapter, the basic concepts of the finite strip method for analysis of rectangular plate and box structures are presented. The energy approach is introduced through an elementary application to a simple beam. This is followed by the derivation of various property matrices used in the analysis. The concept is then extended to the analysis of continuous structures with intermediate supports via the flexibility approach. The stiffness approach, which directly uses eigenfunctions of continuous beams as displacement functions, is also introduced for the analysis of structures that are continuous over rigid line supports. Numerical examples are included throughout this chapter to illustrate the performance of the finite strip method.

4.2 Energy approach for a simple beam

The energy approach as related to the finite strip method is best discussed using a simple beam under a point load at midspan (Fig. 4.2). The boundary conditions at both ends ($y = 0$ and $y = l$) are

$$\begin{aligned} w(y) &= 0 \\ M(y) &= -EI\frac{d^2w}{dy^2} = 0 \end{aligned} \tag{4.1}$$

where w and M are the deflection and bending moments, and EI is the flexural rigidity. The deflection of the beam can be approximately

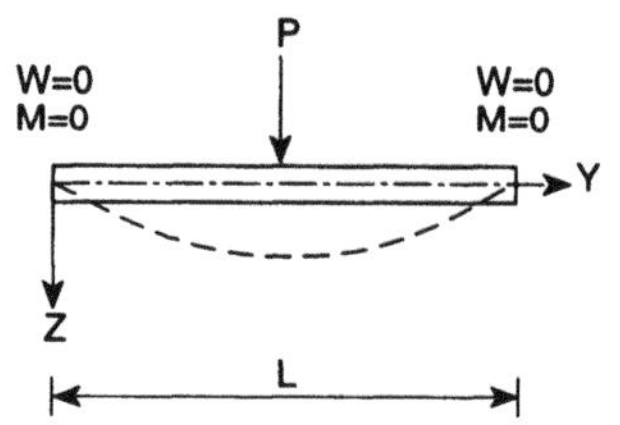

Figure 4.2 *Simple beam under a point load at midspan.*

represented by a half-sine wave:

$$w(y) = \delta \sin\frac{\pi y}{l} \tag{4.2}$$

in which δ is the unknown coefficient.

The assumed displacement function in Equation 4.2 also satisfies the end conditions given by Equation 4.1. In order to obtain the best approximation, the value of δ should make the total potential energy developed in the system become a minimum. For the beam in Fig. 4.2, the total potential energy Π is defined as

$$\Pi = U + W = \frac{EI}{2}\int_0^l (\frac{d^2w}{dy^2})^2 dy - Pw \tag{4.3}$$

where U and W are the strain energy and the potential energy of the load, respectively. Thus, the first derivative of Π with respect to δ must be zero, i.e.

$$\frac{d\Pi}{d\delta} = 0 \tag{4.4}$$

Substitution of Equation 4.2 into 4.3 yields

$$\Pi = \frac{\pi^4 EI\delta^2}{2l^4} \int_0^l \sin^2 \frac{\pi y}{l} dy - P\delta \tag{4.5}$$

or

$$\Pi = \frac{\pi^4 EI\delta^2}{4l^3} - P\delta \tag{4.6}$$

Then, the value of δ is obtained by solving Equation 4.4:

$$\delta = \frac{2Pl^3}{\pi^4 EI} \tag{4.7}$$

The deflection function proposed in Equation 4.2 becomes

$$w(y) = \frac{2Pl^3}{\pi^4 EI} \sin \frac{\pi y}{l} \tag{4.8}$$

from which the bending moment is obtained as

$$M(y) = -EI\frac{d^2 w}{dy^2} = \frac{2Pl}{\pi^2} \sin \frac{\pi y}{l} \tag{4.9}$$

The maximum deflection and bending moments occur at midspan with values of $w_{max} = 0.02053Pl^3/EI$ and $M_{max} = 0.2026Pl$. The exact solutions are $w_{max} = 0.02083Pl^3/EI$ and $M_{max} = 0.2500Pl$. This indicates that the energy approach using one sine function yields a maximum deflection that is in good agreement with the closed-form solution; however, the maximum bending moment is underestimated. This is expected since the exact solution varies linearly from ends to center, which can not be represented by a single half-sine wave.

The accuracy of the analysis can be improved by simulating the deflection curve using a series of sine functions

$$w(y) = \sum_{m=1}^{r} \delta_m \sin \frac{m\pi y}{l} \tag{4.10}$$

where δ_m is the unknown coefficient of the mth series term and r is the number of terms required in the analysis. Applying the energy approach again yields

$$\begin{aligned} w(y) &= \frac{2Pl^3}{\pi^4 EI} \sum_{m=1}^{r} \frac{1}{m^4} \sin \frac{m\pi}{2} \sin \frac{m\pi y}{l} \\ M(y) &= \frac{2Pl}{\pi^2} \sum_{m=1}^{r} \frac{1}{m^2} \sin \frac{m\pi}{2} \sin \frac{m\pi y}{l} \end{aligned} \tag{4.11}$$

Table 4.1 *Deflection and moment at midspan*

m	$w_{max} = \alpha_m \frac{Pl^3}{EI}$ α_m	$M_{max} = \beta_m Pl$ β_m
1	0.02053	0.20264
3	0.02079	0.22516
5	0.02082	0.23326
7	0.02083	0.23740
9	0.02083	0.23990
Beam theory	0.02083	0.25

Since the deformation is symmetrical with respect to the center, the even terms of the series given in Equation 4.11 vanish. The resulting deflection and moment at midspan are listed in Table 4.1. The results illustrate that the convergency of the series for deflection is fast whereas the convergency of the bending moment is still slow. However, the error is only of the order of 4% when five symmetrical terms are used.

4.3 Plate strip

4.3.1 Displacement function

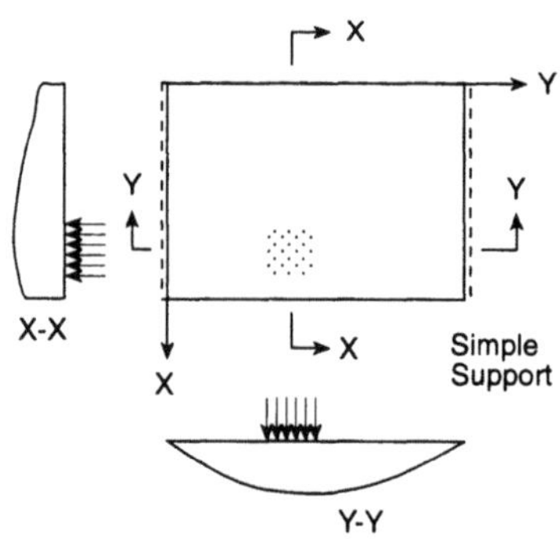

Figure 4.3 *Rectangular plate under transverse loading.*

Figure 4.3 illustrates a simply supported rectangular plate under transverse loading. Although the analysis of the plate as a single element using the energy approach is still possible, it is extremely difficult to postulate a displacement function that can simulate all the possible deformations under any load combinations and boundary conditions. If the plate is divided into a number of longitudinal strips (Fig. 4.4), the deflected surface of each strip

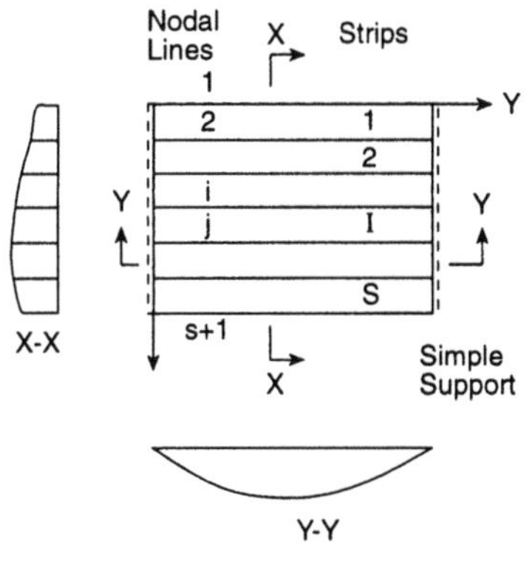

Figure 4.4 *Plate divided into strips.*

may be adequately modeled using simple displacement functions. All the strips are smoothly connected to each other along strip boundaries referred to as nodal lines, and all the displacement fields of the individual strips constitute the deformation pattern of the entire plate. Since the ends of the strip are simply supported, the appropriate displacement function must give zero deflection at both ends. Furthermore, the longitudinal bending moments must also be zero at the support sections. A series of sinusoidal functions can be taken to represent the deformed surface in the longitudinal direction (Fig. 4.4), and polynomial functions in each strip may produce an adequate simulation of the deflected shape in the transverse direction. Thus, the displacement function for a strip becomes a combination of sinusoidal series and polynomial functions and takes the form

$$w(x,y) = \sum_{m=1}^{r} f_m(x) \sin \frac{m\pi y}{l} = \sum_{m=1}^{r} (a_0 + a_1 x + a_2 x^2 + a_3 x^3 + ...) \sin \frac{m\pi y}{l} \tag{4.12}$$

where $a_0, a_2, a_3, \ldots$ are the unknown coefficients, and r is the number of series terms required in the analysis.

To ensure the continuity conditions between adjacent strips more conveniently, $a_0, a_1, a_2, a_3, \ldots$ should be expressed in terms of the displacement parameters. The deflection amplitudes w_{im} and w_{jm} at the two nodal lines i and j of a strip can be chosen as the displacement parameters in the following manner (Fig. 4.5):

$$w_i = \sum_{m=1}^{r} w_{im} \sin \frac{m\pi y}{l}$$

$$w_j = \sum_{m=1}^{r} w_{jm} \sin \frac{m\pi y}{l} \tag{4.13}$$

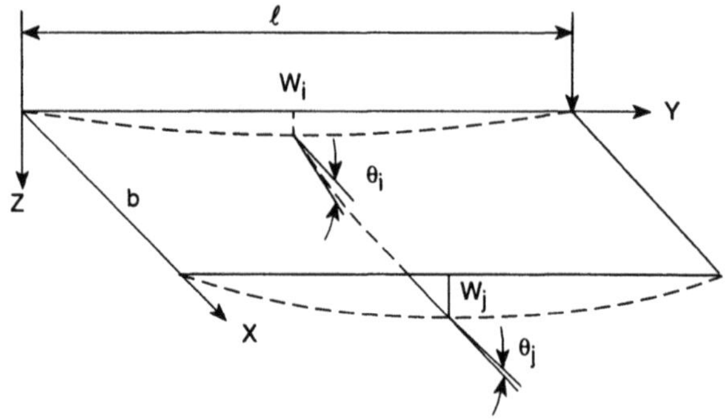

Figure 4.5 *Deflection amplitudes for a plate strip.*

This implies that the two adjacent strips have the same deflection along their common nodal line i since both strips possess the same w_{im}. In order to guarantee a smooth deflected surface for the entire plate, the slope must also be continuous across strip boundaries. Therefore, it is also necessary to take the transverse slope amplitudes θ_{im} and θ_{jm} (Fig. 4.5) at nodal lines i and j as the displacement parameters, which are defined as:

$$\theta_i = (\frac{\partial w}{\partial x})_i = \sum_{m=1}^{r} \theta_{im} \sin \frac{m\pi y}{l}$$

$$\theta_j = (\frac{\partial w}{\partial x})_j = \sum_{m=1}^{r} \theta_{jm} \sin \frac{m\pi y}{l} \tag{4.14}$$

To include the four displacement parameters w_{im}, θ_{im}, w_{jm} and θ_{jm}, a third-order polynomial is required in Equation 4.12:

$$f_m(x) = a_0 + a_1 x + a_2 x^2 + a_3 x^3 \tag{4.15}$$

and $f_m(x)$ must satisfy the conditions

$$at \quad x = 0, \quad f_m(0) = w_{im}, \quad \frac{\partial f_m(0)}{\partial x} = \theta_{im}$$

$$at \quad x = b, \quad f_m(b) = w_{jm}, \quad \frac{\partial f_m(b)}{\partial x} = \theta_{jm}$$

From Equation 4.15, the above conditions lead to

$$\begin{aligned} a_0 &= w_{im} \\ a_1 &= \theta_{im} \\ a_0 + a_1 b + a_2 b^2 + a_3 b^3 &= w_{jm} \\ a_1 + 2a_2 b + 3a_3 b^2 &= \theta_{jm} \end{aligned} \tag{4.16}$$

After solving Equation 4.16 for $a_0, a_1, \ldots$, the displacement function 4.12

can be rewritten as

$$w(x,y) = \sum_{m=1}^{r} (N_1(x)w_{im} + N_2(x)\theta_{im} + N_3(x)w_{jm} + N_4(x)\theta_{jm}) \sin\frac{m\pi y}{l} \tag{4.17}$$

where

$$\begin{aligned} N_1(x) &= (1 - 3X^2 + 2X^3) \\ N_2(x) &= x(1 - 2X + X^2) \\ N_3(x) &= (3X^2 - 2X^3) \\ N_4(x) &= x(X^2 - X) \quad and \\ X &= x/b \end{aligned}$$

or, in a matrix form,

$$w(x,y) = \sum_{m=1}^{r} [N_1, N_2, N_3, N_4] \begin{Bmatrix} w_{im} \\ \theta_{im} \\ w_{jm} \\ \theta_{jm} \end{Bmatrix} \sin\frac{m\pi y}{l} \tag{4.18}$$

More concisely,

$$w(x,y) = \sum_{m=1}^{r} [N]\{\delta\}_m \sin\frac{m\pi y}{l} \tag{4.19}$$

where $\{\delta\}_m$ includes the displacement parameters and $[N]$ is a matrix of transverse shape functions called Hermitian cubic polynomials.

4.3.2 Energy formulation

The strain energy of a plate strip is given by (Timoshenko and Woinowsky-Krieger, 1959)

$$U = \frac{1}{2}\int_0^l \int_0^b (-M_x\frac{\partial^2 w}{\partial x^2} - M_y\frac{\partial^2 w}{\partial y^2} + 2M_{xy}\frac{\partial^2 w}{\partial x \partial y})dxdy \tag{4.20}$$

where M_x, M_y and M_{xy} are the transverse bending moment, the longitudinal bending moment and the twisting moment, respectively. Equation 4.20 can be written in the following matrix form:

$$U = \frac{1}{2}\int_0^l \int_0^b [M_x, M_y, M_{xy}] \begin{Bmatrix} -\dfrac{\partial^2 w}{\partial x^2} \\ -\dfrac{\partial^2 w}{\partial x^2} \\ 2\dfrac{\partial^2 w}{\partial x \partial y} \end{Bmatrix} dxdy = \frac{1}{2}\int_0^l \int_0^b \{M\}^T\{\kappa\}dxdy \tag{4.21}$$

in which $\{M\}$ is the moment vector and $\{\kappa\}$ is the curvature vector.

From Equation 4.19, the curvature vector can be expressed in terms of the displacement parameters as

$$\{\kappa\} = [-\frac{\partial^2 w}{\partial x^2}, -\frac{\partial^2 w}{\partial y^2}, 2\frac{\partial^2 w}{\partial x \partial y}]^T = \sum_{m=1}^{r} [B]_m \{\delta\}_m \tag{4.22}$$

where $[B]_m$ is the strain matrix in the form

$$[B]_m = \begin{bmatrix} -N_1'' \sin k_m y & -N_2'' \sin k_m y & -N_3'' \sin k_m y & -N_4'' \sin k_m y \\ k_m^2 N_1 \sin k_m y & k_m^2 N_2 \sin k_m y & k_m^2 N_3 \sin k_m y & k_m^2 N_4 \sin k_m y \\ 2k_m N_1' \cos k_m y & 2k_m N_2' \cos k_m y & 2k_m N_3' \cos k_m y & 2k_m N_4' \cos k_m y \end{bmatrix} \tag{4.23}$$

in which $N' = \frac{dN}{dx}$, $N'' = \frac{d^2 N}{dx^2}$ and $k_m = m\pi/l$.

The relationships between the moments and the curvatures are as follows:

$$\begin{aligned} M_x &= -(D_x \frac{\partial^2 w}{\partial x^2} + D_1 \frac{\partial^2 w}{\partial y^2}) \\ M_y &= -(D_y \frac{\partial^2 w}{\partial y^2} + D_1 \frac{\partial^2 w}{\partial x^2}) \\ M_{xy} &= 2D_{xy} \frac{\partial^2 w}{\partial x \partial y} \end{aligned} \tag{4.24}$$

where D_x and D_y are the flexural rigidities, D_{xy} is the torsional rigidity, and D_1 is the coupling rigidity. They are defined as

$$\begin{aligned} D_x &= \frac{E_x t^3}{12(1 - \nu_x \nu_y)} \\ D_y &= \frac{E_y t^3}{12(1 - \nu_x \nu_y)} \\ D_{xy} &= \frac{E_{xy} t^3}{12} \\ D_1 &= \nu_x D_x = \nu_y D_y \end{aligned} \tag{4.25}$$

in which E_x, E_y and E_{xy} are the elastic moduli, ν_x is the Poisson's ratio in the x direction (i.e. $\epsilon_x = \nu_x \sigma_y / E_y$ if $\sigma_x = \sigma_z = 0$), ν_y is the Poisson' ratio in the y direction ($\nu_y = \nu_x E_x / E_y$), and t is the thickness of plate strip.

For an isotropic plate, it is only necessary to set $E_x = E_y = E$, $\nu_x = \nu_y = \nu$, and $E_{xy} = G = E/2(1 + \nu)$.

In matrix form, Equation 4.24 becomes

$$\{M\} = \left\{ \begin{array}{c} M_x \\ M_y \\ M_{xy} \end{array} \right\} = \begin{bmatrix} D_x & D_1 & 0 \\ D_1 & D_y & 0 \\ 0 & 0 & D_{xy} \end{bmatrix} \{\kappa\} = [D]\{\kappa\} \tag{4.26}$$

where $[D]$ is referred to as the elasticity matrix.

Substitution of Equation 4.22 into 4.26 yields

$$\{M\} = \sum_{m=1}^{r} [D][B]_m \{\delta\}_m \tag{4.27}$$

Then, by substituting Equations 4.22 and 4.27 into 4.21, the strain energy of the strip can be expressed in terms of the displacement parameters as follows:

$$U = \frac{1}{2} \sum_{m=1}^{r} \sum_{n=1}^{r} \{\delta\}_m^T \int_0^l \int_0^b [B]_m^T [D][B]_n dxdy \{\delta\}_n \tag{4.28}$$

It can be seen from Equation 4.23 that Equation 4.28 contains the following integrations:

$$\int_0^l \sin\frac{m\pi y}{l} \sin\frac{n\pi y}{l} dy = \int_0^l \cos\frac{m\pi y}{l} \cos\frac{n\pi y}{l} dy = \begin{cases} \frac{l}{2} & if \quad m = n \neq 0 \\ 0 & if \quad m \neq n \end{cases}$$

Because of the orthogonal properties of the trigonometrical functions, Equation 4.28 can be simplified as

$$U = \frac{1}{2} \sum_{m=1}^{r} \{\delta\}_m^T \int_0^l \int_0^b [B]_m^T [D][B]_m dxdy \{\delta\}_m \tag{4.29}$$

For the plate strip under distributed transverse load of intensity $q(x, y)$, the potential energy of external loading is defined as

$$W = -\int_0^l \int_0^b q(x, y) w dxdy \tag{4.30}$$

By substituting displacement function 4.19 into the above expression, the potential energy of external loading can also be expressed in terms of the displacement parameters:

$$W = -\sum_{m=1}^{r} \{\delta\}_m^T \int_0^l \int_0^b [N]^T q(x, y) \sin\frac{m\pi y}{l} dxdy \tag{4.31}$$

4.3.3 Stiffness matrix and load vector

After integration, Equations 4.29 and 4.31 are reduced to

$$U = \frac{1}{2} \sum_{m=1}^{r} \{\delta\}_m^T [k]_m \{\delta\}_m \tag{4.32}$$

$$W = -\sum_{m=1}^{r} \{\delta\}_m^T \{p\}_m \tag{4.33}$$

where $[k]_m$ (a 4×4 matrix) and $\{p\}_m$ (a 4×1 vector) are the stiffness matrix and the load vector of the strip, respectively. Their entries are listed in Tables 4.2 and 4.3.

Table 4.2 *Stiffness matrix of plate strip*

$$[k]_m = \begin{bmatrix} [K_{ii}]_b & [K_{ij}]_b \\ [K_{ij}]_b^T & [K_{jj}]_b \end{bmatrix} = \begin{bmatrix} k_1 & \text{symmetrical} & & \\ k_3 & k_2 & & \\ k_4 & k_5 & k_1 & \\ -k_5 & k_6 & -k_3 & k_2 \end{bmatrix}$$

where

$$\begin{aligned} k_1 &= \frac{13lb}{70}k_m^4 D_y + \frac{12l}{5b}k_m^2 D_{xy} + \frac{6l}{5b}k_m^2 D_1 + \frac{6l}{b^3}D_x \\ k_2 &= \frac{lb^3}{210}k_m^4 D_y + \frac{4lb}{15}k_m^2 D_{xy} + \frac{2lb}{15}k_m^2 D_1 + \frac{2l}{b}D_x \\ k_3 &= \frac{11lb^2}{420}k_m^4 D_y + \frac{l}{5}k_m^2 D_{xy} + \frac{3l}{5}k_m^2 D_1 + \frac{3l}{b^2}D_x \\ k_4 &= \frac{9lb}{140}k_m^4 D_y - \frac{12l}{5b}k_m^2 D_{xy} - \frac{6l}{5b}k_m^2 D_1 - \frac{6l}{b^3}D_x \\ k_5 &= \frac{13lb^2}{840}k_m^4 D_y - \frac{l}{5}k_m^2 D_{xy} - \frac{l}{10}k_m^2 D_1 - \frac{3l}{b^2}D_x \\ k_6 &= -\frac{lb^3}{280}k_m^4 D_y - \frac{lb}{15}k_m^2 D_{xy} - \frac{lb}{30}k_m^2 D_1 + \frac{l}{b}D_x \end{aligned}$$

The strain energy of the entire plate and the potential energy of the external loads on the whole structure are obtained by summing the contributions from all strips:

$$U_t = \frac{1}{2}\sum_{I=1}^{S}\sum_{m=1}^{r}\{\delta\}_{Im}^T[k]_{Im}\{\delta\}_{Im} \tag{4.34}$$

$$W_t = -\sum_{I=1}^{S}\sum_{m=1}^{r}\{\delta\}_{Im}^T\{p\}_{Im} \tag{4.35}$$

where S is the number of strips into which the plate is divided, I is the subscript identifying an individual strip, and t is the subscript representing the whole structure.

Exchanging the sequence of summations in Equations 4.34 and 4.35 yields

$$U_t = \frac{1}{2}\sum_{m=1}^{r}\sum_{I=1}^{S}\{\delta\}_{Im}^T[k]_{Im}\{\delta\}_{Im} \tag{4.36}$$

$$W^T = -\sum_{m=1}^{r}\sum_{I=1}^{S}\{\delta\}_{Im}^T\{p\}_{Im} \tag{4.37}$$

In order to carry out the summation over all the strips, a so-called assembling procedure is required.

Table 4.3 *Load vector of plate strip*

1. Point force P_o at (x_o,y_o)

$$\{p\}_m = \begin{Bmatrix} Z_{im} \\ M_{im} \\ Z_{jm} \\ M_{jm} \end{Bmatrix} = \begin{Bmatrix} N_1(x_o) \\ N_2(x_o) \\ N_3(x_o) \\ N_4(x_0) \end{Bmatrix} P_o \sin k_m y_o$$

2. Uniform load of intensity Q_o on the entire strip

$$\{p\}_m = \begin{Bmatrix} Z_{im} \\ M_{im} \\ Z_{jm} \\ M_{jm} \end{Bmatrix} = \begin{Bmatrix} \frac{b}{2} \\ \frac{b^2}{12} \\ \frac{b}{2} \\ -\frac{b^2}{12} \end{Bmatrix} [1-(-1)^m]\frac{Q_o l}{m\pi}$$

3. Patch load of intensity Q_o on area $[x_1,x_2]$ and $[y_1,y_2]$

$$\{p\}_m = \begin{Bmatrix} Z_{im} \\ M_{im} \\ Z_{jm} \\ M_{jm} \end{Bmatrix} = \begin{Bmatrix} \bar{x} - \frac{\bar{x}^3}{b^2} + \frac{\bar{x}^4}{2b^3} \\ \frac{\bar{x}^2}{2} - \frac{2\bar{x}^3}{3b} + \frac{\bar{x}^4}{4b^2} \\ \frac{\bar{x}^3}{b^2} - \frac{\bar{x}^4}{2b^3} \\ -\frac{\bar{x}^3}{3b} + \frac{\bar{x}^4}{4b^2} \end{Bmatrix} Q_o C_m$$

where

$$\begin{aligned} \bar{x}^n &= x_2^n - x_1^n \\ C_m &= \frac{1}{k_m}(\cos k_m y_1 - \cos k_m y_2) \end{aligned}$$

First, for the mth series term, the displacement parameters of the entire plate may be written as an $N \times 1$ vector $\{\delta\}_{tm}$, where N is the total number of parameters. Next, the stiffness matrix $[k]_{Im}$ and the load vector $\{p\}_{Im}$ of strip I can be expanded into an $N \times N$ square matrix $[K]_{Im}$ and an $N \times 1$ vector $\{P\}_{Im}$, respectively. These are constructed by inserting the related stiffness coefficients and loads in their proper locations and filling the remaining locations with zeros as shown in Fig. 4.6. Thus, Equations 4.32 and 4.33 can be written as

$$U_I = \frac{1}{2} \sum_{m=1}^{r} \{\delta\}_{tm}^T [K]_{Im} \{\delta\}_{tm} \tag{4.38}$$

$$W_I = - \sum_{m=1}^{r} \{\delta\}_{tm}^T \{P\}_{Im} \tag{4.39}$$

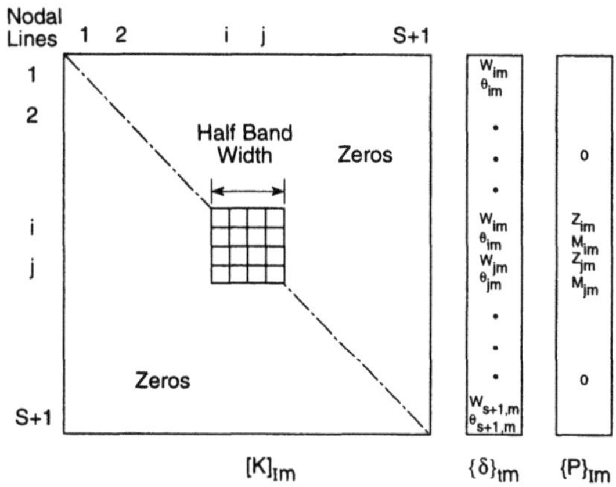

Figure 4.6 *Expanded strip matrices.*

and Equations 4.36 and 4.37 become

$$U_t = \frac{1}{2} \sum_{m=1}^{r} \{\delta\}_{tm}^T (\sum_{I=1}^{S} [K]_{Im}) \{\delta\}_{tm} = \frac{1}{2} \sum_{m=1}^{r} \{\delta\}_{tm}^T [K]_{tm} \{\delta\}_{tm} \tag{4.40}$$

$$W_t = - \sum_{m=1}^{r} \{\delta\}_{tm}^T (\sum_{I=1}^{S} \{P\}_{Im}) = - \sum_{m=1}^{r} \{\delta\}_{tm}^T \{P\}_{tm} \tag{4.41}$$

where $[K]_{tm}$ and $\{P\}_{tm}$ are the overall stiffness matrix and the load vector of the entire structure, respectively.

Since the stiffness matrix of every strip is symmetrical about its diagonal, the overall stiffness matrix is also symmetrical. Moreover, since the strain energy cannot be negative, the overall stiffness matrix is positive-definite.

Furthermore, it can be seen that the overall stiffness matrix has non-zero items in a narrow band symmetrical about the diagonal (for a simply supported plate, the half-bandwidth is four). These properties of the overall stiffness matrix enhance the efficiency of analysis and reduce the requirement for computer storage significantly.

The total potential energy of the entire plate is now expressed as

$$\Pi_t = U_t + W_t = \sum_{m=1}^{r} \left(\frac{1}{2}\{\delta\}_{tm}^T [K]_{tm} \{\delta\}_{tm} - \{\delta\}_{tm}^T \{P\}_{tm}\right) \tag{4.42}$$

4.3.4 Minimization of total potential energy

In order that the total potential energy becomes minimum, its first derivative with respect to every displacement parameter must be zero, i.e.

$$\begin{aligned} \frac{\partial \Pi_t}{\partial w_{im}} &= 0 \\ \frac{\partial \Pi_t}{\partial \theta_{im}} &= 0 \quad ; \quad i = 1, N' \end{aligned} \tag{4.43}$$

where N' is the total number of nodal lines ($N' = S + 1$ for a plate). In matrix form, Equation 4.43 can be written as

$$\frac{\partial \Pi_t}{\partial \{\delta\}_{tm}} = \{0\} \tag{4.44}$$

Substituting Equation 4.42 into 4.44 yields a set of linear algebraic equations for the mth series term:

$$[K]_{tm} \{\delta\}_{tm} = \{P\}_{tm} \tag{4.45}$$

which can be solved by means of standard subprogram for solution of linear equations with positive-definite band symmetric matrix.

After solving Equation 4.45 for the unknown displacement parameters $\{\delta\}_{tm}$ of all the series terms, the displacements and bending moments at any point inside the structure can be obtained from Equations 4.19 and 4.27.

4.3.5 General cases of end support conditions

Displacement functions

In the general case of end support conditions, the displacement function of a plate strip may take the form

$$w(x, y) = \sum_{m=1}^{r} [N] Y_m(y) \{\delta\}_m \tag{4.46}$$

Table 4.4 μ_1 *to* μ_{12} *for clamped ends*

$\mu_1 l$	4.730040744862704	$\mu_7 l$	23.56194490204045
$\mu_2 l$	7.853204624095837	$\mu_8 l$	26.70353755550818
$\mu_3 l$	10.99560783800167	$\mu_9 l$	29.84513020910325
$\mu_4 l$	14.13716549125746	$\mu_{10} l$	32.98672286269282
$\mu_5 l$	17.27875965739948	$\mu_{11} l$	36.12831551628262
$\mu_6 l$	20.42035224562606	$\mu_{12} l$	39.26990816987241

where $[N]$ is the matrix of shape functions of x (see Equation 4.19), and $Y_m(y)$ is the mth term of a series of functions of y which must satisfy the end support conditions *a priori*.

The most commonly used series are the beam eigenfunctions which are derived from the solution of the beam vibration differential equation

$$\frac{d^4Y}{dy^4} = \mu^4 Y \tag{4.47}$$

The general form of the beam eigenfunctions is

$$Y(y) = c_1 \sin(\mu y) + c_2 \cos(\mu y) + c_3 \sinh(\mu y) + c_4 \cosh(\mu y) \tag{4.48}$$

with the coefficients c_i to be determined by the end conditions. These have been worked out explicitly for the various end conditions:

1. Both ends simply supported, i.e. $Y(0) = Y''(0) = 0$ and $Y(l) = Y''(l) = 0$:

$$Y_m(y) = \sin(\mu_m y),$$
$$(\mu_m = \frac{m\pi}{l})$$

2. Both ends clamped, i.e. $Y(0) = Y'(0) = 0$ and $Y(l) = Y'(l) = 0$:

$$Y_m(y) = \sin(\mu_m y) - \sinh(\mu_m y) - \alpha_m[\cos(\mu_m y) - \cosh(\mu_m y)],$$
$$\alpha_m = \frac{\sin(\mu_m l) - \sinh(\mu_m l)}{\cos(\mu_m l) - \cosh(\mu_m l)}$$

   $\mu_m l$ are the solutions of the equation $1 - \cos(\mu l)\cosh(\mu l) = 0$, the first 12 values of which are listed in Table 4.4. For m greater than 12, $\mu_m = (m + 0.5)\pi/l$ can be taken as a very close approximation.

3. One end simply supported and the other end clamped, i.e.

$$Y(0) = Y''(0) = 0 \quad \textit{and} \quad Y(l) = Y'(l) = 0:$$

$$Y_m(y) = \sin(\mu_m y) - \alpha_m \sinh(\mu_m y),$$
$$\alpha_m = \frac{\sin(\mu_m l)}{\sinh(\mu_m l)}$$

Table 4.5 μ_1 *to* μ_{10} *for clamped-free ends*

$\mu_1 l$	1.87510406871196	$\mu_6 l$	17.27875953208823
$\mu_2 l$	4.69409113297417	$\mu_7 l$	20.42035225104125
$\mu_3 l$	7.85475743823761	$\mu_8 l$	23.56194490180644
$\mu_4 l$	10.99554073487547	$\mu_9 l$	26.70353755551830
$\mu_5 l$	14.13716839104647	$\mu_{10} l$	29.84513020910282

$\mu_m l$ are the solutions of the equation $\tan(\mu l) = \tanh(\mu l)$. The first four values are 3.92660231204792, 7.06858274562873, 10.21017612281303 and 13.35176877775409; $\mu_m l = \frac{1}{4}\pi + m\pi$ can be employed for $m \geq 5$.

4. Both ends free, i.e. $Y''(0) = Y'''(0) = 0$, $Y''(l) = Y'''(l) = 0$:

$$Y_1(y) = 1, \mu_1 = 0,$$

$$Y_2(y) = 1 - \frac{2y}{l}, \mu_2 = 1,$$

$$Y_m(y) = \sin(\mu_m y) + \sinh(\mu_m y) - \alpha_m[\cos(\mu_m y) + \cosh(\mu_m y)],$$

$$\alpha_m = \frac{\sin(\mu_m l) - \sinh(\mu_m l)}{\cos(\mu_m l) - \cosh(\mu_m l)}$$

in which μ_m $(m \geq 3)$ is equal to μ_{m-2} in case 2.

5. One end clamped and the other end free, i.e.

$$Y(0) = Y'(0) = 0 \quad and \quad Y''(l) = Y'''(l) = 0 \ :$$

$$Y_m(y) = \sin(\mu_m y) - \sinh(\mu_m y) - \alpha_m[\cos(\mu_m y) - \cosh(\mu_m y)],$$

$$\alpha_m = \frac{\sin(\mu_m l) + \sinh(\mu_m l)}{\cos(\mu_m l) + \cosh(\mu_m l)}$$

$\mu_m l$ are the solutions of the equation $1 + \cos(\mu l)\cosh(\mu l) = 0$, the first 10 values are listed in Table 4.5.

For m greater than 10, $\mu_m l = (m - 0.5)\pi$ can be used with satisfactory accuracy.

6. One end simply supported and the other end free, i.e.

$$Y(0) = Y''(0) = 0 \quad and \quad Y''(l) = Y'''(l) = 0 \ :$$

$$Y_1(y) = \frac{y}{l}, \mu_1 = 1,$$

$$Y_m(y) = \sin(\mu_m y) + \alpha_m \sinh(\mu_m y),$$

$$\alpha_m = \frac{\sin(\mu_m l)}{\sinh(\mu_m l)}$$

μ_m $(m \geq 2)$ are identical to μ_{m-1} in case 3.

It can be proved that the beam eigenfunctions possess the valuable properties of orthogonality, i.e.

$$\int_0^l Y_m Y_n dy = 0 \qquad for \;\; m \neq n$$

$$\int_0^l Y_m'' Y_n'' dy = 0 \qquad for \;\; m \neq n$$

Utilization of these properties will result in a significant saving in computational effort for the calculation of stiffness matrices.

Stiffness matrix and load vector

From the assumed displacement function 4.48, the curvature vector may be expressed in terms of the displacement parameters as

$$\{\kappa\} = [-\frac{\partial^2 w}{\partial x^2}, -\frac{\partial^2 w}{\partial y^2}, 2\frac{\partial^2 w}{\partial x \partial y}]^T = \sum_{m=1}^{r} [B]_m \{\delta\}_m \tag{4.49}$$

where

$$[B]_m = \begin{bmatrix} -N_1''Y_m(y) & -N_2''Y_m(y) & -N_3''Y_m(y) & -N_4''Y_m(y) \\ -N_1Y_m''(y) & -N_2Y_m''(y) & -N_3Y_m''(y) & -N_4Y_m''(y) \\ 2N_1'Y_m'(y) & 2N_2'Y_m'(y) & 2N_3'Y_m'(y) & 2N_4'Y_m'(y) \end{bmatrix} \tag{4.50}$$

The moment vector, the strain energy of the strip and the potential energy of the external loading can then be expressed as

$$\{M\} = \sum_{m=1}^{r} [D][B]_m \{\delta\}_m \tag{4.51}$$

$$U = \frac{1}{2} \sum_{m=1}^{r} \sum_{n=1}^{r} \{\delta\}_m^T \int_0^l \int_0^b [B]_m^T [D][B]_n dx dy \{\delta\}_n \tag{4.52}$$

$$W = -\sum_{m=1}^{r} \{\delta\}_m^T \int_0^l \int_0^b [N]^T q(x,y) Y_m(y) dx dy \tag{4.53}$$

After integration, Equations 4.52 and 4.53 become

$$U = \frac{1}{2} \sum_{m=1}^{r} \sum_{n=1}^{r} \{\delta\}_m^T [k]_{mn} \{\delta\}_n \tag{4.54}$$

$$W = -\sum_{m=1}^{r} \{\delta\}_m^T \{p\}_m \tag{4.55}$$

in which

$$[k]_{mn} = \int_0^l \int_0^b [B]_m^T [D][B]_n dx dy \tag{4.56}$$

and

$$\{p\}_m = \int_0^l \int_0^b [N]^T q(x,y) Y_m(y) dx dy \tag{4.57}$$

From Equation 4.50, it can be seen that Equation 4.56 includes the following integrals:

$$I_1 = \int_0^l Y_m Y_n dy \qquad I_2 = \int_0^l Y_m'' Y_n'' dy$$

$$I_3 = \int_0^l Y_m' Y_n' dy \qquad I_4 = \int_0^l Y_m Y_n'' dy$$

It is already known that integrals I_1 and I_2 are equal to zero for $m \neq n$, whereas integrals I_3 and I_4 will not in general vanish for $m \neq n$. Consequently, the different series terms are coupled, i.e. $[K]_{mn} \neq [0]$ for $m \neq n$, unless the plate is simply supported.

The overall matrix equation for the entire structure and all the series terms takes the form

$$\begin{bmatrix} [K]_{t11} & [K]_{t12} & \dots & [K]_{t1r} \\ [K]_{t21} & [K]_{t22} & \dots & [K]_{t2r} \\ \vdots & \vdots & \dots & \vdots \\ [K]_{tr1} & [K]_{tr2} & \dots & [K]_{trr} \end{bmatrix} \begin{Bmatrix} \{\delta\}_{t1} \\ \{\delta\}_{t2} \\ \vdots \\ \{\delta\}_{tr} \end{Bmatrix} = \begin{Bmatrix} \{P\}_{t1} \\ \{P\}_{t2} \\ \vdots \\ \{P\}_{tr} \end{Bmatrix} \tag{4.58}$$

In programming, in order to take advantage of the banded nature of the stiffness matrix, the displacement parameters should be grouped in the following sequence:

$$\{\delta\}_t = [..., w_{i1}, w_{i2}, ..., w_{ir}, \theta_{i1}, \theta_{i2}, ..., \theta_{ir}, ...]^T \tag{4.59}$$

Once Equation 4.58 has been solved for all the unknown displacement parameters (related to all nodal lines and all series terms), the displacements and moments at any point of the structure can readily be calculated using Equations 4.46 and 4.51.

4.4 Plane stress strip

4.4.1 Displacement functions

For a plate subjected to in-plane forces (Fig. 4.7), all the stresses in the direction of its thickness are negligible, and the plate mid-plane remains plane after deformation. According to the theory of two-dimensional elasticity (Timoshenko and Goodier, 1970), the strain–

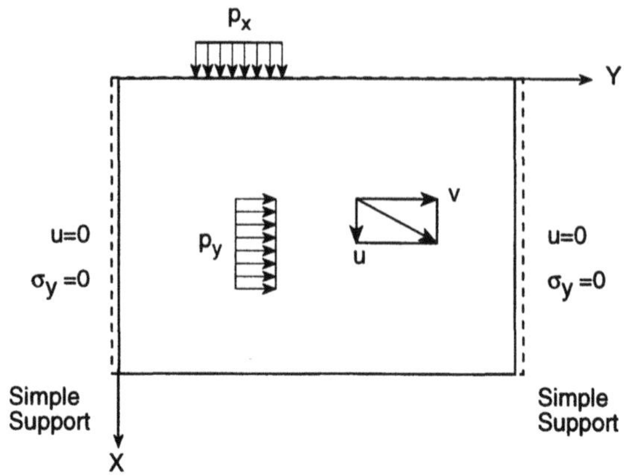

Figure 4.7 *Plate under in-plane loading.*

displacement relationships take the form

$$\{\epsilon\} = \left\{ \begin{array}{c} \epsilon_x \\ \epsilon_y \\ \gamma_{xy} \end{array} \right\} = \left\{ \begin{array}{c} \dfrac{\partial u}{\partial x} \\ \\ \dfrac{\partial v}{\partial y} \\ \\ \dfrac{\partial u}{\partial y} + \dfrac{\partial v}{\partial x} \end{array} \right\} \tag{4.60}$$

where u and v are the displacement components in the x and y directions, respectively. The stresses are related to the strains as

$$\begin{aligned} \{\sigma\} &= \left\{ \begin{array}{c} \sigma_x \\ \sigma_y \\ \tau_{xy} \end{array} \right\} = \frac{1}{1 - \nu_x \nu_y} \left[\begin{array}{ccc} E_x & \nu_x E_x & 0 \\ \nu_y E_y & E_y & 0 \\ 0 & 0 & (1 - \nu_x \nu_y) E_{xy} \end{array} \right] \left\{ \begin{array}{c} \epsilon_x \\ \epsilon_y \\ \gamma_{xy} \end{array} \right\} \\ &= [D_p]\{\epsilon\} \end{aligned} \tag{4.61}$$

where $[D_p]$ is the elasticity matrix for the in-plane stress analysis.

For a rectangular plate with two simply supported ends, the boundary conditions are

$$u = 0 \quad and \quad \sigma_y = 0 \quad at \quad y = 0 \quad and \quad y = l \tag{4.62}$$

In this case, the plate can be analyzed using plane stress strips (Fig. 4.8). At any point in the strip, the displacement components can be represented as

$$\left\{ \begin{array}{c} u \\ v \end{array} \right\} = \sum_{m=1}^{r} \left\{ \begin{array}{c} (c_0 + c_1 x + ...) \sin k_m y \\ (d_0 + d_1 x + ...) \cos k_m y \end{array} \right\} \tag{4.63}$$

in which $k_m = m\pi/l$ and c_0, c_1, d_0, $d_1, \ldots$ are unknown coefficients.

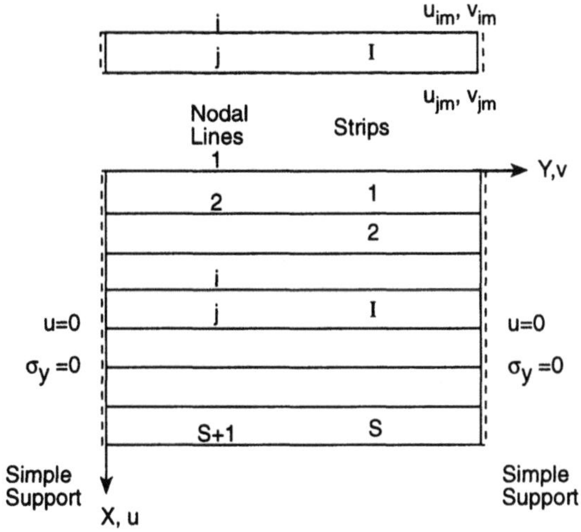

Figure 4.8 *Plane stress strips.*

Derivation shows that Equation 4.63 satisfies the boundary conditions given in Equation 4.62.

By following the procedure used for the plate strip, the coefficients c_0, c_1, d_0, $d_,$ ... can be expressed in terms of the displacement amplitudes u_{im}, v_{im}, u_{jm} and v_{jm} along nodal lines i and j, and Equation 4.63 is rewritten as

$$\left\{ \begin{array}{c} u \\ v \end{array} \right\} = \sum_{m=1}^{r}$$

$$\left[\begin{array}{cccc} (1-X)\sin k_m y & 0 & X \sin k_m y & 0 \\ 0 & (1-X)\cos k_m y & 0 & X\cos k_m y \end{array} \right] \left\{ \begin{array}{c} u_{im} \\ v_{im} \\ u_{jm} \\ v_{jm} \end{array} \right\}$$

$$= \sum_{m=1}^{r} [N]_m \{\delta\}_m \tag{4.64}$$

where $X = x/b$, $[N]_m$ is the matrix of displacement functions, and $\{\delta\}_m = [u_{im}, v_{im}, u_{jm}, v_{jm}]^T$ is the vector of displacement parameters.

4.4.2 Stiffness matrix and load vector

The strain energy in the plane stress strip is defined as (Timoshenko and Goodier, 1970)

$$U = \frac{t}{2}\int_0^l \int_0^b (\sigma_x \epsilon_x + \sigma_y \epsilon_y + \tau_{xy}\gamma_{xy})dxdy = \frac{t}{2}\int_0^l \int_0^b \{\epsilon\}^T \{\sigma\} dxdy \tag{4.65}$$

By substituting the displacement function 4.64 into the strain–displacement relationships 4.60 and then into the stress–strain relationships 4.61, the strain and stress vectors can be expressed in terms of displacement parameters as

$$\{\epsilon\} = \sum_{m=1}^{r} [B]_m \{\delta\}_m \tag{4.66}$$

and

$$\{\sigma\} = \sum_{m=1}^{r} [D_p][B]_m \{\delta\}_m \tag{4.67}$$

in which $[B]_m$ is the strain matrix in the following form:

$$[B]_m = \begin{bmatrix} -\frac{1}{b}\sin k_m y & 0 & \frac{1}{b}\sin k_m y & 0 \\ 0 & -(1-X)k_m \sin k_m y & 0 & -Xk_m \sin k_m y \\ (1-X)k_m \cos k_m y & -\frac{1}{b}\cos k_m y & Xk_m \cos k_m y & \frac{1}{b}\cos k_m y \end{bmatrix} \tag{4.68}$$

By substituting Equation 4.66 and 4.67 into 4.65, the strain energy can be written as

$$U = \frac{t}{2} \sum_{m=1}^{r} \{\delta\}_m^T \int_0^l \int_0^b [B]_m^T [D_p][B]_m dxdy \{\delta\}_m \tag{4.69}$$

Similarly, the potential energy of external loading can also be expressed in terms of the displacement parameters as

$$W = -\int_0^l \int_0^b [u, v] \begin{Bmatrix} p_x \\ p_y \end{Bmatrix} dxdy = -\sum_{m=1}^{r} \{\delta\}_m^T \int_0^l \int_0^b [N]_m^T \begin{Bmatrix} p_x \\ p_y \end{Bmatrix} dxdy \tag{4.70}$$

where p_x and p_y are the components of load distributed in the plane of the strip. Thus, the stiffness matrix and the load vector of the strip are obtained and given by

$$[k]_m = t \int_0^l \int_0^b [B]_m^T [D_p][B]_m dxdy \tag{4.71}$$

and

$$\{p\}_m = \int_0^l \int_0^b [N]_m^T \begin{Bmatrix} p_x \\ p_y \end{Bmatrix} dxdy \tag{4.72}$$

The elements of $[k]_m$ and $\{p\}_m$ are listed in Tables 4.6 and 4.7, respectively.

Following the assembling and minimization procedures described in Section 4.3, a set of linear algebraic equations can be established as follows:

$$[K]_{tm} \{\delta\}_{tm} = \{P\}_{tm} \tag{4.73}$$

Solving the equations simultaneously for all the unknown displacement

Table 4.6 *Stiffness matrix of the plane stress strip*

$$[k]_m = \begin{bmatrix} [K_{ii}]_p & [K_{ij}]_p \\ [K_{ij}]_p^T & [K_{jj}]_p \end{bmatrix} = t \begin{bmatrix} k_1 & \text{symmetrical} & & \\ k_3 & k_2 & & \\ k_4 & -k_6 & k_1 & \\ k_6 & k_5 & -k_3 & k_2 \end{bmatrix}$$

in which

$$k_1 = \frac{l}{2b}E_1 + \frac{lbk_m^2}{6}E_{xy} \qquad k_4 = -\frac{l}{2b}E_1 + \frac{lbk_m^2}{12}E_{xy}$$

$$k_2 = \frac{l}{2b}E_{xy} + \frac{lbk_m^2}{6}E_2 \qquad k_5 = -\frac{l}{2b}E_{xy} + \frac{lbk_m^2}{12}E_2$$

$$k_3 = \frac{lk_m}{4}\nu_x E_1 - \frac{lk_m}{4}E_{xy} \qquad k_6 = \frac{lk_m}{4}\nu_x E_1 + \frac{lk_m}{4}E_{xy}$$

and

$$E_1 = \frac{E_x}{1 - \nu_x \nu_y} \qquad E_2 = \frac{E_y}{1 - \nu_x \nu_y}$$

Table 4.7 *Load vector of the plane stress strip*

1. Point force (F_x, F_y) at (x_o, y_o)

$$\{p\}_m = \begin{Bmatrix} X_{im} \\ Y_{im} \\ X_{jm} \\ Y_{jm} \end{Bmatrix} = \begin{Bmatrix} F_x(1 - x_o/b)\sin k_m y_o \\ F_y(1 - x_o/b)\cos k_m y_o \\ F_x(x_o/b)\sin k_m y_o \\ F_y(x_o/b)\cos k_m y_o \end{Bmatrix}$$

2. Line load of intensity q_x on line $x = x_o$ from y_1 to y_2

$$\{p\}_m = \begin{Bmatrix} X_{im} \\ Y_{im} \\ X_{jm} \\ Y_{jm} \end{Bmatrix} = \begin{Bmatrix} 1 - x_o/b \\ 0 \\ x_o/b \\ 0 \end{Bmatrix} C_m$$

where

$$C_m = \frac{q_x}{k_m}(\cos k_m y_1 - \cos k_m y_2)$$

3. Uniform load of intensity $(p_x, 0)$ in area of the entire strip

$$\{p\}_m = \begin{Bmatrix} X_{im} \\ Y_{im} \\ X_{jm} \\ Y_{jm} \end{Bmatrix} = \begin{Bmatrix} \frac{1}{2} \\ 0 \\ \frac{1}{2} \\ 0 \end{Bmatrix} [1 - (-1)^m]\frac{p_x lb}{m\pi}$$

parameters related to all the nodal lines and all the series terms, and then substituting the results in Equations 4.64 and 4.67, yields the displacements and stresses at any point in the in-plane plate.

4.4.3 General cases of end support conditions

As a general case of end support conditions, the displacement function of an plane stress strip takes the form

$$\begin{Bmatrix} u \\ v \end{Bmatrix} = \sum_{m=1}^{r} \begin{bmatrix} (1-X)Y_m(y) & 0 & XY_m(y) & 0 \\ 0 & (1-X)Y'_m(y)/\mu_m & 0 & XY'_m(y)/\mu_m \end{bmatrix} \begin{Bmatrix} u_{im} \\ v_{im} \\ u_{jm} \\ v_{jm} \end{Bmatrix}$$

$$= \sum_{m=1}^{r} [N]_m \{\delta\}_m \tag{4.74}$$

where $Y_m(y)$ is the beam eigenfunction; its expressions for a variety of end support conditions have been listed in Section 4.3.5.

By substituting the displacement function 4.74 into the strain–displacement relationships, the strain vector, Equation 4.60, can be expressed in terms of displacement parameters as

$$\{\epsilon\} = \sum_{m=1}^{r} [B]_m \{\delta\}_m \tag{4.75}$$

in which

$$[B]_m = \begin{bmatrix} -\frac{1}{b}Y_m(y) & 0 & \frac{1}{b}Y_m(y) & 0 \\ 0 & (1-X)Y''_m(y)/\mu_m & 0 & XY''_m(y)/\mu_m \\ (1-X)Y'_m(y) & -\frac{1}{b}Y'_m(y)/\mu_m & XY'_m(y) & \frac{1}{b}Y'_m(y)/\mu_m \end{bmatrix} \tag{4.76}$$

It can be seen that the shear strain γ_{xy} will vanish in the entire strip if

$$u_{im} = u_{jm} = (v_{jm} - v_{im})/b\mu_m$$

and the longitudinal strain ϵ_y becomes zero on any given straight line $X = x/b = X_0$ (X_0 is any given constant) provided that

$$(1 - X_0)v_{im} + X_0 v_{jm} = 0$$

This implies that the assumed displacement function 4.74 is capable of simulating pure in-plane bending combined with axial elongation. This is necessary to capture the deformation pattern of deep beams and webs of bridges. This complies with beam theory; under this deformation, the shear stress is zero in the entire beam and the longitudinal stress should be zero along the so-called neutral axis.

The displacement function 4.74 has been successfully applied in dynamic analysis, and accurate predictions for frequencies and mode shapes have been obtained for all cases of end support conditions. However, for static analysis, Equation 4.74 is valid only for the simply supported case. For example, the function for the clamped-clamped case requires that $Y_m(y) = 0$ and $Y'_m(y) = 0$ at both ends of the strip. From Equation 4.76 it can be noted that

$$\tau_{xy} = E_{xy}\gamma_{xy} = E_{xy}\sum_{m=1}^{r}(...)Y'_m(y)$$

will be equal to zero at two ends. This means that there is no reaction at the supports to resist the vertical load.

In order to solve this problem, another type of function for the clamped-clamped case was developed and is given by

$$\left\{ \begin{array}{c} u \\ v \end{array} \right\} = \sum_{m=1}^{r}$$

$$\begin{bmatrix} (1-X)\sin k_m y & 0 & X\sin k_m y & 0 \\ 0 & (1-X)\sin k_{m+1} y & 0 & X\sin k_{m+1} y \end{bmatrix} \left\{ \begin{array}{c} u_{im} \\ v_{im} \\ u_{jm} \\ v_{jm} \end{array} \right\}$$

$$= \sum_{m=1}^{r} [N]_m \{\delta\}_m \tag{4.77}$$

in which $k_{m+1} = (m+1)\pi/l$.

From the displacement functions 4.74 or 4.77, the stiffness matrix and the load vector can be obtained as

$$[k]_{mn} = t\int_0^l \int_0^b [B]_m^T [D_p][B]_n dxdy \tag{4.78}$$

$$\{p\}_m = \int_0^l \int_0^b [N]_m^T \left\{ \begin{array}{c} p_x \\ p_y \end{array} \right\} dxdy \tag{4.79}$$

In the general case, $[k]_{mn}$ is not a null matrix for $m \neq n$, unless the two ends of the strip are simply supported. This means the different series terms are coupled in the stiffness matrix.

4.5 Flat shell strip

4.5.1 Stiffness matrix and force vector

A prismatic folded plate structure or box girder bridge (Fig. 4.1) is an assembly of rectangular plates that can offer resistance to both bending and in-plane loadings. This type of structure can be analyzed using the flat shell strip, which is formed by combining the plate strip and the plane stress strip so that it is capable of simulating both the bending and in-plane deformations.

The total potential energy of the flat shell strip is the algebraic sum of those from both the in-plane and bending deformations. For instance, if a strip is simply supported at two opposite ends, the total potential energy can be obtained as

$$\Pi = U + W = \frac{1}{2}\sum_{m=1}^{r}\{\delta\}_m^T[k]_m\{\delta\}_m - \sum_{m=1}^{r}\{\delta\}_m^T\{p\}_m \tag{4.80}$$

where $\{\delta\}_m = [u_{im}, v_{im}, w_{im}, \theta_{im}, u_{jm}, v_{jm}, w_{jm}, \theta_{jm}]^T$ is the vector of the displacement parameters in the strip coordinate system,

$$[k]_m = \begin{bmatrix} [K_{ii}]_p & 0 & [K_{ij}]_p & 0 \\ 0 & [K_{ii}]_b & 0 & [K_{ij}]_b \\ [K_{ij}]_p^T & 0 & [K_{jj}]_p & 0 \\ 0 & [k_{ij}]_b^T & 0 & [K_{jj}]_b \end{bmatrix} \tag{4.81}$$

in which the sub–matrices are given in Tables 4.2 and 4.6, and $\{p\}_m = [X_{im}, Y_{im}, Z_{im}, M_{im}, X_{jm}, Y_{jm}, Z_{jm}, M_{jm}]^T$ is the load vector in the strip coordinate system, the components are listed in Tables 4.3 and 4.7.

4.5.2 Coordinate transformation

The above derivations are carried out in a local coordinate system, wherein the x and y axes coincide with the mid-surface of a strip. In folded plate structures, any two plates will in general meet at an angle and, in order to assemble the stiffness matrices, the displacement vectors and the load vectors of non-coplanar strips, a global coordinate system common to all the strips must be established.

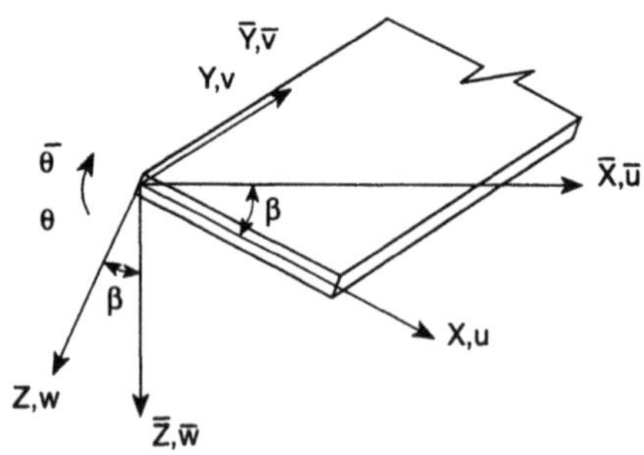

Figure 4.9 *Coordinate transformation.*

In Fig. 4.9 x, y and z are individual coordinates of a strip, and $\bar{x}$, $\bar{y}$ and $\bar{z}$ are the global coordinates; y and $\bar{y}$ coincide with the intersection line of two adjoining strips. The local and global displacement components are related as

$$\begin{aligned} u_{im} &= \bar{u}_{im}\cos\beta + \bar{w}_{im}\sin\beta \\ v_{im} &= \bar{v}_{im} \\ w_{im} &= -\bar{u}_{im}\sin\beta + \bar{w}_{im}\cos\beta \\ \theta_{im} &= \bar{\theta}_{im} \end{aligned} \tag{4.82}$$

This relationship is also applicable for nodal line j. Thus, for a strip, the following transformation is obtained:

$$\{\delta\}_m = \begin{bmatrix} [t] & 0 \\ 0 & [t] \end{bmatrix} \{\bar{\delta}\}_m = [T]\{\bar{\delta}\}_m \tag{4.83}$$

where

$$[t] = \begin{bmatrix} \cos\beta & 0 & \sin\beta & 0 \\ 0 & 1 & 0 & 0 \\ -\sin\beta & 0 & \cos\beta & 0 \\ 0 & 0 & 0 & 1 \end{bmatrix} \tag{4.84}$$

By substituting Equation 4.83 into 4.80, the total potential energy of the strip can be expressed in terms of the global displacement vector and global load vector as

$$\begin{aligned} \Pi &= \frac{1}{2}\sum_{m=1}^{r}\{\bar{\delta}\}_m^T[T]^T[k]_m[T]\{\bar{\delta}\}_m - \sum_{m=1}^{r}\{\bar{\delta}\}_m^T[T]^T\{p\}_m \\ &= \frac{1}{2}\sum_{m=1}^{r}\{\bar{\delta}\}_m^T[\bar{k}]_m\{\bar{\delta}\}_m - \sum_{m=1}^{r}\{\bar{\delta}\}_m^T\{\bar{p}\}_m \end{aligned} \tag{4.85}$$

Thus, the stiffness matrix and the load vector of the strip in the global system are obtained as

$$[\bar{k}_m] = [T]^T[k_m][T] \tag{4.86}$$

$$\{\bar{p}\}_m = [T]^T\{p\}_m \tag{4.87}$$

By following the assembling and minimization procedures described earlier, a set of linear algebraic equations for the entire box structure is established as

$$[\bar{K}]_{tm}\{\bar{\delta}\}_{tm} = \{\bar{P}\}_{tm} \tag{4.88}$$

After the global displacement parameters related to all the nodal lines and all the series terms are obtained from Equation 4.88, Equation 4.83 is used to transform back to the local coordinate system. The displacements, internal moments and stresses at any point in the structure can then be computed using Equations 4.19, 4.27, 4.64 and 4.67.

The above procedure is also valid for flat shell strip with other end support conditions, and the transformation can be carried out by means of Equations 4.83, 4.87 4.89.

$$[\bar{k}_{mn}] = [T]^T[k_{mn}][T] \tag{4.89}$$

4.6 Boundary conditions along nodal lines

In plate or box structures, some longitudinal edges or intermediate lines may be simply supported or clamped. Moreover, in some structures, the geometry and deformation may be symmetrical about a longitudinal center line. In the finite strip simulation of such structures, these edges or lines should be represented by nodal lines; and the corresponding boundary conditions along these lines are imposed.

4.6.1 Simply supported nodal lines

For a simply supported longitudinal edge, the boundary conditions are:

$$\begin{aligned} w &= 0 \\ M_x &= -(D_x\frac{\partial^2 w}{\partial x^2} + D_1\frac{\partial^2 w}{\partial y^2}) = 0 \end{aligned} \tag{4.90}$$

Since w is a constant along this edge, $\frac{\partial^2 w}{\partial y^2}$ vanishes on this line. Thus, Equation 4.90 becomes

$$\begin{aligned} w &= 0 \\ \frac{\partial^2 w}{\partial x^2} &= 0 \end{aligned} \tag{4.91}$$

The first condition in Equation 4.90 is referred to as a displacement boundary condition, which has to be satisfied fully according to the principle of variation. However, the second condition merely reflects the equilibrium requirement on the boundary and is called a natural boundary condition. The natural boundary condition may not be satisfied by assumed displacement field *a priori* but shall be achieved by convergence of the solution through refining the mesh or increasing the order of displacement functions etc.

According to Equation 4.19, along simply supported nodal line i, the deflection is

$$w_i = \sum_{m=1}^{r} w_{im}Y_m(y) \tag{4.92}$$

It is apparent that, if the first condition in Equation 4.91 is to be satisfied, it is only necessary to impose $w_{im} = 0$ for all the series terms. The simplest way to make $w_{im} = 0$ is to assign an infinite value (say 10^{30}) to the item on the leading diagonal of the overall stiffness matrix, which corresponds to w_{im}. Then the solution of the matrix equation automatically leads to $w_{im} = 0$.

4.6.2 Clamped nodal lines

For a clamped nodal line i, the boundary conditions are

$$\begin{aligned} w &= 0 \\ \theta &= \frac{\partial w}{\partial x} = 0 \end{aligned} \tag{4.93}$$

Because the deflection and transverse slope amplitudes (w_{im} and θ_{im} respectively) are chosen as nodal displacement parameters for the plate strip, it is only necessary to assign $w_{im} = 0$ and $\theta_{im} = 0$ when solving the overall matrix equation for any series term m in order to satisfy the clamped conditions in Equation 4.93.

4.6.3 Symmetrical deformation

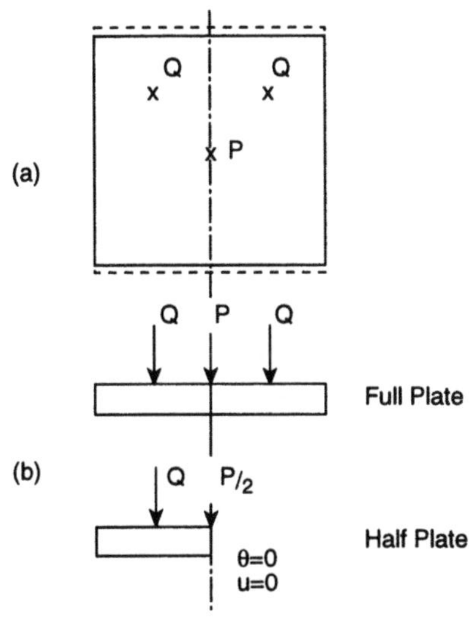

Figure 4.10 *Symmetrical deformation of a plate.*

Some plate structures are symmetrical about the longitudinal center-line as the one shown in Fig. 4.10. If the external loads and the longitudinal supports are also symmetrical with respect to the center-line, then, the deformation of the structure is expected to be symmetrical about this line, and the transverse slope along the center-line will be zero. In this case, only half of the structure needs to be analyzed provided the condition $\theta_{im} = 0$ along the right edge (Fig. 4.10(b)) is imposed.

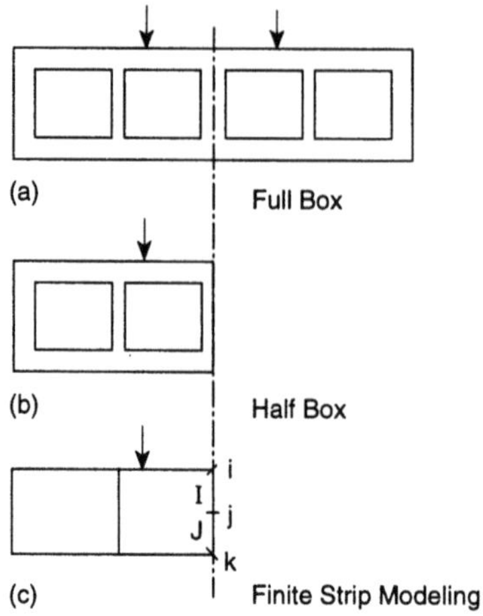

Figure 4.11 *Symmetrical deformation of a box.*

This method is also applicable to the symmetrical deformation of box structures (Fig. 4.11). In this case, only half the thickness of the center web (i.e. strips I and J) should be taken in the analysis. Furthermore, the transverse slope amplitude ($\bar{\theta}_{im}$) and the horizontal displacement amplitude ($\bar{u}_{im}$) should be set to zero along nodal lines i, j and k.

4.7 Continuous structures: flexibility approach

For continuous plate and box structures with intermediate supports, The flexibility approach can be used combined with the finite strip method. In this approach, the continuous structure is first released from all the intermediate supports that are replaced by a set of redundant forces. The structure deforms under both the external loading and the redundant forces. The deflections at all the supported points due to external loads and each unit redundant force are respectively calculated using the finite strip method. The total deflections should be

$$\begin{aligned}
a_1 &= f_{11}r_1 + f_{12}r_2 + ... + f_{1n}r_n + \delta_1 = 0 \\
a_2 &= f_{21}r_1 + f_{22}r_2 + ... + f_{2n}r_n + \delta_2 = 0 \\
&\cdot \\
&\cdot \\
&\cdot \\
a_n &= f_{n1}r_1 + f_{n2}r_2 + ... + f_{nn}r_n + \delta_n = 0
\end{aligned} \tag{4.94}$$

where n is the number of intermediate supports, a_i is the total deflection at supports i, f_{ij} is the deflection at point i due to a unit force at point j, r_i is the redundant force assumed to be the point force or uniformly distributed rectangular patching load centered at point i, and δ_i is the deflection at point i due to external loading.

Thus, the following matrix equation is obtained:

$$\begin{bmatrix} f_{11} & f_{12} & \cdots & f_{1n} \\ f_{21} & f_{22} & \cdots & f_{2n} \\ \cdot & & & \\ f_{n1} & f_{n2} & \cdots & f_{nn} \end{bmatrix} \begin{Bmatrix} r_1 \\ r_2 \\ \cdot \\ r_n \end{Bmatrix} + \begin{Bmatrix} \delta_1 \\ \delta_2 \\ \cdot \\ \delta_n \end{Bmatrix} = \{0\} \tag{4.95}$$

or

$$[f]\{r\} + \{\delta\} = \{0\} \tag{4.96}$$

from which the redundant forces can be solved. It is then possible to evaluate the displacements and stresses at any point of the released structure under both the external loading and redundant forces.

4.8 Continuous structures: stiffness approach

For plate and box structures that are continuous over rigid line supports, the direct stiffness approach has been developed. Using the eigenfunctions of the continuous beam, this approach converges very quickly when evaluating the maximum negative bending moments at intermediate supported sections. However, because different series terms are coupled in the stiffness matrix, this approach is not economical when evaluating the local bending effects of concentrated loads.

4.8.1 Continuous beam eigenfunctions

The free vibration differential equation of the continuous beam is

$$\frac{d^4Y}{dy^4} = \mu^4 Y \tag{4.97}$$

Because the entire beam vibrates at the same frequency, μ is identical for all the spans. The general solution of this equation for the ith span can then be expressed as

$$Y_i(y) = c_{i1}\sin(\mu y) + c_{i2}\cos(\mu y) + c_{i3}\sinh(\mu y) + c_{i4}\cosh(\mu y) \tag{4.98}$$

in which $c_{i1}, \ldots, c_{i4}$ are constants to be determined from the end conditions of the span.

If the first span is simply supported at the end $y_1 = 0$ (Fig. 4.12(a)), the end conditions are $Y_1(0) = Y_1''(0) = 0$, $Y_1(l_1) = 0$. Equation 4.98 then becomes

$$Y_1(y) = A_1\left(\sin(\mu y_1) - \alpha_1 \sinh(\mu y_1)\right) \tag{4.99}$$

and

$$\alpha_1 = \sin(\mu l_1)/\sinh(\mu l_1)$$

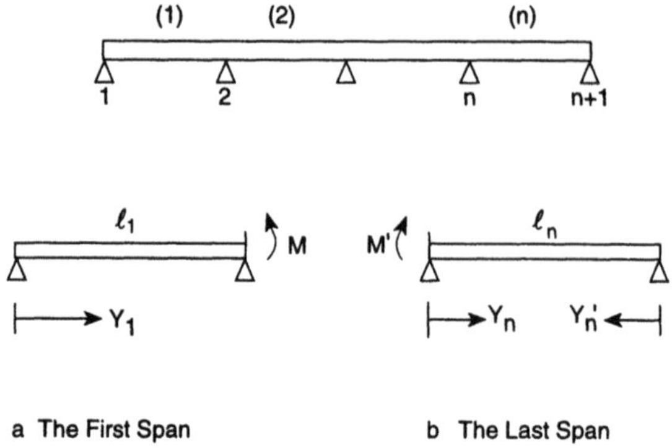

Figure 4.12 *Continuous beam.*

For the last span n (Fig. 4.12(b)), in order to simplify mathematic procedures, it is convenient to use a new coordinate system, $y_n' = l_n - y_n$, to express the general solution for this span. Thus

$$Y_n(y_n') = A_n \left(\sin(\mu y_n') - \alpha_n \sinh(\mu y_n')\right) \tag{4.100}$$

and

$$\alpha_n = \sin(\mu l_n)/\sinh(\mu l_n)$$

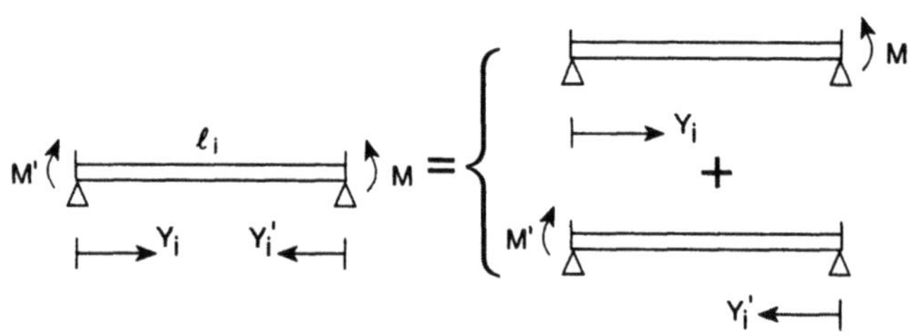

Figure 4.13 *Span i of a continuous beam.*

Each intermediate span i (Fig. 4.13) can be treated as a combination of the above-mentioned two situations, i.e.,

$$Y_i(y) = A_{i1} \left(\sin(\mu y_i) - \alpha_i \sinh(\mu y_i)\right) + A_{i2} \left(\sin(\mu y_i') - \alpha_i \sinh(\mu y_i')\right) \tag{4.101}$$

where

$$\alpha_i = \sin(\mu l_i)/\sinh(\mu l_i)$$

and

$$y_i' = l_i - y_i.$$

Evidently, the first span and the last span of a continuous beam can be considered as particular cases of an intermediate span, with $A_{12} = 0$ and

$A_{n1} = 0$, respectively. Therefore, Equation 4.101 can be used to express the solutions for all spans. The slope, dY_i/dy, and the curvature, d^2Y_i/dy^2, of a continuous beam can then be obtained as

$$\frac{dY_i}{dy} = \mu A_{i1}\left(\cos(\mu y_i) - \alpha_i \cosh(\mu y_i)\right) - \mu A_{i2}\left(\cos(\mu y_i') - \alpha_i \cosh(\mu y_i')\right) \tag{4.102}$$

$$\frac{d^2Y_i}{dy^2} = -\mu^2 A_{i1}\left(\sin(\mu y_i) + \alpha_i \sinh(\mu y_i)\right) - \mu^2 A_{i2}\left(\sin(\mu y_i') + \alpha_i \sinh(\mu y_i')\right) \tag{4.103}$$

For simplicity, the following notation is introduced:

$$SL_i = \cos(\mu l_i) - \alpha_i \cosh(\mu l_i) \tag{4.104}$$

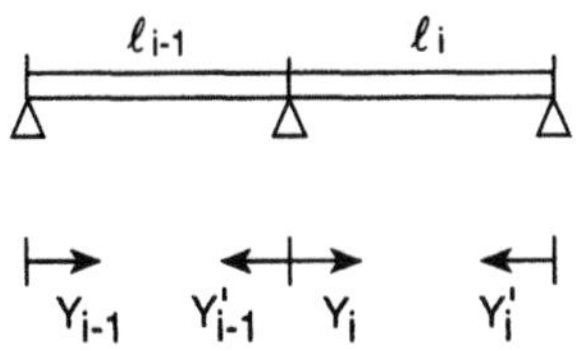

Figure 4.14 *Support i of a continuous beam.*

The continuity conditions between span $(i-1)$ and span i at support i (Fig. 4.14, $y_{i-1} = l_{i-1}$, $y_{i-1}' = 0$, $y_i = 0$, $y_i' = l_i$) are as follows:

1. The slope of adjacent spans must be equal. Therefore, by virtue of Equation 4.102 and 4.104, the following can be obtained:

$$A_{i-1,1}SL_{i-1} - A_{i-1,2}(1-\alpha_{i-1}) = A_{i1}(1-\alpha_i) - A_{i2}SL_i \tag{4.105}$$

2. The bending moment and, consequently, the curvatures must also be continuous over support i. From Equation 4.103 and the expression for α_i, the following equation is obtained:

$$A_{i-1,1}\sin(\mu l_{i-1}) = A_{i2}\sin(\mu l_i) \tag{4.106}$$

The values of μ and A_{ij} (i=1, ..., n, j=1,2) have to satisfy continuity conditions 4.105 and 4.106 at all the intermediate supports. Further, μ must also satisfy the end conditions of the beam:

$$A_{12} = 0 \quad , \quad A_{n1} = 0 \tag{4.107}$$

Equation 4.107 implies that $M_y = 0$ at both ends of the beam.

The A_{ij} represent only the mode shape, not its size; hence the following values may be adopted, without loss of generality:

$$A_{11} = 1, \quad A_{12} = 0 \tag{4.108}$$

For a given μ, A_{i1} and A_{i2} can be evaluated from the known values $A_{i-1,1}$

and $A_{i-1,2}$ by using Equations 4.105 and 4.106. Evaluating a span at a time, all the A_{ij} up to A_{n1} and A_{n2}, which are the functions of μ, can be calculated.

If all the A_{ij} are not simultaneously zero, and the value of μ yields

$$A_{n1}(\mu) = 0 \tag{4.109}$$

then μ and A_{ij} have satisfied both the continuity conditions and the end conditions, therefore these values give a valid solution.

Solutions for μ and A_{ij} can be obtained using existing numerical methods for solving algebraic equation, such as the first-order Reguli–Falsi iteration technique. However, special attention should be given to ensure that no roots are missed (Cheung and Li, 1988A.)

4.8.2 Continuous flat shell strip

To analyze a continuous slab on girder or box girder bridge using the direct stiffness approach, the continuous flat shell strips are required. For the mth series term, the displacement parameters of this strip are chosen as

$$\{\delta\}_m = [u_{im}\ v_{im}\ w_{im}\ \theta_{im}\ u_{jm}\ v_{jm}\ w_{jm}\ \theta_{jm}]^T$$

Within the strip, the displacements are interpolated as

$$\begin{aligned} u &= \sum_{m=1}^{r} ((1-X)u_{im} + Xu_{jm})\, Y_m(y) \\ v &= \sum_{m=1}^{r} ((1-X)v_{im} + Xv_{jm})\, Y'_m(y)/\mu_m \\ w &= \sum_{m=1}^{r} (N_1 w_{im} + N_2\theta_{im} + N_3 w_{jm} + N_4\theta_{jm}) Y_m(y) \end{aligned} \tag{4.110}$$

where Y_m is the eigenfunction of a continuous beam with the span lengths of the actual bridge.

Assuming the displacement functions 4.110, and following the procedure given in Section 4.5, the stiffness matrix and load vector for each flat shell strip can be formed. The final solution is obtained by solving the overall matrix equation

$$[\bar{K}]_t\{\bar{\delta}\}_t = \{\bar{P}\}_t \tag{4.111}$$

which is related to all the nodal lines and all the series terms.

4.9 Numerical examples

1. Simply supported square plate under uniform load

A square plate with side length a is simply supported along the four edges, and is subjected to a uniform load of intensity p. For symmetry,

Table 4.8 *Simply supported plate under uniform load*

No. of strips	No. of terms	No. of DOF	wD/pa^4	M_x/pa^2	M_y/pa^2
1	1	4	0.004140	0.05608	0.05400
	2	8	0.004089	0.05459	0.04945
2	1	6	0.004111	0.05016	0.05197
	2	12	0.004060	0.04867	0.04744
	3	18	0.004065	0.04897	0.04847
3	1	8	0.004110	0.04957	0.05178
	2	16	0.004059	0.04805	0.04724
	3	24	0.004063	0.04836	0.04827
	4	32	0.004062	0.04825	0.04789
Theory			0.00406	0.0479	0.0479

only half the plate needs to be analyzed, and only symmetrical series terms with m being an odd number are included in the finite strip simulation. One, two and three plate strips of equal width are employed in the analysis. The results of maximum deflection and bending moments are presented in Table 4.8, in which $D = Et^3/12(1-\nu^2)$ is the flexural rigidity of the plate, and $\nu = 0.3$.

For comparison, the analytical results (Timoshenko and Woinowsky-Krieger, 1959) are also given in the same table. It is apparent that the finite strip method converges very rapidly especially for the central deflection.

2. Simple beam modeled by plane stress strips

A simple beam of span length $l = 10$ m, depth $h = 1.0$ m and thickness $t = 0.2$ m is subjected to a uniform load of intensity $q = 100$ kN m^{-1} from above. The material properties are $E = 30000$ MPa and $\nu = 0.2$. The beam is modeled using one, two and three plane stress strips of equal width in order to take into consideration the shear effects. Only symmetrical series terms are required in the analysis since the deformation is symmetrical about the midspan. The resulting maximum deflections, longitudinal stresses and shear stresses are given in Table 4.9. The beam theory solution including shear effects is also given in same table. It can be observed that the use of only one strip and one series term is enough to produce satisfactory results for the maximum deflection and longitudinal stress with less than 4% error, whereas three strips and ten series terms are needed to yield a value for the maximum shear stress with an error of less than 5%.

Table 4.9 *Simple beam modeled by plane strips*

No. of strips	No. of terms	w_{max} (m)	σ_{ymax} (MPa)	τ_{yzmax} (MPa)
1	1	0.02562	38.70	2.05
	2	0.02549	37.27	2.29
	3	0.02550	37.58	2.39
	5	0.02550	37.52	2.49
	10	0.02550	37.50	2.60
2	1	0.02639	39.08	2.16
	2	0.02621	37.65	2.40
	3	0.02622	37.95	2.49
	5	0.02627	37.89	2.57
	10	0.02627	37.87	2.61
3	1	0.02651	39.11	2.93
	2	0.02643	37.65	3.26
	3	0.02645	37.97	3.38
	5	0.02644	37.90	3.48
	10	0.02644	37.89	3.57
Theory		0.02664	37.50	3.75

3. Box beam under self-weight

A simple box beam of span length $l = 20.0$ m deforms under self-weight. The outside width and depth are respectively 2.2 m and 1.2 m. The thickness of each wall is 0.2 m. The material properties are $E = 30000$ MPa, $\nu = 0.2$ and $\gamma = 24.0$ kN m^{-3}. By virtue of symmetry, only half of the structure is to be modeled, and only symmetrical series terms are necessary. The flanges and web are divided into strips of identical width. The maximum values of deflection, longitudinal stress and shear stress are obtained as shown in Table 4.10. The solutions based on beam theory are also given in the table. It can be noted that the web can be adequately modeled by only one strip in order to obtain the satisfactory maximum longitudinal stresses (at the bottom surface of midspan). However, in this case, the result for maximum shear stress (at the web center of end section) has an error of 8% when 20 symmetrical series terms are used. If more accurate shear stress is of interest, more strips should be used for each web. For maximum deflection, the finite strip solutions converge to a value larger than the beam theory solution, since the latter does not include transverse shear deformation.

Table 4.10 *Simple box beam under self-weight*

No. of strips for each web	No. of terms	w_{max} (mm)	σ_{ymax} (MPa)	τ_{yzmax} (MPa)
1	1	8.83	3.79	0.577
	2	8.78	3.64	0.640
	5	8.78	3.67	0.680
	10	8.78	3.66	0.691
	20	8.78	3.66	0.696
2	1	8.89	3.81	0.582
	3	8.84	3.65	0.646
	5	8.84	3.69	0.686
	10	8.84	3.68	0.696
	20	8.84	3.68	0.700
3	1	8.90	3.81	0.614
	3	8.85	3.66	0.682
	5	8.85	3.66	0.726
	10	8.85	3.69	0.738
	20	8.85	3.68	0.743
Beam Theory		8.33	3.60	0.756

4. Simply supported continuous plate

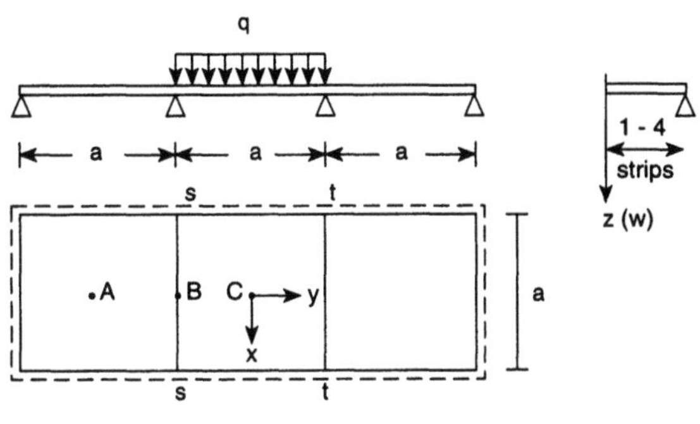

Figure 4.15 *Simply supported continuous plate.*

A rectangular plate of width a and length $3a$, simply supported along four edges and also along the intermediate lines ss and tt forms a simply supported continuous plate over three spans (Fig. 4.15). The middle span

Table 4.11 *Deflections and moments of a continuous plate.*

Number of strips	Number of terms	w_A	w_C	M_{yB}	M_{xC}	M_{yC}
1	3	−0.0612	0.2916	−0.0275	0.0372	0.0417
	9	−0.0586	0.2918	−0.0350	0.0367	0.0378
	15	−0.0585	0.2922	−0.0369	0.0369	0.0394
	21	−0.0585	0.2922	−0.0375	0.0368	0.0368
2	3	−0.0609	0.2892	−0.0273	0.0327	0.0405
	9	−0.0584	0.2893	−0.0347	0.0322	0.0367
	15	−0.0583	0.2898	−0.0365	0.0324	0.0378
	21	−0.0583	0.2897	−0.0372	0.0323	0.0374
4	3	−0.0609	0.2891	−0.0273	0.0324	0.0404
	9	−0.0583	0.2892	−0.0346	0.0346	0.0366
	15	−0.0583	0.2897	−0.0365	0.0320	0.0377
	21	−0.0583	0.2896	−0.0372	0.0319	0.0373
Exact		−0.0583	0.2896	−0.0381	0.0317	0.0375
Multiplier		qa^4/D			qa^2	

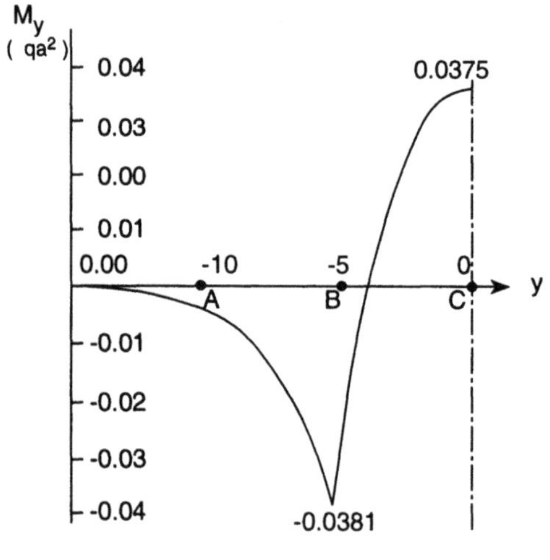

Figure 4.16 *Variation of M_y along the line $x = 0$.*

is uniformly loaded while the two side spans are unloaded. A series of calculations for half a plate were carried out for different meshes, using different number of series terms. The results are shown in Table 4.11 in comparison with the exact solution of thin plate theory (Timoshenko and Woinowsky–Krieger, 1959).

It can be seen that w_A, w_C and M_{xC} require only three terms and two strips in order to obtain reasonable accurate results, while M_{yC}, M_{yB} need 9 and 15 terms, respectively to yield values with errors less than 5%. This implies that more series terms are required to compute the bending moment in the y direction. The distribution of the longitudinal bending moment M_y along the longitudinal center line $x = 0$ is shown in Fig. 4.16.

The outcome of this example dictates that the total number of series terms is more critical than the number of strips for the accuracy of longitudinal bending moment, while the number of strips has greater effect than the number of series terms on the accuracy of M_x.

CHAPTER 5

Higher order finite strips

5.1 Introduction

In Chapter 4, cubic polynomials are used as transverse shape functions for the strips in plate-bending analysis. These shape functions ensure compatibility of deflection and transverse slopes at strip boundaries, but do not guarantee continuity of the curvature between adjacent strips. Therefore, the resulting transverse bending moment is often discontinuous across strip boundaries, and not equal to zero on the free edges of a plate. Similar situations also happen to the strips for in-plane stress analysis.

This problem can be solved by using higher order polynomials as transverse shape functions instead of increasing the number of strips. In this aspect, there exist two independent approaches:

1. increasing the degrees of freedom at each nodal line;
2. introducing an internal nodal line.

These two procedures are described in this chapter. The equations are given in the form for the strips with both ends simply supported only. However, the concepts are also applicable for strips with other boundary conditions.

5.2 Plate strip HO2 with curvature continuity

In this strip (Fig. 5.1(a)), the transverse curvature amplitude is included as a nodal displacement parameter. Thus, the curvatures and consequently the bending moments become continuous across the strip boundaries, and more accurate results are expected. However, for plates with an abrupt change in thickness or material properties, the curvature should not be continuous, and this strip becomes inappropriate. Moreover, this approach is also not applicable for box structures because more than two strips may join together on one nodal line where the transverse curvature and bending moment are both discontinuous. Therefore, the application of this approach is basically limited to constant thickness plates.[1]

[1] In some references e.g. (Brown & Ghali, 1975), this type of strip is also used for analysis of box girder bridges, but the higher-order degrees of freedom at each slab girder joint are eliminated before the individual stiffness matrices are assembled.

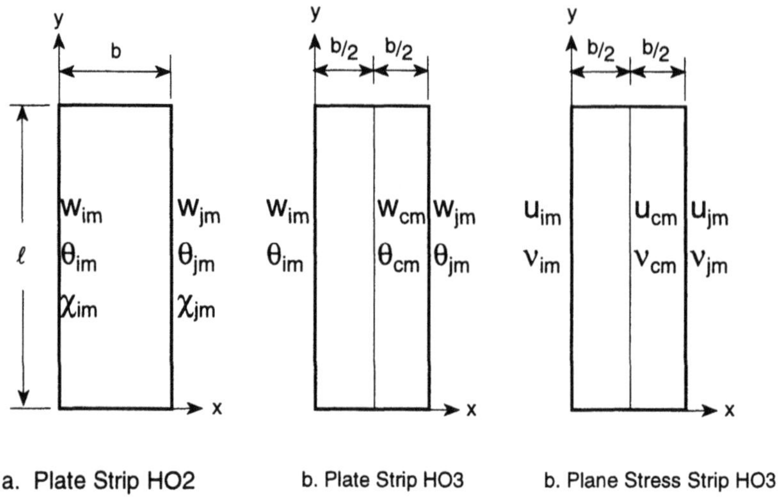

Figure 5.1 *Higher order finite strips: (a) plate strip HO2; (b) plate strip HO3; (c) plane stress strip HO3.*

Within this strip, the deflection at any point is interpolated as follows:

$$w = \sum_{m=1}^{r} [N]_m \{\delta\}_m = \sum_{m=1}^{r} [N_1, N_2, N_3, N_4, N_5, N_6] \{\delta\}_m \sin \frac{m\pi y}{l} \tag{5.1}$$

where

$$\begin{aligned} N_1 &= 1 - 10X^3 + 15X^4 - 6X^5 \\ N_2 &= x(1 - 6X^2 + 8X^3 - 3X^4) \\ N_3 &= 0.5x^2(1 - 3X + 3X^2 - X^3) \\ N_4 &= 10X^3 - 15X^4 + 6X^5 \\ N_5 &= x(-4X^2 + 7X^3 - 3X^4) \\ N_6 &= 0.5x^2(X - 2X^2 + X^3) \quad and \\ X &= x/b \end{aligned} \tag{5.2}$$

$$\{\delta\}_m = [w_{im}, \theta_{im}, \chi_{im}, w_{jm}, \theta_{jm}, \chi_{jm}]^T \tag{5.3}$$

in which $\chi_{im} = (\partial^2 w/\partial x^2)_{im}$ is the curvature amplitude at nodal line i for the m-th series term.

Using the above displacement functions, the curvature vector can be expressed in terms of the displacement parameters as

$$\{\kappa\} = [-\frac{\partial^2 w}{\partial x^2}, -\frac{\partial^2 w}{\partial y^2}, 2\frac{\partial^2 w}{\partial x \partial y}]^T = \sum_{m=1}^{r} [B]_m \{\delta\}_m \tag{5.4}$$

$$[B]_m = \begin{bmatrix} -N_1'' \sin k_m y & -N_2'' \sin k_m y & -N_3'' \sin k_m y \\ k_m^2 N_1 \sin k_m y & k_m^2 N_2 \sin k_m y & k_m^2 N_3 \sin k_m y \\ 2k_m N_1' \cos k_m y & 2k_m N_2' \cos k_m y & 2k_m N_3' \cos k_m y \end{bmatrix.$$
$$\left. \begin{matrix} -N_4'' \sin k_m y & -N_5'' \sin k_m y & -N_6'' \sin k_m y \\ k_m^2 N_4 \sin k_m y & k_m^2 N_5 \sin k_m y & k_m^2 N_6 \sin k_m y \\ 2k_m N_4' \cos k_m y & 2k_m N_5' \cos k_m y & 2k_m N_6' \cos k_m y \end{matrix} \right] \tag{5.5}$$

in which $k_m = m\pi/l$.

The moment vector becomes

$$\{M\} = \sum_{m=1}^{r} [D][B]_m\{\delta\}_m \tag{5.6}$$

Then, the stiffness matrix is obtained as

$$[k]_m = \int_0^l \int_0^b [B]_m^T [D][B]_m dxdy \tag{5.7}$$

The elements of this matrix are listed in Table 5.1.

The load vector $\{p\}_m$ of the strip is calculated according to

$$\{p\}_m = \int_0^l \int_0^b [N]_m^T q(x,y) dxdy \tag{5.8}$$

For a uniform load of intensity Q_o on the entire strip,

$$\{p\}_m = \left[\frac{b}{2}, \frac{b^2}{10}, \frac{b^3}{120}, \frac{b}{2}, -\frac{b^2}{10}, \frac{b^3}{120}\right]^T [1-(-1)^m]\frac{Q_o l}{m\pi} \tag{5.9}$$

For a point load P_o acting at (x_o, y_o)

$$\{p\}_m = [N_1(x_o), N_2(x_o), ..., N_6(x_o)]^T P_o \sin k_m y_o \tag{5.10}$$

5.3 Plate strip HO3 with an internal nodal line

In this strip (Fig. 5.1 (b)), the displacement parameters for each nodal line remain unchanged from the basic plate strip, but one internal nodal line c is introduced and usually placed at midway between the two longitudinal edges. The displacement functions also incorporate a fifth-order polynomial, as follows:

$$w = \sum_{m=1}^{r} [N]\{\delta\}_m = \sum_{m=1}^{r} [N_1, N_2, N_3, N_4, N_5, N_6]\{\delta\}_m \sin k_m y \tag{5.11}$$

where

$$\{\delta\}_m = [w_{im}, \theta_{im}, w_{cm}, \theta_{cm}, w_{jm}, \theta_{jm}]^T \tag{5.12}$$

$$\begin{aligned} N_1 &= 1 - 23X^2 + 66X^3 - 68X^4 + 24X^5 \\ N_2 &= x(1 - 6X + 13X^2 - 12X^3 + 4X^4) \end{aligned}$$

Table 5.1 *Stiffness matrix of higher order plate strip HO2*

$$[k]_m = \begin{bmatrix} k_1 & & & & & \\ k_4 & k_2 & & \text{symmetrical} & & \\ k_5 & k_6 & k_3 & & & \\ k_7 & k_8 & k_9 & k_1 & & \\ -k_8 & k_{10} & k_{11} & -k_4 & k_2 & \\ k_9 & -k_{11} & k_{12} & k_5 & -k_6 & k_3 \end{bmatrix}$$

where

$$\begin{aligned}
k_1 &= \frac{181bA}{462} + \frac{10B}{7b} + \frac{20C}{7b} + \frac{120D}{7b^3} \\
k_2 &= \frac{52b^3A}{3465} + \frac{8bB}{35} + \frac{16bC}{35} + \frac{192D}{35b} \\
k_3 &= \frac{b^5A}{9240} + \frac{b^3B}{630} + \frac{b^3C}{315} + \frac{3bD}{35} \\
k_4 &= \frac{311b^2A}{4620} + \frac{3B}{14} + \frac{10C}{7} + \frac{60D}{7b^2} \\
k_5 &= \frac{281b^3A}{55440} + \frac{bB}{84} + \frac{bC}{42} + \frac{3D}{7b} \\
k_6 &= \frac{23b^4A}{18480} + \frac{b^2B}{60} + \frac{b^2C}{30} + \frac{11D}{35} \\
k_7 &= \frac{25bA}{231} - \frac{10B}{7b} - \frac{20C}{7b} - \frac{120D}{7b^3} \\
k_8 &= \frac{151b^2A}{4620} - \frac{3B}{14} - \frac{3C}{7} - \frac{60D}{7b^2} \\
k_9 &= \frac{181b^3A}{55440} - \frac{bB}{84} - \frac{bC}{42} - \frac{3D}{7b} \\
k_{10} &= -\frac{19b^3A}{1980} - \frac{bB}{70} - \frac{bC}{35} + \frac{108D}{35b} \\
k_{11} &= -\frac{13b^4A}{13860} - \frac{b^2B}{210} - \frac{b^2C}{105} + \frac{4D}{35} \\
k_{12} &= \frac{b^5A}{11088} + \frac{b^3B}{1260} + \frac{b^3C}{630} + \frac{bD}{70}
\end{aligned}$$

where

$$A = \frac{1}{2}lk_m^4 D_y \qquad B = 2lk_m^2 D_{xy}$$

$$C = \frac{1}{2}lk_m^2 D_1 \qquad D = \frac{1}{2}lD_x$$

$$\begin{aligned} N_3 &= 16X^2 - 32X^3 + 16X^4 \\ N_4 &= x(-8X + 32X^2 - 40X^3 + 16X^4) \\ N_5 &= 7X^2 - 34X^3 + 52X^4 - 24X^5 \\ N_6 &= x(-X + 5X^2 - 8X^3 + 4X^4) \quad and \\ X &= x/b \end{aligned} \tag{5.13}$$

For this strip, the strain matrix $[B]_m$ has the same form as given in Equation 5.5, but with new definitions for N_1, $N_2, \ldots$ and N_6. Then, the stiffness matrix $[k]_m$ is readily obtained. Its elements are given explicitly in Table 5.2.

The load vector $\{p\}_m$ of the strip can be calculated using Equation 5.8 with new definition of $N_1, \ldots, N_6$. For a uniform load of intensity Q_o on the entire strip,

$$\{p\}_m = \left[\frac{7b}{30}, \frac{b^2}{60}, \frac{8b}{15}, 0, \frac{7b}{30}, -\frac{b^2}{60}\right]^T [1 - (-1)^m]\frac{Q_o l}{m\pi} \tag{5.14}$$

The compatibility of curvatures at strip boundaries is not guaranteed by using this strip. However, the solution leads to higher order expressions for displacement and stress, and consequently yields an improved accuracy over the basic finite strips.

Since the internal nodal line c is not connected to any other strips, it is possible to eliminate the unknown displacement parameters of this nodal line before assembly through the static condensation technique (Cook et al, 1989). Using the subscripts c and o to identify the inner and outer nodal lines respectively, the matrix equation of the strip is divided as follows:

$$\begin{bmatrix} [k_{oo}] & [k_{oc}] \\ [k_{oc}]^T & [k_{cc}] \end{bmatrix} \begin{Bmatrix} \{\delta_o\} \\ \{\delta_c\} \end{Bmatrix} = \begin{Bmatrix} \{F_o\} \\ \{F_c\} \end{Bmatrix} \tag{5.15}$$

From the lower part of the above equation, we have

$$[k_{cc}]\{\delta_c\} = \{F_c\} - [k_{oc}]^T\{\delta_o\} \tag{5.16}$$

or

$$\{\delta_c\} = [k_{cc}]^{-1}(\{F_c\} - [k_{oc}]^T\{\delta_o\}) \tag{5.17}$$

Substituting this expression into the upper part of the strip matrix equation, the modified matrices are obtained with the external displacement parameters only:

$$([k_{oo}] - [k_{oc}][k_{cc}]^{-1}[k_{oc}]^T)\{\delta_o\} = \{F_o\} - [k_{oc}][k_{cc}]^{-1}\{F_c\} \tag{5.18}$$

Apparently, the condensed stiffness and load matrices have the same volumes of the basic plate strip. After the overall matrix equation is solved, the displacement parameters of the internal nodal line can be obtained as

$$w_{cm} = \frac{1}{k_{11}}(p_3 - k_3 w_{im} - k_8\theta_{im} - k_3 w_{jm} + k_8\theta_{jm})$$

Table 5.2 *Stiffness matrix of higher order plate strip HO3*

$$[k]_m = \begin{bmatrix} k_1 & & & & & \\ k_2 & k_7 & & symmetrical & & \\ k_3 & k_8 & k_{11} & & & \\ k_4 & k_9 & 0 & k_{12} & & \\ k_5 & -k_6 & k_3 & -k_4 & k_1 & \\ k_6 & k_{10} & -k_8 & k_9 & -k_2 & k_7 \end{bmatrix}$$

where

$$\begin{aligned}
k_1 &= \frac{523bA}{3465} + \frac{278B}{105b} + \frac{278C}{105b} + \frac{5092D}{35b^3} \\
k_2 &= \frac{19b^2A}{2310} + \frac{13B}{210} + \frac{59C}{105} + \frac{1138D}{35b^2} \\
k_3 &= \frac{4bA}{63} - \frac{256B}{105b} - \frac{256C}{105b} - \frac{512D}{5b^3} \\
k_4 &= -\frac{8b^2A}{693} + \frac{8B}{21} + \frac{8C}{21} + \frac{384D}{7b^2} \\
k_5 &= \frac{131bA}{6930} - \frac{22B}{105b} - \frac{22C}{105b} - \frac{1508D}{35b^3} \\
k_6 &= -\frac{29b^2A}{13860} - \frac{B}{70} - \frac{C}{70} + \frac{242D}{35b^2} \\
k_7 &= \frac{2b^3A}{3465} + \frac{2bB}{45} + \frac{2bC}{45} + \frac{332D}{35b} \\
k_8 &= \frac{2b^2A}{315} - \frac{8B}{105} - \frac{8C}{105} - \frac{128D}{5b^2} \\
k_9 &= -\frac{b^3A}{1155} - \frac{4bB}{315} - \frac{4bC}{315} + \frac{64D}{7b} \\
k_{10} &= -\frac{b^3A}{4620} - \frac{bB}{126} - \frac{bC}{126} + \frac{38D}{35b} \\
k_{11} &= \frac{128bA}{315} + \frac{512B}{105b} + \frac{512C}{105b} + \frac{1024D}{5b^3} \\
k_{12} &= \frac{32b^3A}{3465} + \frac{128bB}{315} + \frac{128bC}{315} + \frac{256D}{7b}
\end{aligned}$$

where

$$A = \frac{1}{2}lk_m^4 D_y \qquad B = 2lk_m^2 D_{xy}$$

$$C = lk_m^2 D_1 \qquad D = \frac{1}{2}lD_x$$

$$\theta_{cm} = \frac{1}{k_{12}}(p_4 - k_4 w_{im} - k_9\theta_{im} + k_4 w_{jm} - k_9\theta_{jm}) \tag{5.19}$$

The explicit expressions of the condensed stiffness matrix and load vector have been worked out by Loo (1972), and are also given in Loo and Cusens (1978).

5.4 Plane stress strip HO3

In Chapter 4, a linear function of x is employed to represent transverse variation of displacement components in a plane stress strip. This displacement function leads to constant transverse strains and a linear distribution of longitudinal strains across the width of a strip. Although this approach yields satisfactory longitudinal stress, too many narrow strips are often necessary to obtain an adequate accuracy for the transverse stress. This difficulty can also be solved by introducing the higher order strips.

The higher order strip with additional compatibilities at strip boundaries is not applicable to the analysis of box structures because this strip demands that the transverse strain must be continuous across any nodal line. Therefore, only the higher order strip with one internal nodal line is discussed here.

In this strip (Fig. 5.1 (c)), the displacement functions take the form

$$\left\{ \begin{array}{c} u \\ v \end{array} \right\} = \sum_{m=1}^{r} [N_{1m}][N_{2m}][N_{3m}] \left\{ \begin{array}{c} u_{im} \\ v_{im} \\ u_{cm} \\ v_{cm} \\ u_{jm} \\ v_{jm} \end{array} \right\} \tag{5.20}$$

where

$$[N_{1m}] = \left[\begin{array}{cc} N_1 \sin k_m y & 0 \\ 0 & N_1 \cos k_m y \end{array} \right]$$

$$\begin{aligned} N_1 &= 1 - 3X + 2X^2 \\ N_2 &= 4X - 4X^2 \\ N_3 &= -X + 2X^2 \quad \textit{and} \\ X &= x/b \end{aligned} \tag{5.21}$$

By substituting the above displacement functions into the strain–displacement relationships in the plane stress analysis, and then into the stress–strain relationships, the strain and stress vectors can be expressed in terms of displacement parameters as

$$\{\epsilon\} = \sum_{m=1}^{r} [B]_m \{\delta\}_m \tag{5.22}$$

Table 5.3 *Stiffness matrix of higher order plane stress strip HO3*

$$[k]_m = t \begin{bmatrix} k_1 & & & & & \\ k_2 & k_7 & & symmetrical & & \\ k_3 & -k_4 & k_{10} & & & \\ k_4 & k_8 & 0 & K_{11} & & \\ k_5 & -k_6 & k_3 & -k_4 & k_1 & \\ k_6 & k_9 & k_4 & k_8 & -k_2 & k_7 \end{bmatrix}$$

in which

$$k_1 = \frac{7l}{6b}E_1 + \frac{lbk_m^2}{15}E_{xy} \qquad k_2 = \frac{lk_m}{4}\nu_x E_1 - \frac{lk_m}{4}E_{xy}$$

$$k_3 = -\frac{4l}{3b}E_1 + \frac{lbk_m^2}{30}E_{xy} \qquad k_4 = \frac{lk_m}{3}\nu_x E_1 + \frac{lk_m}{3}E_{xy}$$

$$k_5 = \frac{l}{6b}E_1 - \frac{lbk_m^2}{60}E_{xy} \qquad k_6 = -\frac{lk_m}{12}\nu_x E_1 - \frac{lk_m}{12}E_{xy}$$

$$k_7 = \frac{lbk_m^2}{15}E_2 + \frac{7l}{6b}E_{xy} \qquad k_8 = \frac{lbk_m^2}{30}E_2 - \frac{4a}{3b}E_{xy}$$

$$k_9 = -\frac{lbk_m^2}{60}E_2 + \frac{l}{6b}E_{xy} \qquad k_{10} = \frac{8l}{3b}E_1 + \frac{4lbk_m^2}{15}E_{xy}$$

$$k_{11} = \frac{4lbk_m^2}{15}E_2 + \frac{8l}{3b}E_{xy}$$

and

$$E_1 = \frac{E_x}{1-\nu_x\nu_y} \qquad E_2 = \frac{E_y}{1-\nu_x\nu_y}$$

and

$$\{\sigma\} = \sum_{m=1}^{r} [D]_p[B]_m\{\delta\}_m \tag{5.23}$$

in which $[B]_m$ is the strain matrix

$$[B]_m = \left[\begin{matrix} N_1' \sin k_m y & 0 & N_2' \sin k_m y \\ 0 & -k_m N_1 \sin k_m y & 0 \\ k_m N_1 \cos k_m y & N_1' \cos k_m y & k_m N_2 \cos k_m y \end{matrix} \right.$$
$$\left. \begin{matrix} 0 & N_3' \sin k_m y & 0 \\ -k_m N_2 \sin k_m y & 0 & -k_m N_3 \sin k_m y \\ N_2' \cos k_m y & k_m N_3 \cos k_m y & N_3' \cos k_m y \end{matrix} \right] \tag{5.24}$$

Then the stiffness matrix $[k]_m$ and load vector $\{p\}_m$ are readily obtained. The elements of the stiffness matrix are given in Table 5.3.

Eliminating the displacement parameters of the internal nodal line by static condensation, the stiffness matrix and load vector are reduced to a

4×4 matrix and a 4×1 vector, respectively. [2] After the overall matrix equation has been solved, the unknown parameters of the internal nodal line can be obtained as

$$\begin{aligned} u_{cm} &= \frac{1}{k_{10}}(p_3 - k_3 u_{im} + k_4 v_{im} - k_3 u_{jm} - k_4 v_{jm}) \\ v_{cm} &= \frac{1}{k_{11}}(p_4 - k_4 u_{im} - k_8 v_{im} + k_4 u_{jm} - k_8 v_{jm}) \end{aligned} \quad (5.25)$$

5.5 Flat shell strip HO3

For analysis of box girder bridges, the higher order flat shell strip (HO3) with one internal nodal line can be formed by combination of the higher order plate strip and the higher order plane stress strip. After static condensation, the higher order flat shell strip and the basic one use the same displacement parameters. Therefore, the final solution for this strip, including all the displacement and stress components at any points of concern, can be obtained by following the same procedure as described in Section 4.5 for the basic flat shell strips.

5.6 Choice of approach

In the analysis of a bridge structure, if accurate values of transverse in-plane stresses, transverse in-plane shear and transverse bending moments are required, the higher order approach with one internal nodal line is preferable to other approaches, because of its higher accuracy at only a minor extra cost.

If only displacements and longitudinal in-plane stresses are the primary interest in an analysis, the basic lower order procedure can be used with satisfactory accuracy and the best economy.

5.7 Numerical examples

Several problems involving plate flexural analysis are presented here. It should be stressed that the performance of the higher order strip in plate bending is really much superior to that of the lower order strip and that only a small number of equations is used in all the examples given in this section. The examples given here are analyzed using higher order strips with curvature terms, higher order strips with intermediate nodal line are given elsewhere.

[2] Details about reduced matrices can be found in Loo (1972) and Loo and Cusens (1978).

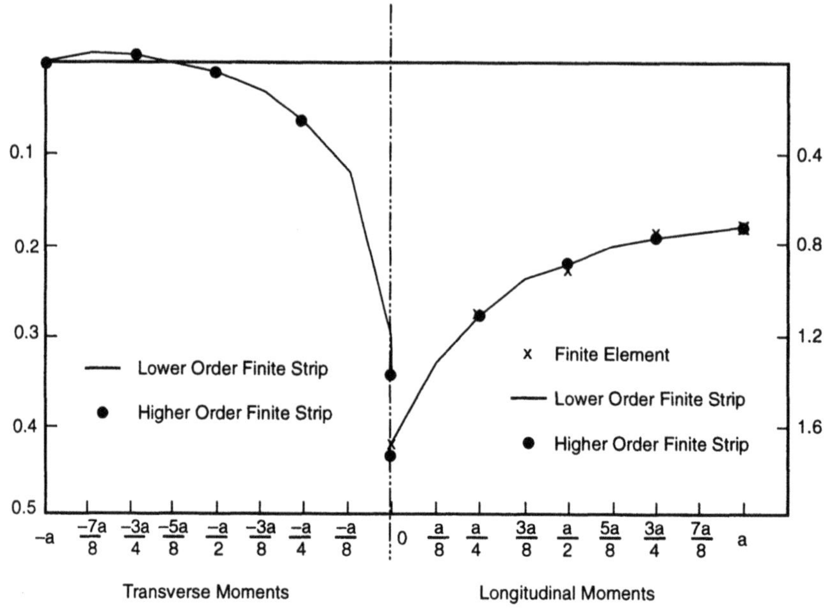

Figure 5.2 *Distribution coefficients for longitudinal and transverse moments due to a unit load at the center of a square bridge.*

Table 5.4 *Clamped slab under concentrated load P : H = higher order finite strip, L = lower order finite strip ($\nu = .3$)*

	Central concentrated load P					
	w_{max}		$-M_{xmax}$		$-M_{ymax}$	
	H	L	H	L	H	L
m=1	0.00511	0.00519	−0.11838	−0.11654	−0.14385	−0.14591
m=3	0.00030	0.00027	−0.00949	−0.01361	0.05122	0.04568
m=5	0.00011	0.00012	+0.00331	+0.00439	−0.04690	−0.05152
m=7	0.00003	0.00003	−0.00198	−0.00182	0.02677	0.02172
$\sum$	0.00555	0.00561	−0.12654	−0.12758	−0.11276	−0.13003
Exact	0.00560	0.00560	−0.12570	−0.12570	−0.12570	−0.12570
Multiplier	$\frac{Pa^2}{D}$			P		

Example 5.1 A square slab bridge simply supported at two ends is analyzed for a central point load $P = 1$, using four strips for half the bridge. The results are compared with those using rectangular finite elements (4×8 mesh) and lower order finite strips (8 strips). The distribution coefficients (Fig. 5.2) computed for longitudinal and transverse moments are practically identical for all three cases.

Example 5.2 A square slab with all edges clamped is analyzed for two load cases a uniformly distributed load q and central concentrated load P. Only two higher order plate strips are used for half the plate. It can be seen from Tables 5.4 and 5.5 that the results are more accurate than those by using five equal lower order strips.

Example 5.3 A series of calculations for different meshes and a different number of terms was carried out for a simply supported isotropic slab. The rate of convergence is compared with that of lower order finite strip, finite element and exact solution in Table 5.6. It reveals that with only two strips and four terms, the higher order strip analysis is identical with the exact solution. The lower order strip will achieve the same accuracy with four strips and four terms.

Table 5.5 *Clamped slab under uniform load* $q(\nu = .3)$. H = *higher order finite strip,* L = *lower order finite strip.*

Parameter	Uniformly distributed load q									
	$W_{\max}$		$M_{x\,\max}$		$M_{y\,\max}$		$-M_{x\,\max}$		$-M_{y\,\max}$	
	H	L	H	L	H	L	H	L	H	L
$m = 1$	0.00131	0.00133	0.02488	0.02554	0.02793	0.02837	−0.05411	−0.03744	−0.03699	−0.05235
$m = 3$	−0.00005	−0.00007	−0.00246	−0.00336	−0.00660	−0.00821	+0.00291	−0.01113	−0.00906	0.00114
$m = 5$	0.00001	0.00001	+0.00072	0.00027	0.0217	0.00348	−0.00003	−0.00523	−0.00445	0.00054
Sum	0.00127	0.00127	0.02314	0.02245	0.02350	0.02364	−0.051232	−0.05380	−0.05050	−0.05067
Exact	0.00126	0.00126	0.02310	0.02310	0.02310	0.02310	−0.05130	−0.05130	−0.05130	−0.05130
Multiplier	qa^4/D					qa^4				

Table 5.6 *Simply supported slab under uniform load* $q(\nu = .3)$.

	Finite strip method							Finite element method		
	Higher order				Lower order					
Number of lines for 1/2 slab	Total number of terms	$W_{\max}$	$M_{x\,\max}$	$M_{y\,\max}$	$W_{\max}$	$M_{x\,\max}$	$M_{y\,\max}$	Mesh size for 1/4 slab	$W_{\max}$	$M_{\max}$
2	1	0.00411	0.0496	0.0518	0.00414	0.0561	0.0540			
	2	0.00406	0.0479	0.0472	0.00409	0.0546	0.0494			
	3	0.00406	0.0482	0.0482	0.00409	0.0549	0.0504	2×2	0.00350	0.0669
	4	0.00406	0.0480	0.0478	0.00409	0.0549	0.0504			
3	1	0.00411	0.0492	0.0517	0.00411	0.0502	0.0520			
	2	0.00406	0.0476	0.0471	0.00406	0.0487	0.0474			
	3	0.00406	0.0479	0.0481	0.00406	0.0490	0.0485	3×3	0.00395	0.0502
	4	0.00406	0.0479	0.0478	0.00406	0.0489	0.0481			
4	1	0.00411	0.0492	0.0516	0.00411	0.0496	0.0518			
	2	0.00406	0.0476	0.0472	0.00406	0.0480	0.0472			
	3	0.00406	0.0479	0.0482	0.00406	0.0484	0.0483	4×4	0.00401	0.0483
	4	0.00406	0.0479	0.0479	0.00406	0.0481	0.0479			
5	1				0.00411	0.0494	0.0517			
	2				0.00406	0.0479	0.0472			
	3				0.00406	0.0482	0.0482	5×5	0.00403	
	4				0.00406	0.0481	0.0478			
Exact		0.00406	0.0479	0.0479	0.00406	0.0479	0.0479		0.00406	0.0479
Multiplier		qa^4/D	qa^2		qa^4/D	qa^2			qa^4/D	qa^2

CHAPTER 6

Spline finite strip method

6.1 Introduction

The semi analytical finite strip method discussed so far is very efficient for analysis of prismatic structures under distributed loading, especially for structures with both ends simply supported. Nevertheless, the use of this method can sometimes lead to difficulties. For instance, because the beam functions are continuously differentiable, it is difficult to use such functions to simulate the abrupt changes of bending moment at internal supports or at loaded cross-sections of point forces. In addition, the beam functions must satisfy the end conditions of a strip *a priori*; hence, these functions cannot deal with discrete supports at strip ends.

In order to overcome these difficulties, a mathematical tool known as the 'B_3 spline' has been used to provide the longitudinal displacement functions to form the spline finite strips for analysis of rectangular plates by Y K Cheung, Fan and Wu (1982).

In this method, each nodal line is divided into a number of sections by equally or unequally spaced knots. Every knot is then taken as the center of a local B_3 spline, which has non-zero values over only four consecutive sections (Fig. 6.1). All the B_3 splines on a nodal line form a series which is utilized to simulate the longitudinal variation of displacements.

The B_3-spline functions can ensure continuity up to the second derivative (the so-called C_2 continuity). By comparison, in order to achieve the same continuity condition, the finite element method needs three times as many unknowns at the element nodes. Hence, when C_2 continuity is required, the use of B_3 splines is computationally much more efficient than the finite element method.

The second derivative of B_3 spline varies linearly in each longitudinal section, as a result of which it can more easily simulate peak values of bending moment at the loaded cross-section or at an intermediate support.

Furthermore, similarly to the finite element method, a spline finite strip can easily take up any prescribed external and internal boundary conditions e.g. by a penalty function approach (see Section 15.3.3). Thus, the spline finite strip method is more flexible than the semianalytical finite strip method in imposing boundary conditions.

At a later stage in its development, the spline finite strip method was applied to the analysis of skew and arbitrary shaped plates, arbitrary

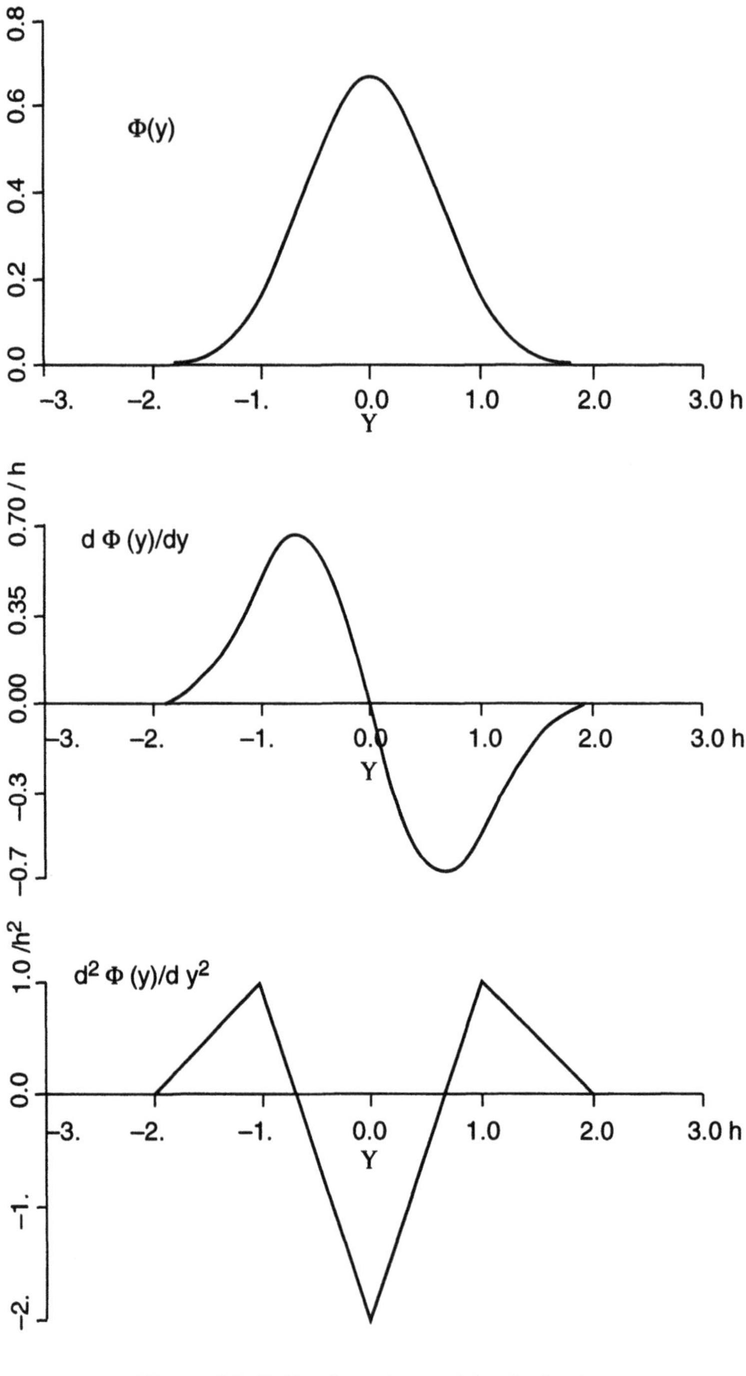

Figure 6.1 *Spline function and its derivatives.*

Table 6.1 *Values of spline function at knots*

Function	y_{m-2}	y_{m-1}	y_m	y_{m+1}	y_{m+2}
$\Phi_m(y)$	0	1/6	2/3	1/6	0
$\Phi'_m(y)$	0	1/2h	0	−1/2h	0
$\Phi''_m(y)$	0	$1/h^2$	$-2/h^2$	$1/h^2$	0

curved slab bridges, rectangular box-girder bridges, continuous haunched bridges and to vibration, stability and postbuckling analysis etc.

This chapter describes the basic procedures in the spline finite strip method and its applications to the static analysis of rectangular plates, arbitrarily shaped slabs and right box-girder bridges. Some more advanced topics will be presented in the following chapters.

6.2 Spline function interpolation

The term Spline was originally the name of a small flexible wooden strip employed by draughtsmen as a tool for drawing a continuous smooth curve segment by segment. Actually, the spline function can be derived with any required continuity (or discontinuity) conditions. In most references, the B_3 spline (see Prenter, 1975) is chosen to represent the longitudinal variation of displacements. This function is a piecewise cubic polynomial with continuity up to the second derivative.

In order to use spline functions to interpolate an arbitrary function f(y) on the interval $a \le y \le b$, the interval may be divided into a number of sections by equally spaced knots y_m ($0 \le m \le r$, $y_0 = a$, $y_r = b$, $y_m = y_0 + m(b-a)/r$). The equally spaced spline function $\Phi_m(y)$ with $y = y_m$ as the center is defined by

$$\Phi_m(y) = \frac{1}{6h^3}\begin{cases} (y-y_{m-2})^3 & y_{m-2} \le y \le y_{m-1} \\ h^3 + 3h^2(y-y_{m-1}) + 3h(y-y_{m-1})^2 - 3(y-y_{m-1})^3 & y_{m-1} \le y \le y_m \\ h^3 + 3h^2(y_{m+1}-y) + 3h(y_{m+1}-y)^2 - 3(y_{m+1}-y)^3 & y_m \le y \le y_{m+1} \\ (y_{m+2}-y)^3 & y_{m+1} \le y \le y_{m+2} \\ 0 & \text{otherwise} \end{cases} \tag{6.1}$$

The function $\Phi_m(y)$ and its first and second derivatives are shown graphically in Fig. 6.1, and their values at knots are given in Table 6.1. It can be seen that the function $\Phi_m(y)$ is twice continuously differentiable over the entire interval, and its second derivative is a linear function of y.

The spline functions centered at all the knots comprise a series $s(y)$ which can be used to interpolate any continuous and differentiable function

$f(y)$. In each section the value of $s(y)$ is related to four splines which are centered at the two ends of this section and the two knots next to those ends, respectively. Therefore, two additional knots y_{-1} and y_{r+1} are needed to interpolate $f(x)$ in the first and the last sections. Thus the series of spline functions can be expressed as

$$s(y) = \sum_{m=-1}^{r+1} c_m \Phi_m(y) \tag{6.2}$$

in which the c_m are coefficients to be determined by interpolation requirements such as

$$\begin{aligned} s'(y_0) &= f'(y_0) \\ s(y_m) &= f(y_m) \quad 0 \le m \le r \\ s'(y_r) &= f'(y_r) \end{aligned} \tag{6.3}$$

Substituting Equation 6.2 into Equation 6.3 and using the values in Table 6.1 yields a set of linear equations with c_m as unknowns:

$$\begin{aligned} \frac{1}{2h}(-c_{-1} + c_1) &= f'(y_0) \\ \frac{1}{6}(c_{m-1} + 4c_m + c_{m+1}) &= f(y_m) \quad 0 \le m \le r \\ \frac{1}{2h}(-c_{r-1} + c_{r+1}) &= f'(y_r) \end{aligned}$$

Solving these equations for c_m gives the required series of spline functions for interpolation:

$$s(y) = \sum_{m=-1}^{r+1} c_m \Phi_m(y) = f(y)$$

In spline finite strip analysis, every loaded cross-section of point force and every supported cross-section should coincide with a longitudinal knot in order to obtain satisfactory accuracy. Therefore, in some cases, unequal spacing between knots must be adopted. The unequally spaced B_3-spline $\Phi_m(y)$ with y_m as center is given by

$$\Phi_m(y) = \begin{cases} 0 & y < y_{m-2} \\ f_1 & y_{m-2} \le y < y_{m-1} \\ f_2 & y_{m-1} \le y < y_m \\ f_3 & y_m \le y < y_{m+1} \\ f_4 & y_{m+1} \le y < y_{m+2} \\ 0 & y_{m+2} \le y \end{cases} \tag{6.4}$$

where

$$f_1 = \frac{(y - y_{m-2})^3}{(y_{m+1} - y_{m-2})(y_m - y_{m-2})(y_{m-1} - y_{m-2})}$$

$$\begin{aligned} f_2 &= f_1 - \frac{(y_{m+2}-y_{m-2})(y-y_{m-1})^3}{(y_{m+2}-y_{m-1})(y_{m+1}-y_{m-1})(y_m-y_{m-1})(y_{m-1}-y_{m-2})} \\ f_3 &= f_4 - \frac{(y_{m+2}-y_{m-2})(y_{m+1}-y)^3}{(y_{m+1}-y_{m-2})(y_{m+1}-y_{m-1})(y_{m+1}-y_m)(y_{m+2}-y_{m+1})} \\ f_4 &= \frac{(y_{m+2}-y)^3}{(y_{m+2}-y_{m-1})(y_{m+2}-y_m)(y_{m+2}-y_{m+1})} \end{aligned} \tag{6.5}$$

6.3 Flat shell spline finite strip

Flat shell spline finite strips (Fig. 6.2) developed by Y. K. Cheung and Fan in 1983 are applicable for analysis of right slab bridges, slab-on-girder bridges and box girder bridges.

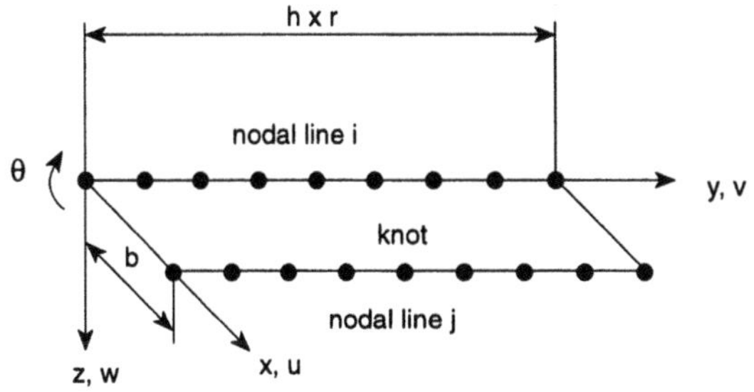

Figure 6.2 *Flat shell spline finite strip.*

The displacements in the strip are expressed as

$$\begin{aligned} u &= \sum_{m=-1}^{r+1} \left((1-X)\Phi^u_{im}(y)u_{im} + X\Phi^u_{jm}(y)u_{jm}\right) \\ v &= \sum_{m=-1}^{r+1} \left((1-X)\Phi^v_{im}(y)v_{im} + X\Phi^v_{jm}(y)v_{jm}\right) \\ w &= \sum_{m=-1}^{r+1} \left(N_{1m}w_{im} + N_{2m}\theta_{im} + N_{3m}w_{jm} + N_{4m}\theta_{jm}\right) \end{aligned} \tag{6.6}$$

where $X = x/b$, r is the total number of longitudinal sections on a nodal line, and

$$\begin{aligned} N_{1m} &= (1-3X^2+2X^3)\Phi^w_{im}(y) \\ N_{2m} &= x(1-2X+X^2)\Phi^\theta_{im}(y) \\ N_{3m} &= (3X^2-2X^3)\Phi^w_{jm}(y) \\ N_{4m} &= x(X^2-X)\Phi^\theta_{jm}(y) \end{aligned} \tag{6.7}$$

In the above equations, $\Phi^u_{im}(y)$, $\Phi^v_{im}(y)$, $\Phi^w_{im}(y)$ and $\Phi^\theta_{im}(y)$ are B_3-spline functions centered at y_m and modified according to boundary conditions of corresponding displacement components at the ends of nodal line i.

In order to adapt to various prescribed boundary conditions at end knot $m = 0$, only three boundary local splines centered at y_{-1}, y_0 and y_1 should be amended, while the standard splines defined by Equations 6.1 or 6.5 are used for other knots. If the knots $m = -1$ through $m = 3$ are equally spaced, the amended schemes for Φ^w_{im} are summarized as follows:

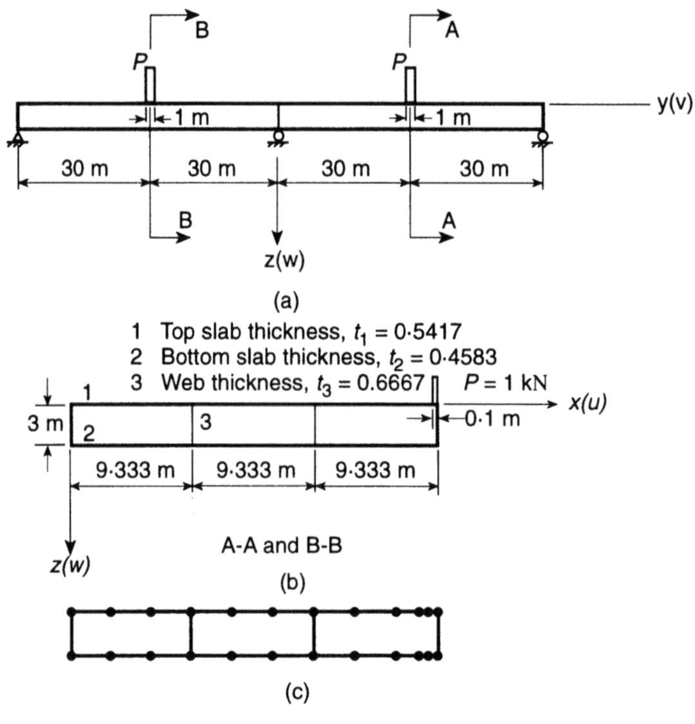

Figure 6.3 *Two-equal span three-cell girder bridge: (a) and (b) geometry and loading; (c) strip division (Y. K. Cheung and Fan 1983).*

1. Free end, i.e. ($w(0) \neq 0$ and $w'(0) \neq 0$):

$$\Phi^w_{-1} = \Phi_{-1}, \qquad \Phi^w_0 = \Phi_0, \qquad \Phi^w_1 = \Phi_1 \tag{6.8}$$

2. Simply supported end, i.e. ($w(0) = 0$ and $w'(0) \neq 0$):

$$\Phi^w_{-1} = 0, \qquad \Phi^w_0 = \Phi_0 - 4\Phi_{-1}, \qquad \Phi^w_1 = \Phi_1 - \Phi_{-1} \tag{6.9}$$

3. Clamped end, i.e. ($w(0) = 0$ and $w'(0) = 0$):

$$\Phi^w_{-1} = 0, \qquad \Phi^w_0 = 0, \qquad \Phi^w_1 = \Phi_1 - \tfrac{1}{2}\Phi_0 + \Phi_{-1} \tag{6.10}$$

4. Sliding clamped end, i.e. $(w(0) \neq 0$ and $w'(0) = 0)$:

$$\Phi^w_{-1} = 0, \qquad \Phi^w_0 = \Phi_0, \qquad \Phi^w_1 = \Phi_1 - \tfrac{1}{2}\Phi_0 + \Phi_{-1} \tag{6.11}$$

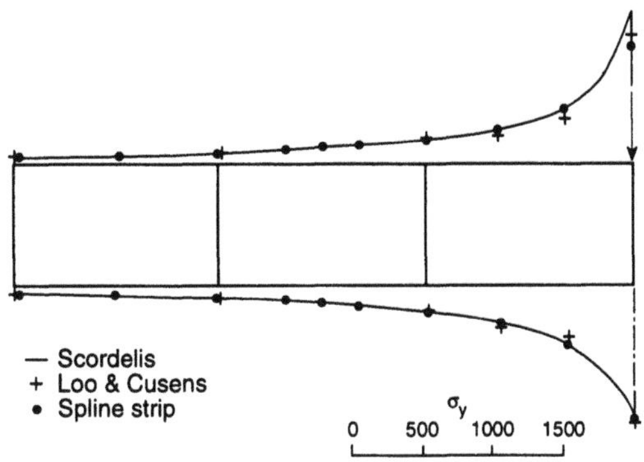

Figure 6.4 *Distribution of longitudinal stress σ_y at a loaded section (Y. K. Cheung and Fan 1983).*

If there are any intermediate supports, they can be treated by the penalty function method introduced in the next section.

The splines for other displacement components may also be modified in a similar minor.

Once the displacement functions for the strip have been chosen, the strip characteristics including stiffness, load and mass matrices can be obtained in line with the standard finite strip formulation as described earlier.

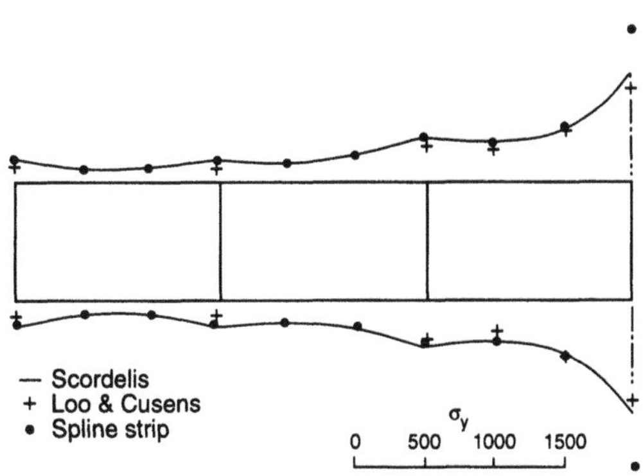

Figure 6.5 *Distribution of longitudinal stress σ_y at an intermediate support (Y. K. Cheung and Fan 1983).*

The present method was applied by Y.K. Cheung and Fan (1983) to the analysis of a two-equal span continuous three-cell box girder bridge under eccentric concentrated loads (Fig. 6.3). By symmetry, analyses were performed only on one span using 26 strips with subdivision of four sections to seven sections. The small patch loads, however, have not been replaced by concentrated loads as was originally done by Scordelis and Davis (1969) and Scordelis (1971). Results of transverse distributions of longitudinal stresses at loaded cross-section and intermediate support are compared with elasticity solutions of Scordelis et al. and also with finite strip solutions (in conjunction with the flexibility approach) given by Loo and Cusens (1971) as shown in Figs. 6.4 and 6.5. Agreement is good at all positions except in the vicinity of the loads where the present method yields lower values. This is presumably due to the replacement of the patch loads by point loads in the analysis carried out by other workers.

6.4 Analysis of arbitrarily shaped plates

6.4.1 Introduction

Incorporating the concept of subparametric mapping (Cook et al, 1989), the spline finite strip method has been extended to the analysis of arbitrarily shaped plate structures and arbitrarily curved slab bridges by Y. K. Cheung et al (1986).

In this analysis, the plate is first mapped into a rectangular domain in the natural coordinate plane ($\xi - \eta$ plane) by subparametric mapping, and the 'mapped' rectangular plate is then discretized into a number of strips (Fig. 6.6). The displacements of the strip are expressed in terms of the natural coordinate variables ξ and η. The formulation and solution procedure can then follow those of the standard finite strip method.

6.4.2 Subparametric transformation

The analyses of arbitrarily shaped plates are carried out on the ξ–η plane. For convenience of description, let the transformation be

$$\begin{aligned} x &= \sum_{i=1}^{12} N_i(\xi,\eta)x_i = x(\xi,\eta) \\ y &= \sum_{i=1}^{12} N_i(\xi,\eta)y_i = y(\xi,\eta) \end{aligned} \qquad (6.12)$$

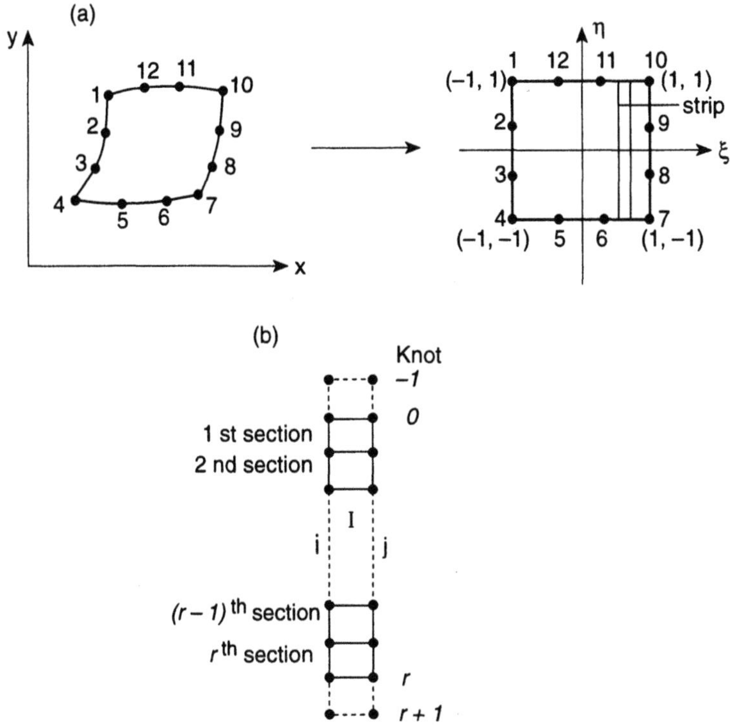

Figure 6.6 *(a) Subparametric mapping and (b) typical spline strip.*

where (x_i, y_i) are the coordinates of ith boundary node of the plate and $N_i(\xi, \eta)$ are the shape functions

$$N_i(\xi,\eta) = \begin{cases} \frac{1}{32}(1+\xi_i\xi)(1+\eta_i\eta)[9(\xi^2+\eta^2)-10] \\ \qquad\qquad for\ \ i = 1,4,7,10 \ \ (\xi_i = \pm 1, \eta_i = \pm 1) \\ \frac{9}{32}(1+\xi_i\xi)(1-\eta^2)(1+9\eta_i\eta) \\ \qquad\qquad for\ \ i = 2,3,9,8 \ \ (\xi_i = \pm 1, \eta_i = \pm\frac{1}{3}) \\ \frac{9}{32}(1+\eta_i\eta)(1-\xi^2)(1+9\xi_i\xi) \\ \qquad\qquad for\ \ i = 5,6,11,12 \ \ (\xi_i = \pm\frac{1}{3}, \eta_i = \pm 1) \end{cases} \tag{6.13}$$

From the above transformation, we have

$$\frac{\partial x}{\partial \xi} = \sum_{i=1}^{12} \frac{\partial N_i}{\partial \xi} x_i$$

$$\frac{\partial y}{\partial \xi} = \sum_{i=1}^{12} \frac{\partial N_i}{\partial \xi} y_i$$

$$\frac{\partial x}{\partial \eta} = \sum_{i=1}^{12} \frac{\partial N_i}{\partial \eta} x_i$$
$$\frac{\partial y}{\partial \eta} = \sum_{i=1}^{12} \frac{\partial N_i}{\partial \eta} y_i \tag{6.14}$$

For any function $f(\xi, \eta)$, the following relationships hold:

$$\left\{ \begin{array}{c} \dfrac{\partial f}{\partial \xi} \\ \\ \dfrac{\partial f}{\partial \eta} \end{array} \right\} = \left[\begin{array}{cc} \dfrac{\partial x}{\partial \xi} & \dfrac{\partial y}{\partial \xi} \\ \\ \dfrac{\partial x}{\partial \eta} & \dfrac{\partial y}{\partial \eta} \end{array} \right] \left\{ \begin{array}{c} \dfrac{\partial f}{\partial x} \\ \\ \dfrac{\partial f}{\partial y} \end{array} \right\} = [J] \left\{ \begin{array}{c} \dfrac{\partial f}{\partial x} \\ \\ \dfrac{\partial f}{\partial y} \end{array} \right\} \tag{6.15}$$

where $[J]$ is the so-called Jacobian matrix.

Thus, the derivatives of $f(\xi, \eta)$ with respect to x and y are obtained as

$$\left\{ \begin{array}{c} \dfrac{\partial f}{\partial x} \\ \\ \dfrac{\partial f}{\partial y} \end{array} \right\} = [J]^{-1} \left\{ \begin{array}{c} \dfrac{\partial f}{\partial \xi} \\ \\ \dfrac{\partial f}{\partial \eta} \end{array} \right\} = \frac{1}{\mid J \mid} \left[\begin{array}{cc} \dfrac{\partial y}{\partial \eta} & -\dfrac{\partial y}{\partial \xi} \\ \\ -\dfrac{\partial x}{\partial \eta} & \dfrac{\partial x}{\partial \xi} \end{array} \right] \left\{ \begin{array}{c} \dfrac{\partial f}{\partial \xi} \\ \\ \dfrac{\partial f}{\partial \eta} \end{array} \right\} \tag{6.16}$$

where $\mid J \mid$ is the Jacobian determinant written as

$$\mid J \mid = \left| \begin{array}{cc} \dfrac{\partial x}{\partial \xi} & \dfrac{\partial y}{\partial \xi} \\ \\ \dfrac{\partial x}{\partial \eta} & \dfrac{\partial y}{\partial \eta} \end{array} \right| = \frac{\partial x}{\partial \xi}\frac{\partial y}{\partial \eta} - \frac{\partial y}{\partial \xi}\frac{\partial x}{\partial \eta} \tag{6.17}$$

If $f(\xi, \eta) = \xi$, then $\frac{\partial f}{\partial \xi} = 1$ and $\frac{\partial f}{\partial \eta} = 0$, thus from Equation 6.16 we obtain

$$\frac{\partial \xi}{\partial x} = \frac{1}{|J|}\frac{\partial y}{\partial \eta}$$
$$\frac{\partial \xi}{\partial y} = -\frac{1}{|J|}\frac{\partial x}{\partial \eta} \tag{6.18}$$

Similarly, by assuming $f(\xi, \eta) = \eta$ we get

$$\frac{\partial \eta}{\partial x} = -\frac{1}{|J|}\frac{\partial y}{\partial \xi}$$
$$\frac{\partial \eta}{\partial y} = \frac{1}{|J|}\frac{\partial x}{\partial \xi} \tag{6.19}$$

By substituting $f(\xi, \eta) = w$ into Equation 6.16, the transformation for the slopes is obtained as

$$\left\{\begin{array}{c}\dfrac{\partial w}{\partial x}\\[2ex]\dfrac{\partial w}{\partial y}\end{array}\right\}=[J]^{-1}\left\{\begin{array}{c}\dfrac{\partial w}{\partial \xi}\\[2ex]\dfrac{\partial w}{\partial \eta}\end{array}\right\}=\frac{1}{\mid J\mid}\left[\begin{array}{cc}\dfrac{\partial y}{\partial \eta} & -\dfrac{\partial y}{\partial \xi}\\[2ex]-\dfrac{\partial x}{\partial \eta} & \dfrac{\partial x}{\partial \xi}\end{array}\right]\left\{\begin{array}{c}\dfrac{\partial w}{\partial \xi}\\[2ex]\dfrac{\partial w}{\partial \eta}\end{array}\right\} \tag{6.20}$$

Further, the transformation for curvatures can be derived as

$$\left\{\begin{array}{c}\dfrac{\partial^2 w}{\partial x^2}\\[2ex]\dfrac{\partial^2 w}{\partial y^2}\\[2ex]2\dfrac{\partial^2 w}{\partial x\partial y}\end{array}\right\}=[T_B]\left\{\begin{array}{c}\dfrac{\partial^2 w}{\partial \xi^2}\\[2ex]\dfrac{\partial w}{\partial \xi}\\[2ex]\dfrac{\partial^2 w}{\partial \eta^2}\\[2ex]\dfrac{\partial w}{\partial \eta}\\[2ex]2\dfrac{\partial^2 w}{\partial \xi\partial \eta}\end{array}\right\} \tag{6.21}$$

where

$$[T_B]=\left[\begin{array}{ccccc}(\dfrac{\partial \xi}{\partial x})^2 & \dfrac{\partial^2 \xi}{\partial x^2} & (\dfrac{\partial \eta}{\partial x})^2 & \dfrac{\partial^2 \eta}{\partial x^2} & \dfrac{\partial \eta}{\partial x}\dfrac{\partial \xi}{\partial x}\\[2ex](\dfrac{\partial \xi}{\partial y})^2 & \dfrac{\partial^2 \xi}{\partial y^2} & (\dfrac{\partial \eta}{\partial y})^2 & \dfrac{\partial^2 \eta}{\partial y^2} & \dfrac{\partial \eta}{\partial y}\dfrac{\partial \xi}{\partial y}\\[2ex]2\dfrac{\partial \xi}{\partial x}\dfrac{\partial \xi}{\partial y} & 2\dfrac{\partial^2 \xi}{\partial x\partial y} & 2\dfrac{\partial \eta}{\partial x}\dfrac{\partial \eta}{\partial y} & 2\dfrac{\partial^2 \eta}{\partial x\partial y} & (\dfrac{\partial \eta}{\partial x}\dfrac{\partial \xi}{\partial y}+\dfrac{\partial \xi}{\partial x}\dfrac{\partial \eta}{\partial y})\end{array}\right] \tag{6.22}$$

in which

$$\begin{aligned}\frac{\partial^2 \xi}{\partial x^2} &= \frac{\partial \xi}{\partial x}\frac{\partial}{\partial \xi}(\frac{\partial \xi}{\partial x})+\frac{\partial \eta}{\partial x}\frac{\partial}{\partial \eta}(\frac{\partial \xi}{\partial x})\\ \frac{\partial^2 \xi}{\partial y^2} &= \frac{\partial \xi}{\partial y}\frac{\partial}{\partial \xi}(\frac{\partial \xi}{\partial y})+\frac{\partial \eta}{\partial y}\frac{\partial}{\partial \eta}(\frac{\partial \xi}{\partial y})\\ \frac{\partial^2 \xi}{\partial x\partial y} &= \frac{\partial \xi}{\partial y}\frac{\partial}{\partial \xi}(\frac{\partial \xi}{\partial x})+\frac{\partial \eta}{\partial y}\frac{\partial}{\partial \eta}(\frac{\partial \xi}{\partial x})\\ \frac{\partial^2 \eta}{\partial x^2} &= \frac{\partial \xi}{\partial x}\frac{\partial}{\partial \xi}(\frac{\partial \eta}{\partial x})+\frac{\partial \eta}{\partial x}\frac{\partial}{\partial \eta}(\frac{\partial \eta}{\partial x})\end{aligned}$$

$$\begin{aligned}\frac{\partial^2\eta}{\partial y^2} &= \frac{\partial\xi}{\partial y}\frac{\partial}{\partial\xi}\left(\frac{\partial\eta}{\partial y}\right)+\frac{\partial\eta}{\partial y}\frac{\partial}{\partial\eta}\left(\frac{\partial\eta}{\partial y}\right)\\ \frac{\partial^2\eta}{\partial x\partial y} &= \frac{\partial\xi}{\partial y}\frac{\partial}{\partial\xi}\left(\frac{\partial\eta}{\partial x}\right)+\frac{\partial\eta}{\partial y}\frac{\partial}{\partial\eta}\left(\frac{\partial\eta}{\partial x}\right)\end{aligned} \tag{6.23}$$

6.4.3 Displacement functions

In the context of the spline finite strip, the whole domain in the $\xi-\eta$ plane is partitioned into n strips along ξ-axis and r sections along the η-axis, so that there are $n\times r$ subdomains and $(r+3)\times(n+1)$ knots for computation. Note that it is not necessary for the strips and the sections to have equal width or length, respectively.

In the rectangular strip I with nodal lines i and j as boundaries (Fig. 6.6(b)), the deflection w is expressed as

$$w=\sum_{m=-1}^{r+1}(N_{1m}w_{im}+N_{2m}\theta_{im}+N_{3m}w_{jm}+N_{4m}\theta_{jm}) \tag{6.24}$$

where

$$\begin{aligned}N_{1m} &= (1-3\Xi^2+2\Xi^3)\Phi^w_{im}(\eta)\\ N_{2m} &= b\Xi(1-2\Xi+\Xi^2)\Phi^\theta_{im}(\eta)\\ N_{3m} &= (3\Xi^2-2\Xi^3)\Phi^w_{jm}(\eta)\\ N_{4m} &= b\Xi(\Xi^2-\Xi)\Phi^\theta_{jm}(\eta)\end{aligned} \tag{6.25}$$

in which $\Xi=(\xi-\xi_i)/b$ and $b=\xi_j-\xi_i$, r is the total number of longitudinal sections on a nodal line, $\Phi^w_{im}(\eta)$ and $\Phi^\theta_{im}(\eta)$ are B_3-spline functions centered at η_m and modified according to boundary conditions of corresponding displacement components at the ends of nodal line i (see Section 6.3).

Intermediate supports are included by penalty function method (Section 15.3.3). The end boundary conditions may also be treated by that method with modifications of standard splines being avoided.

The stiffness matrix for each section is

$$[k^s]=\int[B]^T[D][B]|J|d\xi d\eta \tag{6.26}$$

where $[B]$ is the strain matrix, and $[D]$ is the elasticity matrix.

6.4.4 Numerical examples

1. Lobley Hill South overbridge

Lobley Hill South overbridge is a four-span continuous bridge. Its epoxy–sand model was analyzed using the spline finite strip method by Cheung, Tham and Li (1986). The model was 1:30 of full size. It had a width of

448.7 mm and there were five rows of column support, as shown in Fig. 6.7. The stiffnesses of the columns are given in Table 6.2.

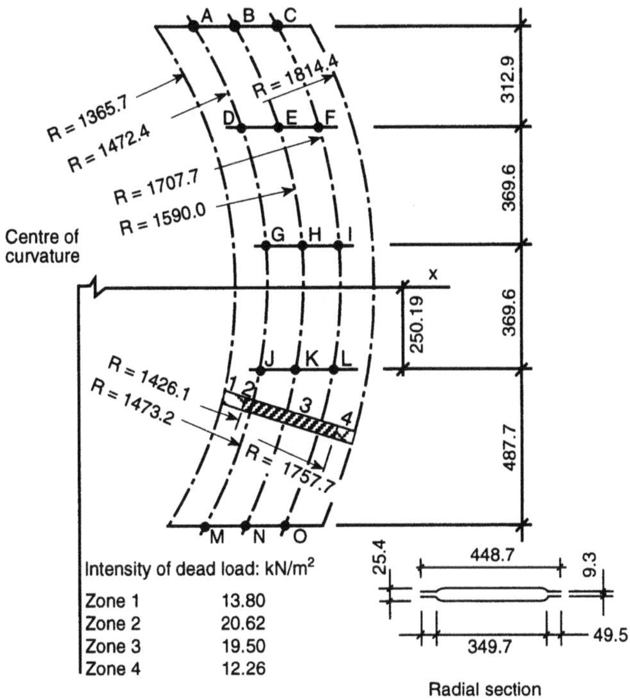

Figure 6.7 *Detail of idealization of Lobley Hill South overbridge model (dimensions in mm); (Y. K. Cheung, Tham and Li, 1986).*

Table 6.2 *Elastic stiffness of column supports of Lobley Hill South overbridge model (KN mm^{-1}; (Y. K. Cheung, Tham and Li, 1986).*

Column	A	B	C	D	E	F	G	H
Elastic Stiffness	33.80	33.62	33.45	33.62	33.27	32.75	32.22	31.87
Column	I	J	K	L	M	N	O	
Elastic Stiffness	31.52	31.00	30.65	30.47	33.97	33.80	33.80	

Only the dead load case had been considered, and the loading can be divided into four zones of different intensities (Fig. 6.7) varying from 12.26 KN m^{-2} to 20.62 KN m^{-2}. The values of the modulus of elasticity and Poisson's ratio were 1.65×10^7 KN m^{-2} and 0.22, respectively. Both deflection and longitudinal bending moment were depicted in Fig. 6.8, in comparison with those obtained from finite element analysis and experiments by Lim and Moffatt (1971). All results were in good agreement. Lim's computed results were based on a 9×26 mesh of elements with a total of 720 degrees of freedom, whereas the present mesh is 9×20 but with only 360 degrees of freedom.

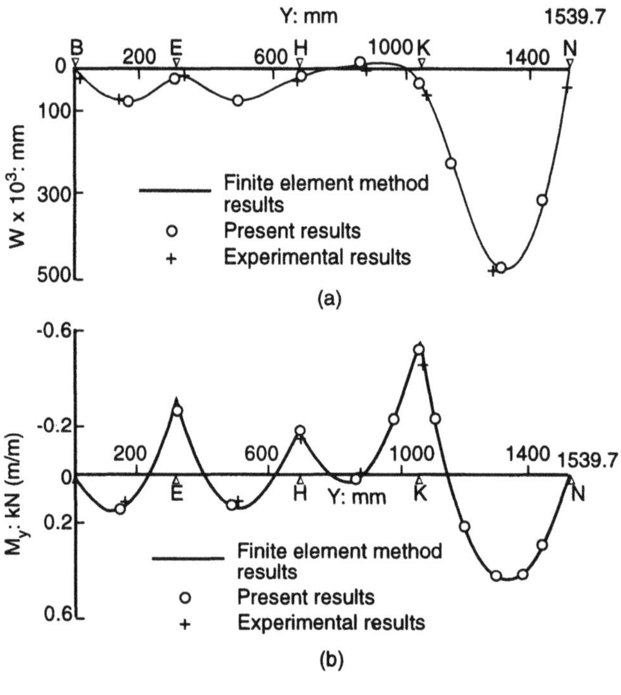

Figure 6.8 *Lobley Hill South overbridge: (a) central longitudinal deflexion profile for model under dead load; (b) distribution of longitudinal bending moments along central line for model under dead load (Y. K. Cheung, Tham and Li, 1986).*

2. S-shaped continuous two-span bridge

A two-span S-shaped continuous slab bridge was analyzed by Cheung et al. (1986) to demonstrate the versatility of the spline finite strip method. Four control points per curved side were used to define the shape of each S-curve. The expressions of the two S-shaped sides were defined by cubic polynomials in terms of a parameter $t(-1 \leq t \leq 1)$ (Fig. 6.9):

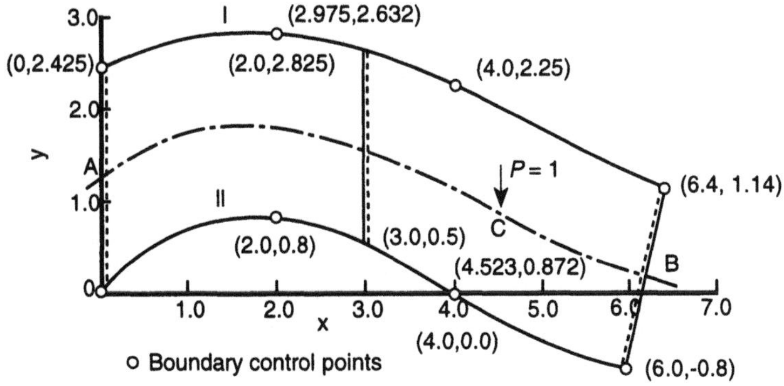

Figure 6.9 *S-shaped slab bridge.*

Curve I

$$
\begin{aligned}
x &= 0.225t^3 + 0.225t^2 + 2.975t + 2.975 \\
y &= -0.6638t^3 - 0.5456t^2 + 0.0213t + 2.3281
\end{aligned}
$$

Curve II

$$
\begin{aligned}
x &= 3t + 3 \\
y &= 0.9t^3 - 0.9t^2 - 1.3t + 0.5
\end{aligned}
$$

Line supports were assumed for both ends and for the intermediate one. Analyses were carried out for both a uniformly distributed load and a concentrated load at point C (Fig. 6.9); the results are given in Fig. 6.10 and compared with those computed using the standard finite elements by Bazeley et al. (1966). In this example, the results of both the coarse mesh (110 unknowns) and the fine mesh (378 unknowns) compare favorably with the finite element results based on 525 degrees of freedom.

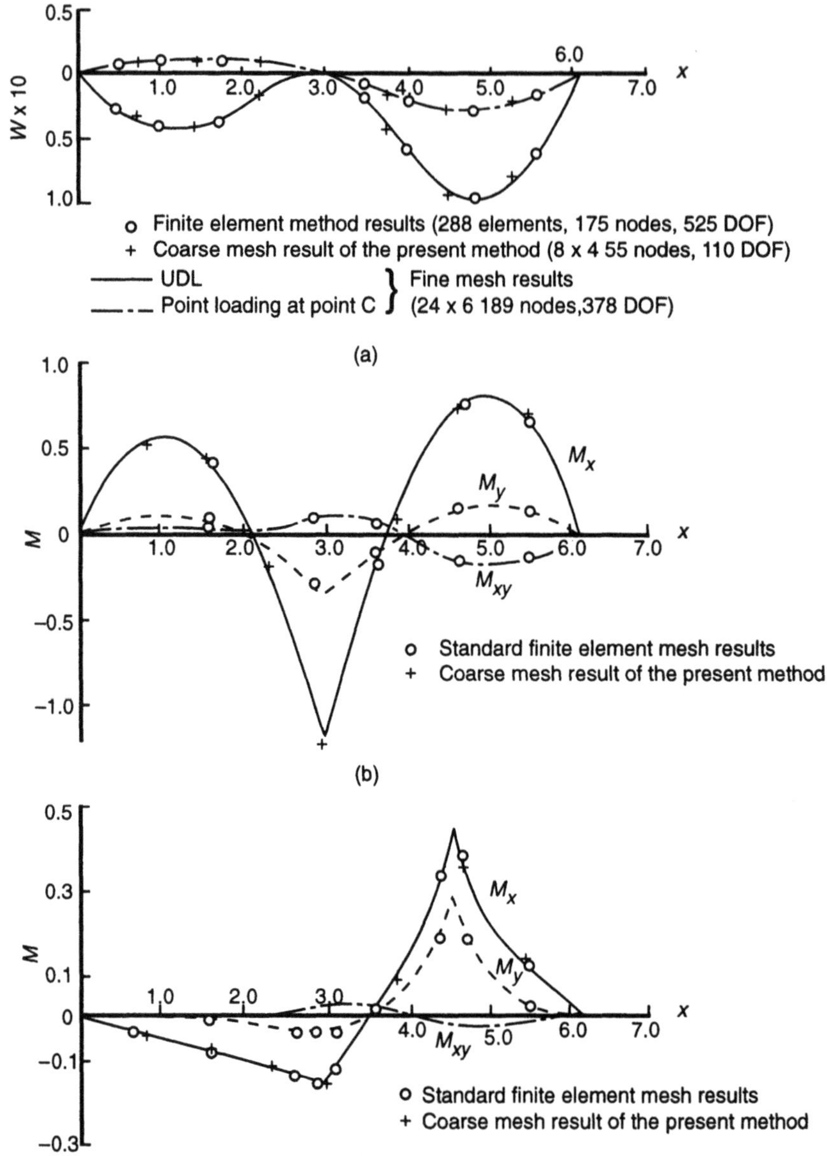

Figure 6.10 *Deflexions and bending moments of S-shaped continuous slab bridge: (a) deflexion along central line AB; (b) moments along central line AB (UDL); (c) moments along central line AB (point loading at point C) (Y. K. Cheung, Tham and Li 1986).*

CHAPTER 7

Compound strip method

7.1 Introduction

In order to analyze plate structures, which are continuous over a number of flexible beams and columns or stiffened by many longitudinal and transverse beams, a compound strip method has been developed by Puckett and Gutkowski (1983). This method incorporates the effects of the support elements in a direct stiffness methodology so that the solution procedure becomes more straightforward than the flexibility approach.

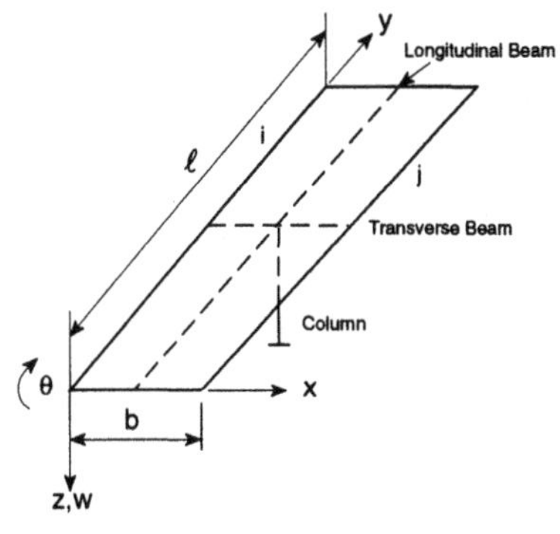

Figure 7.1 *Rectangular compound strip.*

The so-called compound strip is a substructure which allows beam and column elements to be attached to a plate strip or a flat shell strip (Fig. 7.1). The plate strip or flat shell strip and the attached elements deform monolithically. Therefore, the displacements of the attached elements can be expressed in terms of the nodal displacement parameters of the strip by means of its displacement functions. In this case, the strain energy of the compound strip includes not only the strain energy of the plate strip or flat shell strip, but also the strain energy of all the attached elements. Therefore, the stiffness matrix of the compound strip can be formed by adding the stiffness matrices of the attached elements directly to that of the plate strip or flat shell strip prior to assembly:

$$K^e = K^p + \sum_l K^l + \sum_t K^t + \sum_c K^c \tag{7.1}$$

where K^p is stiffness matrix of the plate strip or flat shell strip, K^l is stiffness matrix of longitudinal beam in a strip, K^t is stiffness matrix of transverse beam in a strip, K^c is stiffness matrix of column in a strip and K^e is total stiffness matrix of a compound strip.

The subsequent assembly of the K^e into the global stiffness matrix K, and solution of the structure matrix equation are similar to the procedure in the ordinary finite strip analysis.

This method has been successfully employed to model right and curved slab-girder bridges (Puckett and Lang, 1986; Puckett, 1986), box girders and folded plates (Maleki, 1987 and 1991). The concept is also extended to spline compound strip analysis by using the B_3 splines as the longitudinal displacement functions (Chen, Gutkowski and Puckett, 1990).

This chapter presents the basic procedures in forming the compound strip and the spline compound strip. Their application to the right and curved slab-girder bridges will be described in Chapters 12 and 13, respectively.

7.2 Rectangular compound strip

A rectangular compound strip may include a rectangular plate strip or flat shell strip, longitudinal beams, transverse beams and columns (Fig. 7.1).

The stiffness matrix of the plate strip or flat shell strip, $[K^p]$, is given in Chapter 4. Therefore, only the stiffness matrices of the attached elements need to be derived. In a compound strip, the deflection at any point can be written as

$$w = \sum_{m=1}^{r} [N_1, N_2, N_3, N_4] Y_m(y) \{\delta\}_m = \sum_{m=1}^{r} [N]_m \{\delta\}_m \tag{7.2}$$

where r is the number of series terms employed in the analysis, and $\{\delta\}_m$ is the vector of nodal displacement parameters corresponding to the mth series term and defined as

$$\{\delta\}_m = [w_{im}, \theta_{im}, w_{jm}, \theta_{jm}]^T \tag{7.3}$$

The terms N_1 to N_4 are the transverse displacement functions as

$$\begin{aligned} N_1 &= 1 - 3X^2 + 2X^3 \\ N_2 &= x(1 - 2X + X^2) \\ N_3 &= 3X^2 - 2X^3 \\ N_4 &= x(X^2 - X) \\ X &= x/b \end{aligned} \tag{7.4}$$

$Y_m(y)$ is the mth series term of beam eigenfunctions which satisfy the boundary conditions of the strip at both ends. $N_1, \ldots, N_4$ and $Y_m(y)$ form the matrix $[N]$:

$$[N]_m = [N_1, N_2, N_3, N_4] Y_m(y) \tag{7.5}$$

Thus, the stiffness matrix for each type of the attached element can be obtained as follows:

1. Flexural stiffness of a longitudinal beam
   The flexural strain energy for a longitudinal beam is

$$U^{fl} = \frac{EI}{2}\int_0^l (\frac{\partial^2 w}{\partial y^2})^2 dy \tag{7.6}$$

where EI is the flexural rigidity of the beam.

By substituting Equation 7.2 into 7.6, the flexural strain energy can be expressed in terms of the nodal displacement parameters as

$$\begin{aligned} U^{fl} &= \frac{EI}{2}\int_0^l \frac{\partial^2}{\partial y^2}(\sum_{m=1}^{r}[N]_m\{\delta\}_m)\frac{\partial^2}{\partial y^2}(\sum_{n=1}^{r}[N]_n\{\delta\}_n)dy \\ &= \frac{1}{2}\sum_{m=1}^{r}\sum_{n=1}^{r}\{\delta\}_m^T EI\int_0^l (\frac{\partial^2 [N]_m}{\partial y^2})^T \frac{\partial^2 [N]_n}{\partial y^2}dy\{\delta\}_n \\ &= \frac{1}{2}\sum_{m=1}^{r}\sum_{n=1}^{r}\{\delta\}_m^T[K_{mn}^{fl}]\{\delta\}_n \end{aligned} \tag{7.7}$$

where $[K_{mn}^{fl}]$ is the flexural stiffness matrix of the longitudinal beam corresponding to the mth and nth series terms, and can be written as

$$[K_{mn}^{fl}] = EII_4 \begin{bmatrix} N_1N_1 & N_1N_2 & N_1N_3 & N_1N_4 \\ & N_2N_2 & N_2N_3 & N_2N_4 \\ & & N_3N_3 & N_3N_4 \\ \text{symmetrical} & & & N_4N_4 \end{bmatrix} \tag{7.8}$$

in which $N_1, N_2, \ldots$ are evaluated at the local x coordinate of the beam, and I_4 is the integral

$$I_4 = \int_0^l Y_m''(y)Y_n''(y)dy \tag{7.9}$$

2. Torsional stiffness of a longitudinal beam
   The torsional strain energy of a longitudinal beam is

$$U^{tl} = \frac{GJ}{2}\int (\frac{\partial^2 w}{\partial x \partial y})^2 dy \tag{7.10}$$

where GJ is the torsional rigidity of the beam.

Following the same procedure as in type 1 yields the torsional stiffness matrix of the longitudinal beam as

$$[K_{mn}^{tl}] = GJ\int_0^l (\frac{\partial^2[N]_m}{\partial x\partial y})^T \frac{\partial^2[N]_n}{\partial x\partial y} dy$$

$$= GJI_5 \begin{bmatrix} N_1'N_1' & N_1'N_2' & N_1'N_3' & N_1'N_4' \\ & N_2'N_2' & N_2'N_3' & N_2'N_4' \\ & & N_3'N_3' & N_3'N_4' \\ \text{symmetrical} & & & N_4'N_4' \end{bmatrix} \quad (7.11)$$

where the prime denotes the first derivative with respect to x and

$$I_5 = \int_0^l Y_m'(y)Y_n'(y)dy \quad (7.12)$$

3. Flexural stiffness of a transverse beam
The flexural strain energy for a transverse beam is

$$U^{ft} = \frac{EI}{2}\int (\frac{\partial^2 w}{\partial x^2})^2 dx \quad (7.13)$$

Thus, the flexural stiffness matrix of the beam can be obtained as

$$[K_{mn}^{ft}] = EI\int_0^b (\frac{\partial^2[N]_m}{\partial x^2})^T \frac{\partial^2[N]_n}{\partial x^2} dx$$

$$= EIY_mY_n \begin{bmatrix} 12/b^3 & 6/b^2 & -12/b^3 & 6/b^2 \\ & 4/b & -6/b^2 & 2/b \\ & & 12/b^3 & -6/b^2 \\ \text{symmetrical} & & & 4/b \end{bmatrix} \quad (7.14)$$

in which Y_m and Y_n are evaluated at the local y coordinate of the beam, while b is the width of the strip.

4. Torsional stiffness of a transverse beam
The torsional strain energy of a transverse beam is

$$U^{tt} = \frac{GJ}{2}\int (\frac{\partial^2 w}{\partial x\partial y})^2 dx \quad (7.15)$$

The torsional stiffness matrix of the transverse beams can then be derived as

$$[K_{mn}^{tt}] = GJ\int_0^b (\frac{\partial^2[N]_m}{\partial x\partial y})^T \frac{\partial^2[N]_n}{\partial x\partial y} dx$$

$$= \frac{GJ}{30b}Y_m'Y_n' \begin{bmatrix} 36 & 3b & -36 & 3b \\ & 4b^2 & -3b & b^2 \\ & & 36 & 3b \\ \text{symmetrical} & & & 4b^2 \end{bmatrix} \quad (7.16)$$

5. Axial stiffness of a column
The axial strain energy of a support column is

$$U^{ac} = \frac{1}{2}K_a w_c^2 \tag{7.17}$$

where K_a is the axial stiffness of the column, and w_c is the axial deflection which can be obtained by evaluating the displacement function $[N_m]$ at the local coordinates x_c, y_c of the column.

Substitution of Equation 7.2 into 7.17 gives the corresponding stiffness matrix as

$$\begin{aligned}[K^{ac}_{mn}] &= K_a[N]^T_m[N]_n \\ &= K_a Y_m Y_n \begin{bmatrix} N_1N_1 & N_1N_2 & N_1N_3 & N_1N_4 \\ & N_2N_2 & N_2N_3 & N_2N_4 \\ & & N_3N_3 & N_3N_4 \\ \text{symmetrical} & & & N_4N_4 \end{bmatrix}\end{aligned} \tag{7.18}$$

where N_1 to N_4 are evaluated at x_c, while Y_m and Y_n are calculated at y_c.

6. Flexural stiffness of a column
The flexural strain energy in a column has two components, which are related to rotations of the column head in the transverse and longitudinal directions of the strip, respectively.

The strain energy for the transverse rotation is

$$U^{xc} = \frac{1}{2}K_x(\frac{\partial w_c}{\partial x})^2 \tag{7.19}$$

where K_x is the transverse rotation stiffness of the column head. From Equation 7.2, the corresponding stiffness matrix can be formed as

$$\begin{aligned}[K^{xc}_{mn}] &= K_x(\frac{\partial[N]_m}{\partial x})^T\frac{\partial[N]_n}{\partial x} \\ &= K_x Y_m Y_n \begin{bmatrix} N_1'N_1' & N_1'N_2' & N_1'N_3' & N_1'N_4' \\ & N_2'N_2' & N_2'N_3' & N_2'N_4' \\ & & N_3'N_3' & N_3'N_4' \\ \text{symmetrical} & & & N_4'N_4' \end{bmatrix}\end{aligned} \tag{7.20}$$

The strain energy for the longitudinal rotation of column head is

$$U^{yc} = \frac{1}{2}K_y(\frac{\partial w_c}{\partial y})^2 \tag{7.21}$$

where K_y is the longitudinal rotation stiffness of the column head.

In the same manner, the related stiffness matrix can be given as

$$\begin{aligned}[K^{yc}_{mn}] &= K_y(\frac{\partial[N]_m}{\partial y})^T\frac{\partial[N]_n}{\partial y}\\ &= K_yY'_mY'_n\begin{bmatrix} N_1N_1 & N_1N_2 & N_1N_3 & N_1N_4\\ & N_2N_2 & N_2N_3 & N_2N_4\\ & & N_3N_3 & N_3N_4\\ \text{symmetrical} & & & N_4N_4\end{bmatrix}\end{aligned} \tag{7.22}$$

After forming the stiffness matrices of plate and all the support elements in the strip, the stiffness submatrix of the compound strip $[K^e_{mn}]$ can readily be obtained as

$$\begin{aligned}[K^e_{mn}] &= [K^p_{mn}] + \sum_1^{nl}([K^{fl}_{mn}] + [K^{tl}_{mn}]) + \sum_1^{nt}([K^{ft}_{mn}] + [K^{tt}_{mn}])\\ &\quad + \sum_1^{nc}([K^{ac}_{mn}] + [K^{xc}_{mn}] + [K^{yc}_{mn}])\end{aligned} \tag{7.23}$$

in which nl is the number of longitudinal beams attached to the strip, nt is the number of transverse beams in the strip, and nc is the number of columns supporting this strip.

From the expressions for the stiffness matrices, it can be noted that the series terms are generally coupled, i.e. the $[K^e_{mn}]$ does not vanish for $m \neq n$, unless the strip includes plate and longitudinal beams only.

7.3 Rectangular B-spline compound strip

Chen *et al.* (1990) developed the rectangular B-spline compound strip by using the series of B_3-spline functions (see Chapter 6) as the longitudinal displacement functions in order to simulate abrupt changes of bending moments at intermediate supports and loaded sections more efficiently. Both the bending and in-plane deformations are included in this strip, the eccentricity of the stiffening member is also taken into consideration. The displacement functions given in Equation 6.6 are employed for the present strip, and K^p in Equation 7.1 represents the stiffness matrix of the flat shell spline finite strip (see Section 6.3).

The strain energy of a longitudinal beam is defined as

$$U^l = \frac{1}{2}\int_0^l \left[EA(\frac{\partial(v - e^l w_y)}{\partial y})^2 + EI(\frac{\partial^2 w}{\partial y^2})^2 + GJ(\frac{\partial^2 w}{\partial x \partial y})^2\right] dy \tag{7.24}$$

where e^l is the eccentricity of the longitudinal beam, and EA is the axial stiffness of the beam.

The strain energy of a transverse beam can be calculated as

$$U^t = \frac{1}{2}\int_0^b \left[EA(\frac{\partial(u - e^t w_x)}{\partial y})^2 + EI(\frac{\partial^2 w}{\partial x^2})^2 + GJ(\frac{\partial^2 w}{\partial x \partial y})^2 \right] dx \quad (7.25)$$

in which e^t is the eccentricity of the transverse beam.

The strain energy of a column is given by

$$U^c = \frac{1}{2}[K_a w^2 + K_y(\frac{\partial w}{\partial y})^2 + K_x(\frac{\partial w}{\partial_x})^2] \quad (7.26)$$

Substituting displacement functions 6.6 into the expressions for strain energy and following a procedure similar to that used in the previous section, yields the stiffness matrices of the stiffeners and columns. The stiffness matrix of the compound strip is then readily obtained from Equation 7.1.

7.4 Numerical examples

1. Single panel plate

A square plate with sides of length 10.0 in (254 mm) is supported as shown

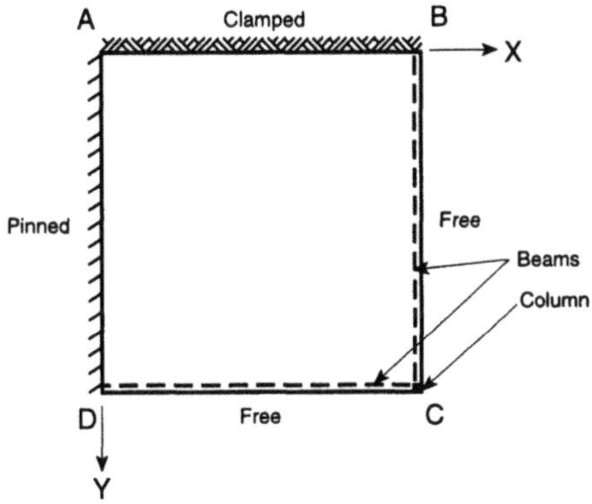

Figure 7.2 *Single panel plate example (Puckett and Gutkowski, 1986, courtesy of ASCE).*

in Fig. 7.2, and subjected to uniform load 1.0 ksi (6.90 MPa). The plate rigidities are $D_x = D_y = 1.0$ ksi (6.90 MPa), $D_1 = 0.30$ ksi (2.07 MPa), and $D_{xy} = 0.35$ ksi (2.41 MPa). The beam rigidities are $EI = GJ = 1.0$ kin^2 (2.870 kN mm^2). The axial stiffness of the column is 10 000 k in^{-1} (1750 kN mm^{-1}).

The plate was simulated by nine compound strips with three, five and seven harmonics (Puckett and Gutkowski, 1986). The strips are oriented in the y direction; therefore, the eigenfunctions of a cantilever beam

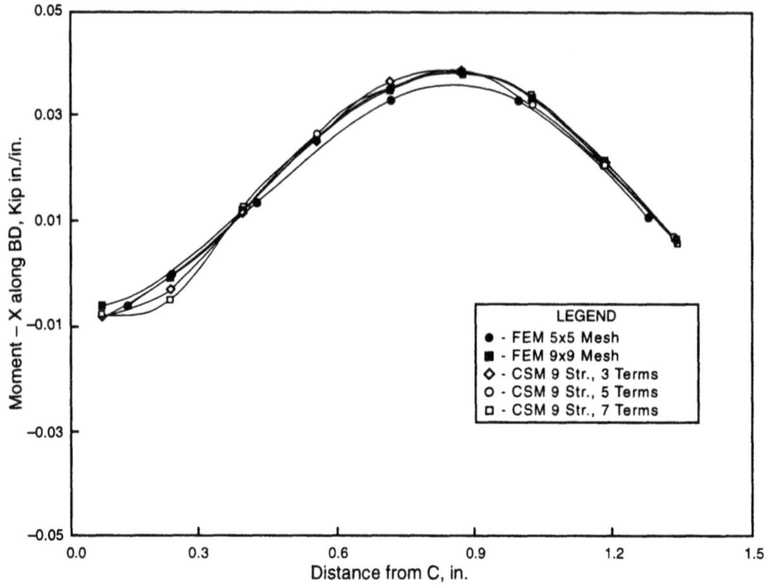

Figure 7.3 *Moment – X along BD (Puckett and Gutkowski, 1986, courtesy of ASCE). 1 in = 0.0254 m, 1 Kip = 4.4498 KN.*

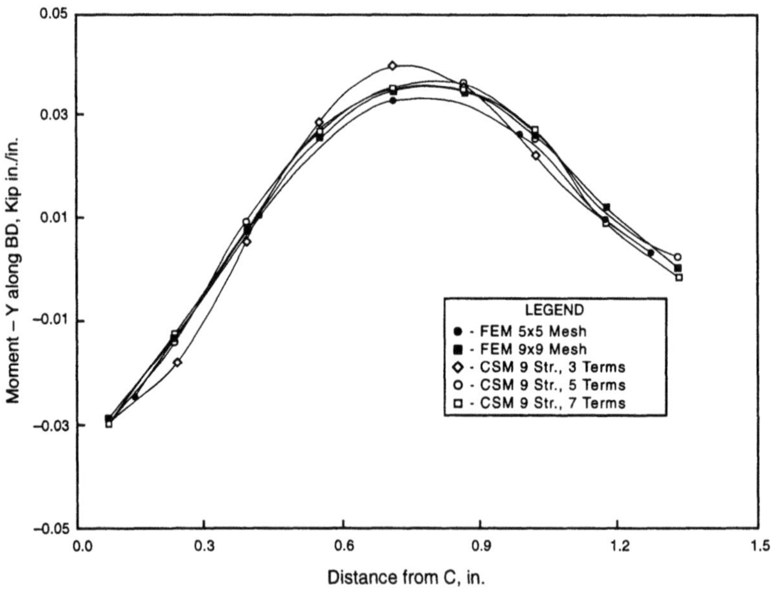

Figure 7.4 *Moment – Y along BD (Puckett and Gutkowski, 1986, courtesy of ASCE). 1 in = 0.0254 m, 1 Kip = 4.4498 KN.*

Table 7.1 *Deflection 1 in from the beam center (inches) (Chen et al., 1990, courtesy of Elsevier Science).*

d (in)	Load on each beam (lb)	Theoretical	Experimental	BCSM mesh c
0.752	39.7	0.0160	0.0178	0.0151
0.600	36.2	0.0279	0.0255	0.0260
0.450	13.8	0.0233	0.0206	0.0213
Degrees of freedom				196

Table 7.2 *Stress at the bottom of the beam center (psi) (Chen et al., 1990, courtesy of Elsevier Science).*

d (in)	Load on each beam (lb)	Theoretical	Experimental	BCSM mesh c
0.752	39.7	1929	2224	2139
0.600	36.2	2667	2909	2992
0.450	13.8	1665	1882	1868
Degrees of freedom				196

was used. For comparison, the plate was also analyzed using the finite element program SAP IV. The resulting bending moment distributions along the diagonal BD are given in Figs. 7.3 and 7.4. The results from both the methods compare favorably for moment with the difference in the maximums of less than three percent. The deflection plots for the two methods are virtually indistinguishable and are omitted for brevity.

2. Rectangular slab with two edge beams

Figure 7.5 shows a rectangular slab with two edge beams. A vertical point load is applied at the central cross-section of each beam. The structure was analyzed by Chen *et al.*, (1990) using the B-spline compound strip method (BSCM) with various meshes as shown in Fig. 7.5(b). With a 6×6 mesh for the whole plate, the resulting deflection at the beam center (Table 7.1) and the stress at bottom of the beam center (Table 7.2) are in a good agreement with the theoretical solution and the experimental results obtained by Allen and Severn in 1961. The results from different meshes are compared in Table 7.3. It reveals that the unequally spaced mesh c yields significant improvement in the results over the equally spaced meshes a and b. In Table 7.1 to Table 7.3, 1 in = 0.0254 m, 1 lb = 4.45 N, 1 psi = 6.90 KN m^{-2}.

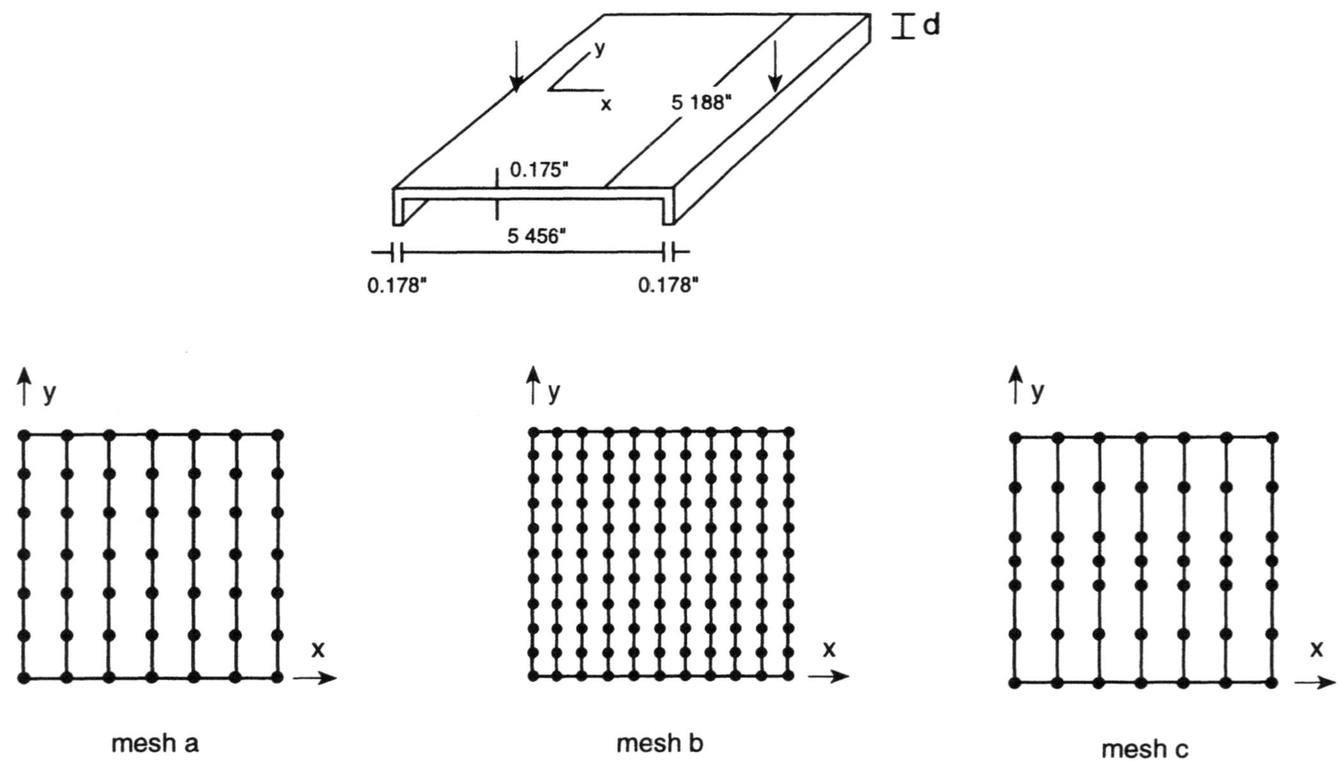

Figure 7.5 *(a) Rectangular slab with edge beams (dimensional in inches). (b) Various meshes (Chen et al., 1990, courtesy of Elsevier Science). (1 in = 0.0254 m)*

Table 7.3 *Results comparison for various meshes for BCSM (Chen et al., 1990, courtesy of Elsevier Science).*

d (in)	Load on each beam (lb)	Deflection i in from beam centre (in)			Stress at the bottom of beam center (psi)		
		Mesh a	Mesh b	Mesh c	Mesh a	Mesh b	Mesh c
0.752	39.7	0.0172	0.0152	0.0151	2163	2135	2139
0.600	36.2	0.0300	0.0260	0.0260	3047	2981	2992
0.450	13.8	0.0253	0.0213	0.0213	1931	1854	1868
Degrees of freedom		196	484	196	196	484	196

CHAPTER 8

Finite layer method; finite prism method

8.1 Introduction

For a very thick plate, any line normal to the mid-plane does not remain straight or normal to the deformed surface during deformation. Therefore, the thin plate theory is no longer applicable, and three-dimensional analysis is required. If a rectangular plate is supported only on its edges, the displacements can be interpolated by beam eigenfunctions in both the x and y directions. As a result, it is only necessary to divide the plate into several layers and to use polynomial shape functions in the z direction (Fig. 8.1). Thus, the three-dimensional elastic analysis reduces to a one-dimensional analysis. This method is referred to as the finite layer method.

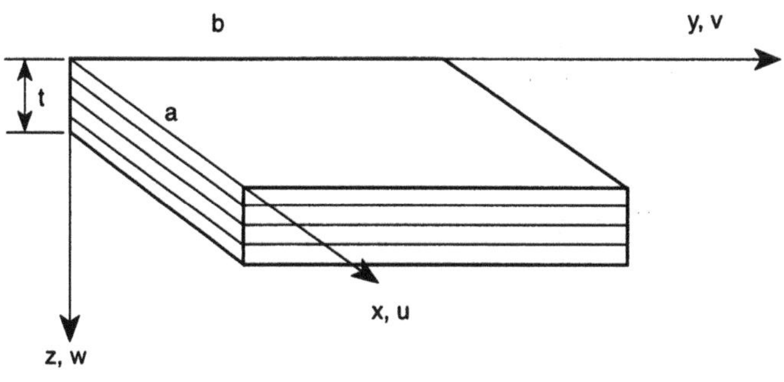

Figure 8.1 *Finite layer simulation.*

In thick slab bridges with variable thickness and holes in their cross-section, and in concrete box girder bridges with very thick flanges and webs (Fig. 8.2), the local stresses such as those at the junctions between the flanges and webs often induce undesired failure.To solve this problem, a three-dimensional analysis is also needed. As an economical solution, the bridge can be simulated by a number of longitudinal prisms using the shape functions of in-plane finite element in the $x-z$ plane and the beam eigenfunctions in the y direction (Fig. 8.2). This method is called the finite prism method.

Both methods are applicable not only for the static analysis, but also for the dynamic and buckling analyses (Y. K. Cheung and Chakrabarti, 1971; Y. K. Cheung and Chakrabarti, 1972; Y. K. Cheung, Tham and Chong,

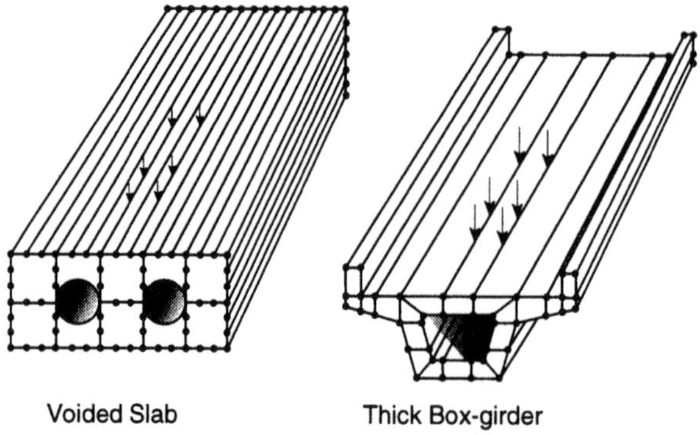

Figure 8.2 *Finite prism simulation (M. S. Cheung and Chan, 1978B).*

1982; Zienkiewicz and Too, 1972; M. S. Cheung and Chan, 1978B; M. S. Cheung and Chan, 1981).

The concepts and applications of these methods are introduced in this chapter.

8.2 Finite layer method

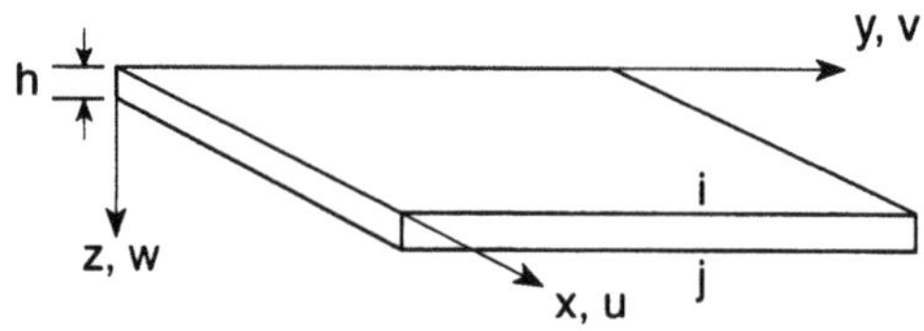

Figure 8.3 *A finite layer.*

A finite layer has two nodal surfaces labeled i and j (Fig. 8.3). The displacement parameters of the finite layer corresponding to series terms m and n are

$$\{\delta\}_{mn} = [u_{imn}, v_{imn}, w_{imn}, u_{jmn}, v_{jmn}, w_{jmn}]^T \tag{8.1}$$

where u, v and w represent the displacement components in the x, y and z directions, respectively.

The displacement field within the finite layer is interpolated as

$$u = \sum_{m=1}^{M}\sum_{n=1}^{N}[(1-\bar{z})u_{imn} + \bar{z}u_{jmn}]X'_n(x)Y_m(y)$$

$$v = \sum_{m=1}^{M}\sum_{n=1}^{N}[(1-\bar{z})v_{imn} + \bar{z}v_{jmn}]X_n(x)Y'_m(y)$$

$$w = \sum_{m=1}^{M}\sum_{n=1}^{N}[(1-\bar{z})w_{imn} + \bar{z}w_{jmn}]X_n(x)Y_m(y) \tag{8.2}$$

and

$$\bar{z} = z/h$$

where $X_n(x)$ and $Y_m(y)$ are the beam eigenfunctions in the x and y directions, respectively (see Section 4.5), M and N are the numbers of eigenfunctions required in the analysis, and h is the thickness of the layer. The first derivatives $X'_n(x)$ and $Y'_m(y)$ are respectively used for the shape functions of u and v because in the most typical deformation, i.e. plate bending, u is proportional to $\partial w/\partial x$ and v to $\partial w/\partial y$.

For the three-dimensional solid, the strain–displacement relationships are

$$\epsilon_x = \frac{\partial u}{\partial x} \qquad \epsilon_y = \frac{\partial v}{\partial y} \qquad \epsilon_z = \frac{\partial w}{\partial z}$$

$$\gamma_{xy} = \frac{\partial u}{\partial y} + \frac{\partial v}{\partial x} \qquad \gamma_{yz} = \frac{\partial v}{\partial z} + \frac{\partial w}{\partial y} \qquad \gamma_{zx} = \frac{\partial u}{\partial z} + \frac{\partial w}{\partial x} \tag{8.3}$$

By substituting Equation 8.2 into 8.3, the strain vector can be expressed in terms of the nodal displacement parameters as

$$\{\epsilon\} = [\epsilon_x, \epsilon_y, \epsilon_z, \gamma_{xy}, \gamma_{yz}, \gamma_{zx}]^T = \sum_{m=1}^{M}\sum_{n=1}^{N}[B]_{mn}\{\delta\}_{mn} = [B]\{\delta\} \tag{8.4}$$

For the isotropic material, the stresses are related to the strains by

$$\begin{Bmatrix} \sigma_x \\ \sigma_y \\ \sigma_z \\ \tau_{xy} \\ \tau_{yz} \\ \tau_{zx} \end{Bmatrix} = \frac{E}{(1+\nu)(1-2\nu)}$$

$$\times \begin{bmatrix} 1-\nu & \nu & \nu & 0 & 0 & 0 \\ \nu & 1-\nu & \nu & 0 & 0 & 0 \\ \nu & \nu & 1-\nu & 0 & 0 & 0 \\ 0 & 0 & 0 & \frac{1-2\nu}{2} & 0 & 0 \\ 0 & 0 & 0 & 0 & \frac{1-2\nu}{2} & 0 \\ 0 & 0 & 0 & 0 & 0 & \frac{1-2\nu}{2} \end{bmatrix} \begin{Bmatrix} \epsilon_x \\ \epsilon_y \\ \epsilon_z \\ \gamma_{xy} \\ \gamma_{yz} \\ \gamma_{zx} \end{Bmatrix} \tag{8.5}$$

or, more concisely

$$\{\sigma\} = [D]\{\epsilon\} \tag{8.6}$$

Then, the stiffness matrix and the load vector of the finite layer can be formed by following the standard procedures given in Chapter 4, and the stiffness matrix takes the following form:

$$[k] = \int [B]^T [D][B] dv$$

$$= \int \begin{bmatrix} [B]^T_{11}[D][B]_{11} & [B]^T_{11}[D][B]_{12} & \cdots & [B]^T_{11}[D][B]_{1N} & [B]^T_{11}[D][B]_{21} & \cdots & [B]^T_{11}[D][B]_{2N} & \cdots & [B]^T_{11}[D][B]_{MN} \\ [B]^T_{12}[D][B]_{11} & [B]^T_{12}[D][B]_{12} & \cdots & [B]^T_{12}[D][B]_{1N} & [B]^T_{12}[D][B]_{21} & \cdots & [B]^T_{12}[D][B]_{2N} & \cdots & [B]^T_{12}[D][B]_{MN} \\ \cdots & \cdots & \cdots & \cdots & \cdots & \cdots & \cdots & \cdots & \cdots \\ [B]^T_{1N}[D][B]_{11} & [B]^T_{1N}[D][B]_{12} & \cdots & [B]^T_{1N}[D][B]_{1N} & [B]^T_{1N}[D][B]_{21} & \cdots & [B]^T_{1N}[D][B]_{2N} & \cdots & [B]^T_{1N}[D][B]_{MN} \\ [B]^T_{21}[D][B]_{11} & [B]^T_{21}[D][B]_{12} & \cdots & [B]^T_{21}[D][B]_{1N} & [B]^T_{21}[D][B]_{21} & \cdots & [B]^T_{21}[D][B]_{2N} & \cdots & [B]^T_{21}[D][B]_{MN} \\ \cdots & \cdots & \cdots & \cdots & \cdots & \cdots & \cdots & \cdots & \cdots \\ [B]^T_{2N}[D][B]_{11} & [B]^T_{2N}[D][B]_{12} & \cdots & [B]^T_{2N}[D][B]_{1N} & [B]^T_{2N}[D][B]_{21} & \cdots & [B]^T_{2N}[D][B]_{2N} & \cdots & [B]^T_{2N}[D][B]_{MN} \\ \cdots & \cdots & \cdots & \cdots & \cdots & \cdots & \cdots & \cdots & \cdots \\ [B]^T_{MN}[D][B]_{11} & [B]^T_{MN}[D][B]_{12} & \cdots & [B]^T_{MN}[D][B]_{1N} & [B]^T_{MN}[D][B]_{21} & \cdots & [B]^T_{MN}[D][B]_{2N} & \cdots & [B]^T_{MN}[D][B]_{MN} \end{bmatrix} dv \tag{8.7}$$

If all the sides of a plate are simply supported, because of the orthogonal properties of the trigonometric series, all the off-diagonal stiffness submatrices vanish, i.e.

$$\int [B]^T_{mn}[D][B]_{rs} dv = 0 \quad for \quad mn \neq rs \tag{8.8}$$

and the different series terms are uncoupled in the analysis.

8.3 Finite prism method

In a finite prism, the displacement parameters for nodal line i and series term m are

$$\{\delta\}_{im} = \{u_{im}, v_{im}, w_{im}\}^T \tag{8.9}$$

The displacement components at any point in the finite prism can be expressed in terms of displacement parameter as:

$$u = \sum_{m=1}^{r} \sum_{i=1}^{n} N_i u_{im} Y_m(y)$$

$$v = \sum_{m=1}^{r}\sum_{i=1}^{n} N_i v_{im} Y'_m(y)$$

$$w = \sum_{m=1}^{r}\sum_{i=1}^{n} N_i w_{im} Y_m(y) \tag{8.10}$$

or simply

$$\{f\} = \{u, v, w\}^T = \sum_{m=1}^{r}\sum_{i=1}^{n} [N]_{im}\{\delta\}_{im} \tag{8.11}$$

where n is the number of nodal lines in each finite prism, N_i is the isoparametric shape function for nodal line i, and $[N]_{im}$ is the displacement submatrix in the form

$$[N]_{im} = \begin{bmatrix} N_i Y_m(y) & 0 & 0 \\ 0 & N_i Y'_m(y) & 0 \\ 0 & 0 & N_i Y_m(y) \end{bmatrix} \tag{8.12}$$

If both ends are simply supported, $Y_m(y) = \sin m\pi y/l$, and Equation 8.10 satisfies the following end conditions:

$$u = 0, \quad w = 0 \quad and \quad \partial v/\partial y = 0 \quad for \quad y = 0 \quad and \quad y = l \tag{8.13}$$

This implies that the in-plane displacements are prevented at the ends, while the longitudinal strains are zero which leads to zero longitudinal stresses at the ends for isotropic and orthotropic materials.

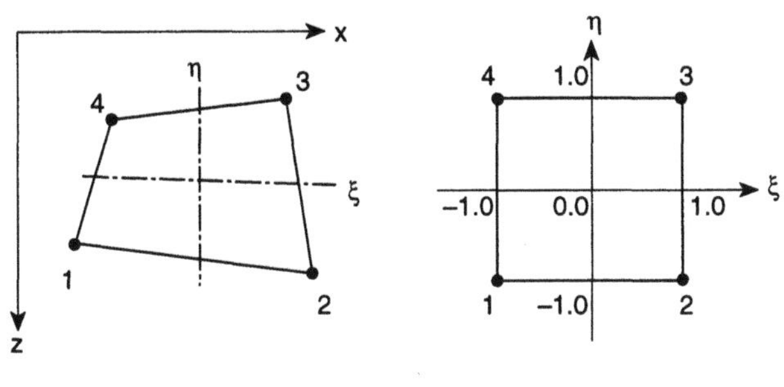

Figure 8.4 *Coordinate transformation.*

The isoparametric shape function N_i is defined in terms of natural coordinates ξ and η (Fig. 8.4). Depending on the number of nodal lines in each prism. N_i can be expressed as follows:

1. $n = 4$ (linear interpolation)

$$N_i = \frac{1}{4}(1 + \xi\xi_i)(1 + \eta\eta_i) \tag{8.14}$$

where ξ_i and η_i are the coordinates of nodal line i.

2. $n = 8$ (quadratic interpolation)

$$N_i = \frac{1}{4}[(1+\xi\xi_i)(1+\eta\eta_i) - (1-\xi^2)(1+\eta\eta_i) - (1+\xi\xi_i)(1-\eta^2)] \quad (8.15)$$

for the corner nodes and

$$N_i = \frac{1}{2}(1-\xi^2)(1+\eta\eta_i) \quad when \quad \xi_i = 0 \quad (8.16)$$

$$N_i = \frac{1}{2}(1+\xi\xi_i)(1-\eta^2) \quad when \quad \eta_i = 0 \quad (8.17)$$

3. $n = 12$ (cubic interpolation)

$$N_i = \frac{1}{32}(1+\xi\xi_i)(1+\eta\eta_i)[9(\xi^2+\eta^2) - 10] \quad (8.18)$$

for the corner nodes,

$$N_i = \frac{9}{32}(1+\xi\xi_i)(1-\eta^2)(1+9\eta\eta_i) \quad (8.19)$$

for the nodes on the sides $\xi_i = \pm 1$ with $\eta_i = \pm 1/3$, and

$$N_i = \frac{9}{32}(1-\xi^2)(1+9\xi\xi_i)(1+\eta\eta_i) \quad (8.20)$$

for the nodes on the sides $\eta_i = \pm 1$ with $\xi_i = \pm 1/3$.

For a given point, the $x - z$ coordinates and the ξ–η coordinates are related by the transformation

$$\begin{aligned} x &= \sum_{i=1}^{n} N_i(\xi,\eta)x_i \\ z &= \sum_{i=1}^{n} N_i(\xi,\eta)z_i \end{aligned} \quad (8.21)$$

where x_i and z_i are the coordinates of nodal line i.

Differentiating both sides of Equation 8.21 with respect to ξ and η yields

$$\begin{aligned} \frac{\partial x}{\partial \xi} &= \sum_{i=1}^{n} \frac{\partial N_i}{\partial \xi} x_i \\ \frac{\partial x}{\partial \eta} &= \sum_{i=1}^{n} \frac{\partial N_i}{\partial \eta} x_i \\ \frac{\partial z}{\partial \xi} &= \sum_{i=1}^{n} \frac{\partial N_i}{\partial \xi} z_i \\ \frac{\partial z}{\partial \eta} &= \sum_{i=1}^{n} \frac{\partial N_i}{\partial \eta} z_i \end{aligned} \quad (8.22)$$

In the finite prism formulation, $\frac{\partial N_i}{\partial x}$ and $\frac{\partial N_i}{\partial z}$ are often required. These derivatives can be derived from the above transformation. First, from the definition of N_i, $\frac{\partial N_i}{\partial \xi}$ and $\frac{\partial N_i}{\partial \eta}$ are readily obtained. $\frac{\partial N_i}{\partial \xi}$ and $\frac{\partial N_i}{\partial \eta}$ are related to $\frac{\partial N_i}{\partial x}$ and $\frac{\partial N_i}{\partial z}$ as

$$\begin{Bmatrix} \dfrac{\partial N_i}{\partial \xi} \\ \dfrac{\partial N_i}{\partial \eta} \end{Bmatrix} = \begin{bmatrix} \dfrac{\partial x}{\partial \xi} & \dfrac{\partial y}{\partial \xi} \\ \dfrac{\partial x}{\partial \eta} & \dfrac{\partial y}{\partial \eta} \end{bmatrix} \begin{Bmatrix} \dfrac{\partial N_i}{\partial x} \\ \dfrac{\partial N_i}{\partial y} \end{Bmatrix} = [J] \begin{Bmatrix} \dfrac{\partial N_i}{\partial x} \\ \dfrac{\partial N_i}{\partial y} \end{Bmatrix} \tag{8.23}$$

where [J] is the Jacobian matrix.

Thus, the derivatives of N_i with respect to x and z can be obtained as

$$\begin{Bmatrix} \dfrac{\partial N_i}{\partial x} \\ \dfrac{\partial N_i}{\partial z} \end{Bmatrix} = [J]^{-1} \begin{Bmatrix} \dfrac{\partial N_i}{\partial \xi} \\ \dfrac{\partial N_i}{\partial \eta} \end{Bmatrix} = \frac{1}{|J|} \begin{bmatrix} \dfrac{\partial y}{\partial \eta} & -\dfrac{\partial y}{\partial \xi} \\ -\dfrac{\partial x}{\partial \eta} & \dfrac{\partial x}{\partial \xi} \end{bmatrix} \begin{Bmatrix} \dfrac{\partial N_i}{\partial \xi} \\ \dfrac{\partial N_i}{\partial \eta} \end{Bmatrix} \tag{8.24}$$

in which $|J|$ is the Jacobian determinant

$$|J| = \frac{\partial x}{\partial \xi}\frac{\partial y}{\partial \eta} - \frac{\partial y}{\partial \xi}\frac{\partial x}{\partial \eta} \tag{8.25}$$

From the strain–displacement relationship 8.3, the strain vector can be expressed in terms of the nodal displacement parameters as

$$\{\epsilon\} = [\epsilon_x, \epsilon_y, \epsilon_z, \gamma_{xy}, \gamma_{yz}, \gamma_{zx}]^T = \sum_{m=1}^{r}\sum_{i=1}^{n}[B]_{im}\{\delta\}_{im} \tag{8.26}$$

where $[B]_{im}$ is the strain matrix,

$$[B]_{im} = \begin{bmatrix} N_{i,x}Y_m & 0 & 0 \\ 0 & N_iY''_m & 0 \\ 0 & 0 & N_{i,z}Y_m \\ N_iY'_m & N_{i,x}Y'_m & 0 \\ 0 & N_{i,z}Y'_m & N_iY'_m \\ N_{i,z}Y_m & 0 & N_{i,x}Y_m \end{bmatrix} \tag{8.27}$$

in which $N_{i,x} = \frac{\partial N_i}{\partial x}$ and $N_{i,z} = \frac{\partial N_i}{\partial z}$.

Then, the stiffness matrix and the load vector of the finite prism can be formed by following the standard procedures given in Chapter 4. The submatrices corresponding to nodal lines i and j can be evaluated as

$$[K]_{ijmn} = \int [B]^T_{im}[D][B]_{jn}dv \tag{8.28}$$

The integration in the above equations can be carried out analytically in the y direction. However, in the x and z directions the Gaussian quadrature commonly used in isoparametric in-plane finite element analysis is preferred (Cook et al., 1989).

By following the standard procedure in Chapter 4, the load vector can be calculated as:

$$\{p\}_{im} = \left\{ \begin{array}{c} X_{im} \\ Y_{im} \\ Z_{im} \end{array} \right\} = \int_{\Omega} [N]_{im}^{T} \left\{ \begin{array}{c} p_x \\ p_y \\ p_y \end{array} \right\} dv \tag{8.29}$$

where p_x, p_y and p_z are the components of load distributed in the region Ω within the prism. For some typical loading conditions, Equation 8.29 can be simplified as follows:

1. Point force (P_x, P_y, P_z) at $y = b$ on nodal line i:

$$\{p\}_{im} = \left\{ \begin{array}{c} X_{im} \\ Y_{im} \\ Z_{im} \end{array} \right\} = \left\{ \begin{array}{c} P_x Y_m(b) \\ P_y Y'_m(b) \\ P_z Y_m(b) \end{array} \right\} \tag{8.30}$$

   where P_x, P_y and P_z are the components of the load in x, y and z directions respectively.

2. Line load of intensity (q_x, q_y, q_z) on nodal line i:

$$\{p\}_{im} = \left\{ \begin{array}{c} X_{im} \\ Y_{im} \\ Z_{im} \end{array} \right\} = \int_l \left\{ \begin{array}{c} q_x Y_m(y) \\ q_y Y'_m(y) \\ q_z Y_m(y) \end{array} \right\} dy \tag{8.31}$$

3. Surface load of intensity (p_x, p_y, p_z) on surface $\eta = 1$:

$$\{p\}_{im} = \left\{ \begin{array}{c} X_{im} \\ Y_{im} \\ Z_{im} \end{array} \right\} = \int_l \int_s \left\{ \begin{array}{c} p_x Y_m(y) \\ p_y Y'_m(y) \\ p_z Y_m(y) \end{array} \right\} N_i ds dy \tag{8.32}$$

   where s represents the side $\eta = 1$ on the cross-section of the strip,

$$ds = \sqrt{dx^2 + dz^2} = \sqrt{(\frac{\partial x}{\partial \xi})^2 + (\frac{\partial z}{\partial \xi})^2} d\xi \tag{8.33}$$

After assembling the above strip matrices over the entire structure, the displacement parameters can be obtained by solving the following matrix equations using standard computer subroutines:

$$[K]\{\delta\} = \{P\} \tag{8.34}$$

Then, the stresses at any point are calculated by means from Equations 8.6 and 8.26.

8.4 Circularly curved finite prism

In the analysis of circularly curved bridges with simply supported ends, the cylindrical coordinate system (Fig. 8.5) is more convenient. The displacement functions for the corresponding finite strip are written as

$$u = \sum_{m=1}^{r}\sum_{i=1}^{n} N_i u_{im} \sin\frac{m\pi\phi}{\alpha} \tag{8.35}$$
$$v = \sum_{m=1}^{r}\sum_{i=1}^{n} N_i v_{im} \cos\frac{m\pi\phi}{\alpha}$$
$$w = \sum_{m=1}^{r}\sum_{i=1}^{n} N_i w_{im} \sin\frac{m\pi\phi}{\alpha}$$

where u, v and w are the displacement components in the r, ϕ and z directions respectively, and α is the spanwise subtended angle in the $r-\phi$ plane.

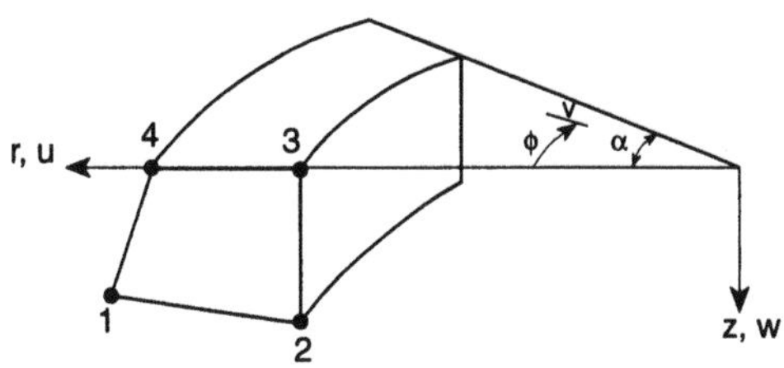

Figure 8.5 *Curved finite prism.*

In the cylindrical coordinate system, the strain–displacement relationships take the form

$$\begin{aligned}
\epsilon_r &= \frac{\partial u}{\partial r}\\
\epsilon_\phi &= \frac{u}{r} + \frac{1}{r}\frac{\partial v}{\partial \phi}\\
\epsilon_z &= \frac{\partial w}{\partial z}\\
\gamma_{r\phi} &= \frac{1}{r}\frac{\partial u}{\partial \phi} + \frac{\partial v}{\partial r} - \frac{v}{r}\\
\gamma_{\phi z} &= \frac{\partial v}{\partial z} + \frac{1}{r}\frac{\partial w}{\partial \phi}\\
\gamma_{zr} &= \frac{\partial u}{\partial z} + \frac{\partial w}{\partial r}
\end{aligned} \tag{8.36}$$

Then, the stiffness matrix and the load vector of the prism can be obtained by following the procedures used in the previous section.

Table 8.1 *Stresses and deflections at the center of rectangular plates under uniform load* $(a = 10, E = 1, \nu = 0.15, q = 1)$ *(Y. K. Cheung, 1976).*

b/a		t/a					
		0.10		0.25		0.50	
		Top	Bottom	Top	Bottom	Top	Bottom
1.0	w	491.9639	491.4200	37.7139	36.4321	9.0208	6.0001
	σ_x	−26.0287	25.9095	−4.2806	4.1699	−1.2427	1.0938
	σ_y	−26.0287	25.9095	−4.2806	4.1699	−1.2427	1.0938
1.5	w	924.2100	923.6643	67.4901	66.1891	13.0846	11.0972
	σ_x	−46.7853	46.7066	−7.6061	−7.5305	−2.0725	1.9684
	σ_y	−24.1373	23.9917	−4.0081	3.8476	−1.1594	0.9990
2.0	w	1207.1999	1206.6533	86.6447	85.3310	16.5533	13.9514
	σ_x	−60.0407	59.9828	−9.7299	9.6724	−2.6056	2.5186
	σ_y	−20.0641	19.9060	−3.3642	3.1857	−1.0032	0.8134

8.5 Numerical examples

1. Simply supported thick plate under uniform load

A rectangular thick plate is simply supported on the four edges, and subjected to uniform load on the top surface. The plate was simulated by five equal finite layers for t=0.5a or three equal finite layers for t/a of 0.1 and 0.25, and the 25 terms of the series (m =1, 3, 5, 7, 9 and n =1, 3, 5, 7, 9) were employed (Cheung and Chakrabarti, 1971). The results are shown in Table 8.1 for $a = 10, E = 1, \nu = 0.15, q = 1$ and a variety of aspect ratios.

It can be noticed that for $t/a = 0.1$ the results are similar to those from thin plate theory, and the deflections and stresses are almost identical at the top and bottom of the plate. For higher t/a ratios the deflections and stresses at these two positions have obvious differences in magnitude.

2. Thick concrete box-girder bridge

A simply supported thick concrete box girder bridge (Fig. 8.6) was analyzed by the finite prism method (Chan and Cheung, 1978). Two loading cases, i.e. the self-weight of the bridge and a AASHTO HS 25 vehicle load as shown in Fig. 8.6, were considered in the analysis. By taking advantage of symmetry of the cross-section and loads, only half the bridge needs to be simulated. Eleven quadratic finite prisms (with eight nodal lines each) and ten series terms were used in each case.

The results obtained for a section under the second axle of the vehicle is given in the same figure. It can be seen that the variation of the longitudinal stresses σ_y over the depth of the cross-section is in accordance with the linear stress distribution assumed in simple beam theory. For the self-weight load case, it is also possible to see the high concentration of transverse stress σ_x at the bottom of cantilever slab.

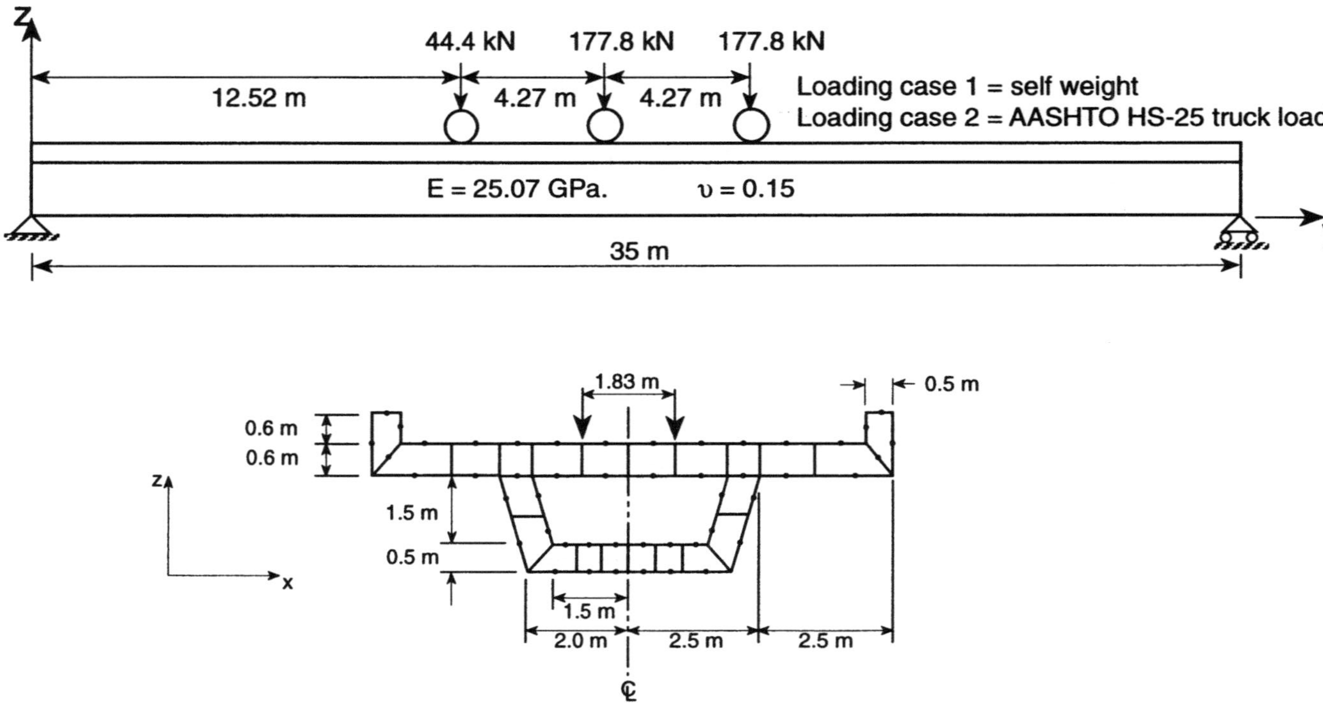

Figure 8.6 *Analysis of thick concrete box girder bridge using finite prisms (M. S. Cheung and Chan, 1978B).*

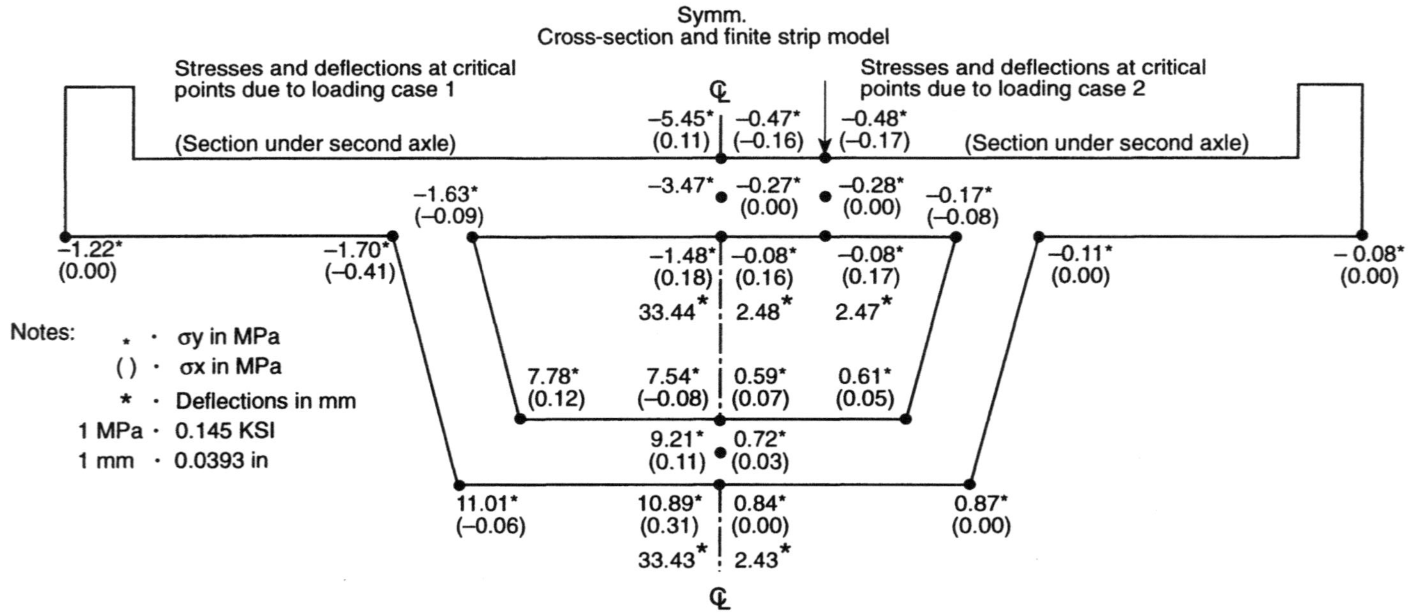

Figure 8.6 *(Continued)*

3. Fan shaped plate under uniform load

A fan shaped plate of $\alpha = 60°$ and $\nu = 0.3$ is subjected to uniform load. First, the plate is analyzed using eight curved plate strips (Fig. 8.7) (M. S. Cheung and Chan, 1981). The maximum moments and deflections along the radial center line are presented in Table 8.2. Then, the plate is analyzed by 8 × 2 curved quadratic finite prisms. The resulting maximum deflections and stresses (on the top surface) along the radial center line are given in Table 8.3 for $a = 12, R_I/R_E = 0.25, E = 10000$ and $q = 1$. Table 8.3 indicates that for relatively thin plates, the finite prism results are not significantly different from those of the finite strip. However, with increasing plate thickness, finite prism results start to deviate from the thin plate values due to the effect of shear deformation. This effect influences the deflection results more than the stress results, and is most pronounced when all the edges of a plate are fully clamped.

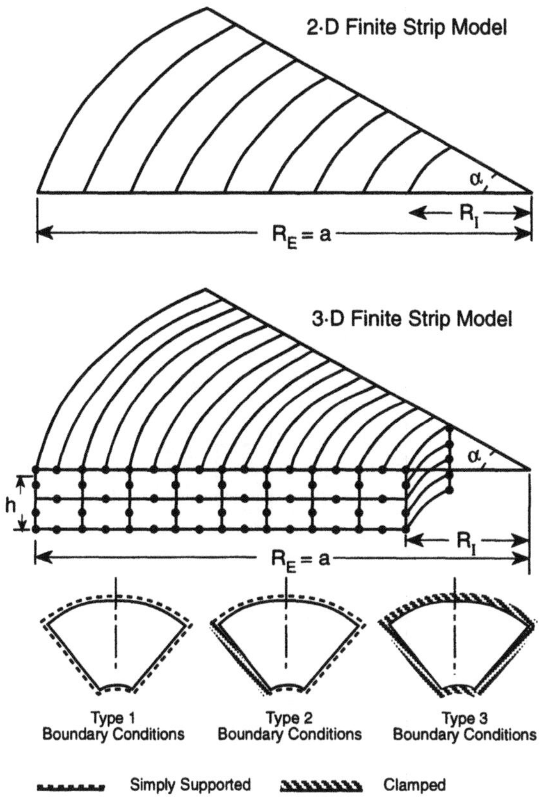

Figure 8.7 *Fan shaped plates (M. S. Cheung and Chan, 1981).*

Table 8.2 *Maximum deflections and moments along the radial center line of fan shaped plates (two-dimensional finite strip solution, $\alpha = 60°, \nu = 0.3$); (M. S. Cheung and Chan, 1981).*

R_I/R_E	Type 1 boundary conditions			Type 2 boundary conditions				Type 3 boundary conditions			
	$W_{\max}$	$M_{r\max}$	$M_{\theta\max}$	$W_{\max}$	$M_{r\max}$	$M_{\theta\max}$	$-M_{\theta\max}$	$W_{\max}$	$M_{r\max}$	$M_{\theta\max}$	$-M_{\theta\max}$
0.00	0.000984	0.024944	0.025060	0.000600	0.019424	0.018935	−0.033250	0.000282	0.011287	0.013011	−0.021597
0.25	0.000927	0.024846	0.023556	0.000612	0.018416	0.016743	−0.032801	0.000276	0.011643	0.012638	−0.021123
0.50	0.000500	0.021236	0.012204	0.000419	0.018720	0.013162	−0.018143	0.000141	0.010173	0.006203	−0.008721
0.75	0.000490	0.007885	0.002438	0.000046	0.007671	0.002687	−0.002484	0.000009	0.002912	0.000907	−0.001335
Multiplier	qa^4/D	qa^2		qa^4/D		qa^2		qa^4/D		qa^2	

Table 8.3 *Maximum deflections ($w_{\max}$) and stresses ($\sigma_{r\max}$ and $\sigma_{\theta\max}$) along the radial center line of fan shaped plates (M. S. Cheung and Chan, 1981)*

Type 1 boundary condition

a/h	$w_{max} \times 10^2$		σ_{rmax}		$\sigma_{\theta max}$	
	3-D	2-D	3-D	2-D	3-D	2-D
15.0	4.3258	4.0998	−33.6134	−33.5421	−31.9316	−31.8006
10.0	1.3633	1.2147	−14.9879	−14.9076	−14.4806	−14.1336
7.5	0.6122	0.5125	−8.5169	−8.3855	−8.2701	−7.9502
5.0	0.2123	0.1518	−3.8765	−3.7269	−3.7176	−3.5334

Type 2 boundary condition

a/h	$w_{max} \times 10^2$		σ_{rmax}		$\sigma_{\theta max}$	
	3-D	2-D	3-D	2-D	3-D	2-D
15.0	2.7938	2.7066	−24.8726	−24.8616	−25.4728	−25.3031
10.0	0.9021	0.8019	−11.1245	−11.0496	−11.2641	−11.2458
7.5	0.4221	0.3383	−6.4172	−6.2154	−6.5121	−6.3258
5.0	0.1594	0.1002	−3.0593	−2.7624	−3.1134	−2.8115

Type 3 boundary condition

a/h	$w_{max} \times 10^2$		σ_{rmax}		$\sigma_{\theta max}$	
	3-D	2-D	3-D	2-D	3-D	2-D
15.0	1.3896	1.2206	−15.9782	−15.7181	−17.1121	−17.0613
10.0	0.4841	0.3617	−7.5896	−6.9858	−7.6132	−7.5828
7.5	0.2457	0.1526	−4.6200	−3.9295	−4.7287	−4.2653
5.0	0.1097	0.0452	−2.4641	−1.7465	−2.5649	−1.8957

Notes: (1) $E = 10\,000, \nu = 0.3, \text{UDL} = 1.0, a = 12.0, \alpha = 60°, R_I/R_E = 0.25$.

(2) Stresses shown are top fiber stresses corresponding to maximum positive bending moments.

CHAPTER 9

Vibration and stability analyses

9.1 Introduction

Soon after their application to the static analysis of a variety of bridge structures, the semianalytical finite strip method and the spline finite strip method extended to the analysis of vibration and stability. This chapter describes the basic procedures for vibration and stability finite strip analyses. Explicit expressions for the mass matrices of some typical finite strips are also given. This is followed by numerical examples.

9.2 Vibration finite strip analysis

In the static finite strip analysis of a structure, after assembling the overall stiffness matrix and load vector from individual finite strip matrices, the following equilibrium matrix equation can be obtained:

$$[K]\{\delta\} = \{P\} \tag{9.1}$$

If the structure is moving, the dynamic analysis can be reduced to the static case by following D'Alembert's principle, i.e. by applying inertia forces on the structure. The inertia force is equal to the product of the mass and the acceleration, and acting in the direction of negative acceleration. The load vector of the inertia forces can be obtained in the same method as for external loading. Thus, for a structure in free vibration, at any instant of time, the overall matrix equation takes the form

$$[K]\{\delta(t)\} = -[M]\frac{\partial^2}{\partial t^2}\{\delta(t)\} \tag{9.2}$$

where $\delta(t)$ is now a function of time, and $[M]$ is the overall mass matrix of the structure assembled from individual strip mass matrices $[M^I]$. The assembly procedure is identical to that for stiffness matrices (see Chapter 4).

For free vibration analysis, it can be assumed that the structure is vibrating as

$$\begin{aligned} \{\delta(t)\} &= \{\delta\}\sin\omega t \\ \frac{\partial^2}{\partial t^2}\{\delta(t)\} &= -\omega^2\{\delta\}\sin\omega t \end{aligned} \tag{9.3}$$

in which ω is the natural frequency.

Substituting Equation 9.3 into 9.2 and cancelling out $\sin\omega t$ on both sides yields the matrix equation

$$[K]\{\delta\} = \omega^2[M]\{\delta\} \tag{9.4}$$

This is a generalized eigenvalue problem in which $[K]$ and $[M]$ are both band symmetrical and positive definite matrices for actual bridge structures. Nowadays, many standard computer programs are available and ready to use for computing the frequency ω and corresponding mode shape $\{\delta\}$ from Equation 9.4.

9.3 Mass matrix of a finite strip

The displacement at any point within a finite strip can be expressed in terms of displacement parameters as

$$\{f\} = [N]\{\delta\} = \sum_{m=1}^{r}[N]_m\{\delta\}_m \tag{9.5}$$

where $\{f\}$ and $\{\delta\}$ are both functions of time.

For a thin plate of thickness h, the inertia force per unit area of mid-plane is

$$\{q\} = -\rho h\frac{\partial^2\{f\}}{dt^2} \tag{9.6}$$

where ρ is the mass per unit volume.

By substituting Equation 9.5 into 9.6 and considering 9.3, the inertia force can be expressed as

$$\begin{aligned}\{q\} &= -\rho h[N]\frac{\partial^2}{\partial t^2}\{\delta(t)\} \\ &= \rho h\omega^2[N]\{\delta\}\sin\omega t\end{aligned} \tag{9.7}$$

Because $\sin\omega t$ will be cancelled out, it can be excluded in subsequent derivations. Thus, Equation 9.7 becomes

$$\{q\} = \rho h\omega^2[N]\{\delta\} \tag{9.8}$$

From the principle of minimum total potential energy, the equivalent nodal forces of the distributed inertia loading can be calculated as

$$\begin{aligned}\{P\} &= \int [N]^T\{q\}dxdy \\ &= \int \rho h\omega^2[N]^T[N]\{\delta\}dxdy \\ &= \int \rho h\omega^2[[N]_1[N]_2...[N]_r]^T[[N]_1[N]_2...[N]_r]\{\delta\}dxdy \\ &= \omega^2[M^I]\{\delta\}\end{aligned} \tag{9.9}$$

in which r is the number of series terms required for the analysis.

Thus, the submatrix in the mass matrix of strip I corresponding to series terms m and n can be expressed as

$$[M^I]_{mn} = \int \rho h [N]_m^T [N]_n dxdy \tag{9.10}$$

If a concentrated mass M_c is located at a point $x = x_c$, $y = y_c$, it is only necessary to substitute the coordinates of the mass into $[N]_m$ and $[N]_n$ in Equation 9.10 and also to dispense with the integration. Therefore, the corresponding mass matrix becomes

$$[M^I]_{mn} = M_c [N]_m^T [N]_n \tag{9.11}$$

in which $[N]_m$ and $[N]_n$ are evaluated at $x = x_c$, $y = y_c$.

9.4 Mass matrix of a plate strip

For the plate strip introduced in Section 4.3, the displacement matrix $[N]_m$ is defined as

$$\begin{aligned} [N]_m &= [(1 - 3X^2 + 2X^3), x(1 - 2X + X^2), (3X^2 - 2X^3), x(X^2 - X)]Y_m \\ &= [N]Y_m \end{aligned} \tag{9.12}$$

If the product ρh in Equation 9.10 is constant in the transverse direction, the following equation can be obtained:

$$[M^I]_{mn} = \int_0^b [N]^T [N] dx \int_0^l \rho h Y_m Y_n dy \tag{9.13}$$

After integration in the x direction, the entries of the mass matrix are shown in Fig. 9.1.

$$[M^{Ib}]_{mn} = \rho h I_l \begin{bmatrix} \frac{13b}{35} & & & \text{Symmetrical} \\ \frac{11b^2}{210} & \frac{b^3}{105} & & \\ \frac{9b}{70} & \frac{13b^2}{420} & \frac{13b}{35} & \\ -\frac{13b^2}{420} & -\frac{3b^3}{420} & -\frac{11b^2}{210} & \frac{b^3}{105} \end{bmatrix}$$

$I_l = \int_0^l Ym\ Yn\ dy$

Figure 9.1 *Mass matrix of a plate strip.*

If ρh is constant within the entire strip, the integral $\int_0^l \rho h Y_m Y_n dy$ in

Equation 9.13 is zero for $m \neq n$ because of the orthogonal properties of beam eigenfunctions. This means that all off-diagonal submatrices vanish and, as a result, the mass matrix can be greatly simplified.

For a simply supported plate strip with constant thickness and uniform density, the integral in the y direction becomes very simple:

$$\int_0^l \rho h Y_m Y_n dy = \rho h \int_0^l \sin^2 k_m y dy = \begin{cases} \rho h l/2 & for \quad m = n \\ 0 & for \quad m \neq n \end{cases} \tag{9.14}$$

Equation 9.13 is also applicable for higher order rectangular plate strip if the matrix $[N]$ contains the higher order transverse displacement functions.

If Y_m represent the eigenfunctions of a continuous beam, Equation 9.13 can also be used for the dynamic analysis of continuous plates.

9.5 Mass matrix of a plane stress strip

In the general case of end support conditions, the displacement function of a plane stress strip takes the following form (Chapter 4):

$$\begin{Bmatrix} u \\ v \end{Bmatrix} = \sum_{m=1}^{r}$$

$$\begin{bmatrix} (1-X)Y_m(y) & 0 & XY_m(y) & 0 \\ 0 & (1-X)Y'_m(y)/\mu_m & 0 & XY'_m(y)/\mu_m \end{bmatrix} \begin{Bmatrix} u_{im} \\ v_{im} \\ u_{jm} \\ v_{jm} \end{Bmatrix} \tag{9.15}$$

Therefore

$$[N]_m = \begin{bmatrix} (1-X)Y_m(y) & 0 & XY_m(y) & 0 \\ 0 & (1-X)Y'_m(y)/\mu_m & 0 & XY'_m(y)/\mu_m \end{bmatrix} \tag{9.16}$$

The mass matrix can then be developed according to Equation 9.10. The explicit form is given in Fig. 9.2. Because the integral $\int Y'_m Y'_n dy$ is generally not zero for $m \neq n$ even for constant ρh, the terms in the mass matrix are coupled except for the simple support conditions.

9.6 Mass matrix of a flat shell strip

The flat shell strip is subjected to both the plane stresses and bending forces. The displacement parameter vector of a strip may take the form

$$\{\delta\}_m = [u_{im}, v_{im}, w_{im}, \theta_{im}, u_{jm}, v_{jm}, w_{jm}, \theta_{jm}]^T \tag{9.17}$$

From Equation 9.10 it can be concluded that the mass matrix of the flat shell can be made up from appropriate elements of the $[M^{Ip}]_{mn}$ and

$$[M^{Ip}]_m = \rho h \begin{bmatrix} \frac{b}{3} I_1 & & & \text{Symmetrical} \\ 0 & \frac{b}{3\mu_m\mu_n} I_2 & & \\ \frac{b}{6} I_1 & 0 & \frac{b}{3} I_1 & \\ 0 & \frac{b}{6\mu_m\mu_n} I_2 & 0 & \frac{b}{3\mu_m\mu_n} I_2 \end{bmatrix}$$

$$I_1 = \int_0^{\ell} Y_m Y_n \, dy \qquad\qquad I_2 = \int_0^{\ell} Y'_m Y'_n \, dy$$

Figure 9.2 *Mass matrix of an in-plane strip.*

$[M^{Ib}]_{mn}$ matrices, where the superscripts p and b refer to in-plane stress and bending, respectively. Then

$$[M^I]_{mn} = \begin{bmatrix} [M^I_{ii}]_{mn} & [M^I_{ij}]_{mn} \\ [M^I_{ji}]_{mn} & [M^I_{jj}]_{mn} \end{bmatrix} \tag{9.18}$$

and

$$[M^I_{ij}]_{mn} = \begin{bmatrix} [M^{Ip}_{ij}]_{mn} & [0] \\ [0] & [M^{Ib}_{ij}]_{mn} \end{bmatrix} \tag{9.19}$$

The coordinate transformation procedure presented in Chapter 4 can be used here for the coordinate transformation of mass matrices.

9.7 Bending and in-plane interaction

So far, the bending and in-plane deformation of plates are assumed to be uncoupled. In fact, if the deflection of a plate is large in comparison with the plate thickness, the deflection will cause in-plane strains, and interaction begins. In the linear stability analysis, only initial buckling is considered. Moreover, the existing in-plane stresses $\{\sigma^\circ\}$ have significant magnitude, and therefore it is possible to assume the in-plane stresses are not affected by bending deformation.

During the initial buckling, if points A and B (Fig. 9.3) in the midplane move vertically, the original length dx becomes

$$dx\sqrt{1 + (\frac{\partial w}{\partial x})^2} = dx\{1 + \frac{1}{2}(\frac{\partial w}{\partial x})^2 + ...\} \tag{9.20}$$

Thus, by neglecting the higher order term, the in-plane strain in the x

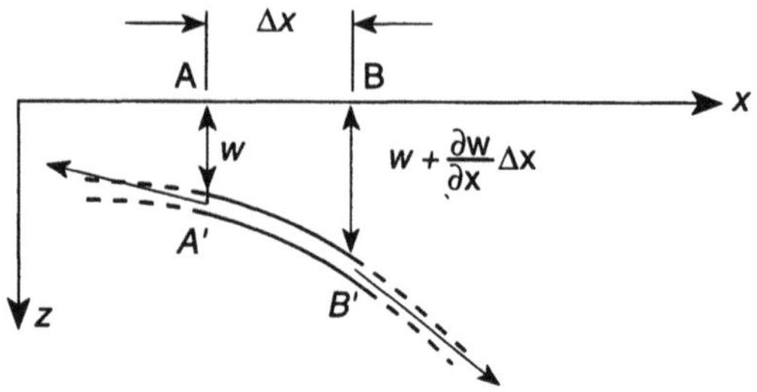

Figure 9.3 *In-plane strain due to lateral deflection (1).*

direction, due to bending, can be obtained as

$$\epsilon_x^\circ = \frac{1}{2}(\frac{\partial w}{\partial x})^2 \tag{9.21}$$

Similarly, it can be shown (Timoshenko and Gere, 1961) that

$$\epsilon_y^\circ = \frac{1}{2}(\frac{\partial w}{\partial y})^2 \tag{9.22}$$

and

$$\gamma_{xy}^\circ = (\frac{\partial w}{\partial x})(\frac{\partial w}{\partial y}) \tag{9.23}$$

During bending, the initial stresses σ_x°, σ_y° and τ_{xy}° perform work on the additional in-plane strains, and this work is transformed into strain energy. Therefore, the strain energy per unit volume of the plate will increase by

$$\sigma_x^\circ \frac{1}{2}(\frac{\partial w}{\partial x})^2 + \sigma_y^\circ \frac{1}{2}(\frac{\partial w}{\partial y})^2 + \tau_{xy}^\circ (\frac{\partial w}{\partial x})(\frac{\partial w}{\partial y}) \tag{9.24}$$

or, in a matrix form

$$\frac{1}{2}\left\{\begin{array}{c} \dfrac{\partial w}{\partial x} \\ \\ \dfrac{\partial w}{\partial y} \end{array}\right\}^T \left[\begin{array}{cc} \sigma_x^\circ & \tau_{xy}^\circ \\ \tau_{xy}^\circ & \sigma_y^\circ \end{array}\right] \left\{\begin{array}{c} \dfrac{\partial w}{\partial x} \\ \\ \dfrac{\partial w}{\partial y} \end{array}\right\} \tag{9.25}$$

In a finite plate strip, by using displacement functions, the slopes can be expressed in terms of displacement parameters:

$$\left\{\begin{array}{c} \dfrac{\partial w}{\partial x} \\ \\ \dfrac{\partial w}{\partial y} \end{array}\right\} = \sum_{m=1}^{r} [G]_m \{\delta\}_m \tag{9.26}$$

Then, the additional strain energy for entire strip becomes

$$\frac{1}{2}\sum_{m=1}^{r}\sum_{n=1}^{r}\{\delta\}_m^T[K_G]_{mn}\{\delta\}_n \tag{9.27}$$

with

$$[K_G]_{mn} = \int [G]_m^T \begin{bmatrix} \sigma_x^\circ & \tau_{xy}^\circ \\ \tau_{xy}^\circ & \sigma_y^\circ \end{bmatrix} [G]_n h dx dy \tag{9.28}$$

This additional strain energy must be taken into consideration in the assembling and minimization procedures described in Chapter 4. The derivatives of this energy with respect to the displacement parameters have the expression

$$[K_G]\{\delta\} \tag{9.29}$$

Thus, the interaction between bending and in-plane stresses can be included by adding an additional stiffness $[K_G]$ to the original $[K]$, so that the structural matrix equation becomes

$$[K]\{\delta\} + [K_G]\{\delta\} = \{P\} \tag{9.30}$$

This additional stiffness matrix $[K_G]$ is symmetrical, as seen from Equation 9.28, and is referred to as the geometric stiffness matrix or initial stress stiffness matrix.

9.8 Stability analysis of plates

In the linear stability analysis of flat plates, in order to find critical load, it can be assumed that the initial in-plane stresses are increasing proportionally by a scaling factor λ.

When the initial buckling occurs, the in-plane forces alone can keep the plate in bending without any lateral loads ({P}={0}). Thus, the following eigenvalue equation can be obtained

$$([K] + \lambda[K_G])\{\delta\} = \{0\} \tag{9.31}$$

After the above equation is solved for the lowest eigenvalue λ_1, the critical loading will be obtained as the system of $\lambda_1\sigma_x^\circ$, $\lambda_1\sigma_y^\circ$ and $\lambda_1\tau_{xy}^\circ$. The buckling pattern can then be found from the corresponding eigenvector $\{\delta\}$ and displacement functions.

9.9 Stability analysis of thin-walled structures

Thin flat-walled structures such as a thin plate stiffened by thin-walled beams can be simulated by the flat shell strips, which include not only the lateral bending but also the in-plane deformation. In the local coordinate system of any individual strip, deflections due to lateral bending of the structure may occur in both z and x directions.

In the case of long flat-walled structures mainly subjected to compression in the y direction, the strain ϵ_y is the most significant strain component, because its accuracy will affect the final results of the analysis much more than other strain components, it therefore deserves more attention in formulation.

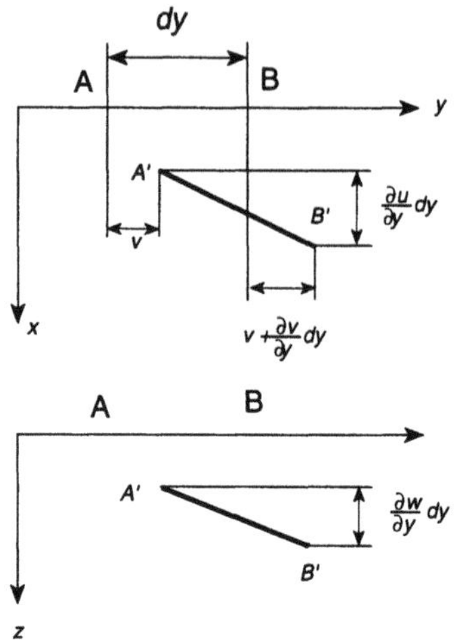

Figure 9.4 *In-plane strain due to lateral deflection (2).*

As seen in Fig. 9.4, if points A and B in the midplane of a strip move to A' and B' respectively, the original length dy becomes

$$dy\sqrt{(1+\frac{\partial v}{\partial y})^2+(\frac{\partial u}{\partial y})^2+(\frac{\partial w}{\partial y})^2}$$
$$= dy\{1+\frac{\partial v}{\partial y}+\frac{1}{2}[(\frac{\partial u}{\partial y})^2+(\frac{\partial v}{\partial y})^2+(\frac{\partial w}{\partial y})^2]+....\} \quad (9.32)$$

Thus, the additional strain in the y direction can be obtained, by neglecting higher order terms, as

$$\epsilon_y^{\circ}=\frac{1}{2}[(\frac{\partial u}{\partial y})^2+(\frac{\partial v}{\partial y})^2+(\frac{\partial w}{\partial y})^2] \quad (9.33)$$

Furthermore, taking into consideration Equations 9.21 and 9.23, all the additional strain components can be written as

$$\begin{aligned}\epsilon_x^{\circ} &= \frac{1}{2}(\frac{\partial w}{\partial x})^2\\ \epsilon_y^{\circ} &= \frac{1}{2}[(\frac{\partial u}{\partial y})^2+(\frac{\partial v}{\partial y})^2+(\frac{\partial w}{\partial y})^2]\\ \gamma_{xy}^{\circ} &= (\frac{\partial w}{\partial x})(\frac{\partial w}{\partial y})\end{aligned} \quad (9.34)$$

Following the same procedure as described in the previous section yields the geometric stiffness matrix

$$[K_G]_{mn} = \int\int \{[G_u]_m^T[S][G_u]_n + [G_v]_m^T[S][G_v]_n + [G_w]_m^T[S][G_w]_n\} h dx dy \tag{9.35}$$

in which

$$[S] = \begin{bmatrix} \sigma_x^\circ & \tau_{xy}^\circ \\ \tau_{xy}^\circ & \sigma_y^\circ \end{bmatrix}$$

$$\left\{ \begin{array}{c} 0 \\ \dfrac{\partial u}{\partial y} \end{array} \right\} = \sum_{m=1}^{r} [G_u]_m \{\delta\}_m$$

$$\left\{ \begin{array}{c} 0 \\ \dfrac{\partial v}{\partial y} \end{array} \right\} = \sum_{m=1}^{r} [G_v]_m \{\delta\}_m$$

$$\left\{ \begin{array}{c} \dfrac{\partial w}{\partial x} \\ \dfrac{\partial w}{\partial y} \end{array} \right\} = \sum_{m=1}^{r} [G_w]_m \{\delta\}_m$$

The stresses σ_x°, σ_y° and τ_{xy}° can be obtained from the axial loading P_0 by static analysis.

The coordinate transformation process presented in Chapter 4 is also applicable for the coordinate transformation of the geometric stiffness matrices.

After the overall matrix of the structure is assembled, the following eigenvalue equation can be obtained:

$$([K] + \lambda[K_G])\{\delta\} = \{0\} \tag{9.36}$$

The smallest eigenvalue λ_1 will give either the critical load $P_{cr} = \lambda_1 P_0$ or the critical stresses $\sigma_{ycr} = \lambda_1 \sigma_y^\circ$. The corresponding eigenvector $\{\delta\}$ combined with the displacement functions represents the buckling pattern.

9.10 Numerical examples

1. Vibration analysis of square plates

The natural frequencies of several square isotropic plates with different boundary conditions were obtained by using eight lower order strips (LO2) with four terms of the series (M. S. Cheung and Y. K. Cheung, 1971) and two higher order strips (HO2) with two terms (M. S. Cheung and Y. K. Cheung, 1972), respectively. The results are compared in Fig. 9.5 with those due to Warburton (1954). Very little difference can be found between the three sets of results.

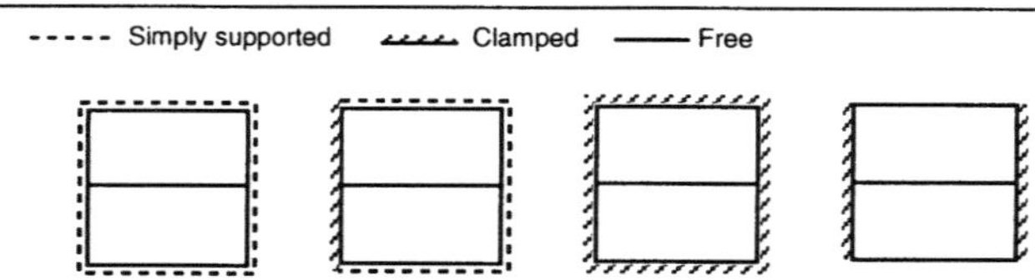

Frequencies	Higher order	Lower order	Warburton (1954)	Higher order	Lower order	Warburton (1954)	Higher order	Lower order	Warburton (1954)	Higher order	Lower order	Warburton (1954)
ω_1	19.74	19.74	19.74	23.66	23.62	23.77	36.05	36.01	36.00	22.33	22.29	22.38
ω_2	49.35	49.32	49.35	51.79	51.62	52.00	73.44	73.48	73.41	27.19	27.08	27.33
ω_3	49.36	49.34	49.35	58.66	58.65	58.65	73.75	73.96	73.41	45.19	44.76	45.49
ω_4	78.97	78.91	78.95	86.41	86.16	86.26	108.53	108.91	108.24	61.55	61.53	61.68
ω_5	98.94	98.94	98.69	100.89	100.35	100.84	133.53	132.09	131.90	68.33	68.29	68.73

Figure 9.5 *Circular frequencies of slabs with various boundary conditions (M. S. Cheung and Y. K. Cheung, 1972).*

2. Vibration of four span continuous plate

A four-span continuous plate with three sides simply supported and the fourth side clamped (Fig. 9.6) was analyzed by using 12 lower order strips (LO2) and three higher-order strips (HO2), respectively. The results from the two analyses are essentially identical.

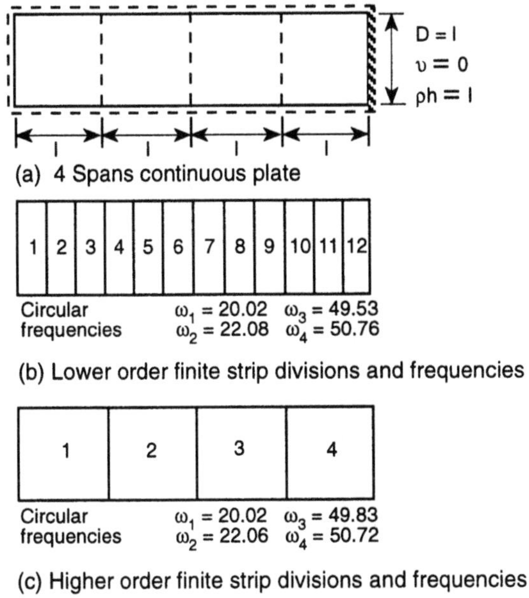

Figure 9.6 *(a) Four-span continuous plate; (b) lower order finite strip divisions and frequencies; (c) higher order finite strip divisions and frequencies (M. S. Cheung and Y. K. Cheung, 1972).*

3. Vibration of stiffened panel

A four-stiffener panel (Fig. 9.7) with two ends simply supported and the other two edges free was analyzed by M. S. Cheung and Y. K. Cheung (1971A), with Poisson's ratio ν assumed to be 0.3 throughout. The lowest 20 natural frequencies are presented in Table 9.1, in terms of a dimensionless parameter $\bar{n}$, defined by

$$\bar{n} = \omega l / \sqrt{E/\rho}$$

In the same table comparison is also made with the results obtained by Wittrick and Williams (1970). It can be seen that a very good agreement is obtained.

The shapes of the modes corresponding to eight natural frequencies listed in Table 9.1 for which the wave number is unity are sketched in Fig. 9.8. The data from which these sketches were prepared consist of displacements and rotations at the top and bottom of all four stiffeners. It should be noted

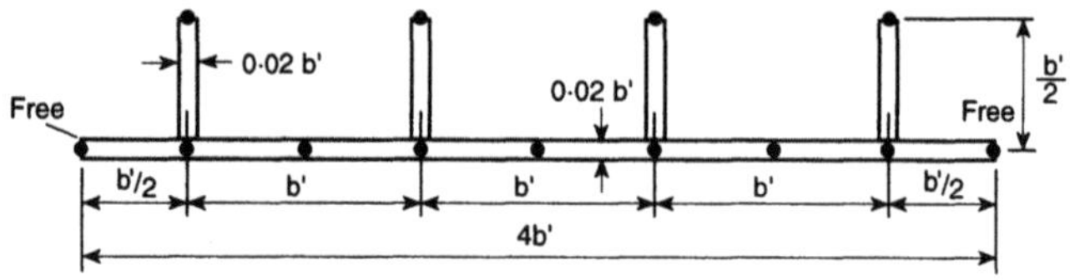

Figure 9.7 *A simply supported four-stiffener panel ($l = 6b'$) (M. S. Cheung and Y. K. Cheung, 1971A).*

Table 9.1 *Natural frequencies of a four-stiffener panel in the range* $0 < \bar{n} < 0.081$ *(12 strips) (M. S. Cheung and Y. K. Cheung, 1971A).*

Mode number	$\bar{n} = \frac{\omega l}{\sqrt{E/\rho}}$ Finite strip	Wittrick and Williams	Wave number (m)	Type of geometry
1	0.0287	0.0286	1	A
2	0.0292	0.0291	1	S
3	0.0365	0.0359	1	S
4	0.0366	0.0362	1	A
5	0.0394	0.0391	1	S
6	0.0396	0.0395	2	S
7	0.0411	0.0410	2	A
8	0.0504	0.0504	3	S
9	0.0521	0.0519	3	A
10	0.0557	0.0555	2	S
11	0.0639	0.0636	1	A
12	0.0643	0.0641	4	S
13	0.0647	0.0642	1	S
14	0.0648	0.0645	2	A
15	0.0653	0.0647	1	A
16	0.0653	0.0647	1	A
17	0.0656	0.0654	4	A
18	0.0736	0.0731	3	A
19	0.0773	0.0768	4	S
20	0.0808	0.0804	5	S

that only half of the cross-section is shown in Fig. 9.8 because the mode is always either symmetrical or antisymmetrical.

4. Vibration of continuous slab bridge with variable thickness
A continuous, variable thickness deck is shown in Fig. 9.9(a). The deck was divided into strips in the transverse direction by averaging the thickness

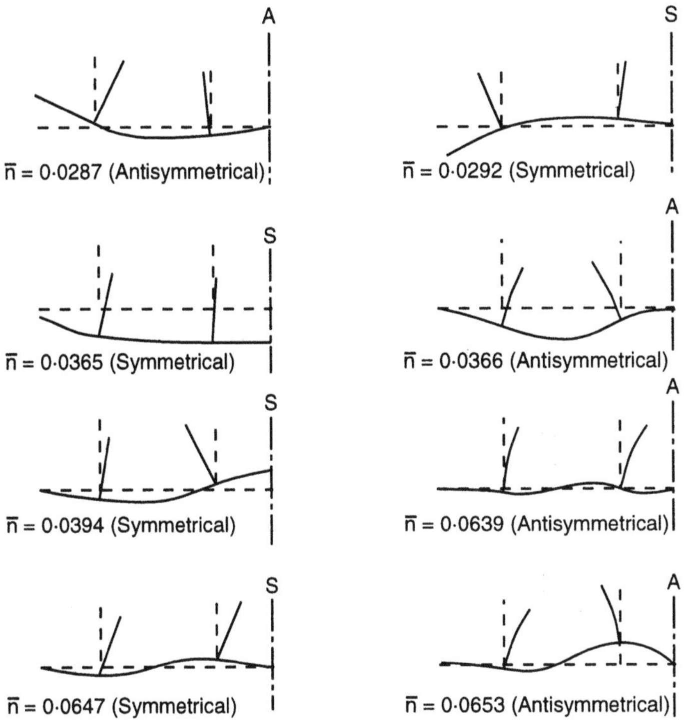

Figure 9.8 *Modal shapes for the first eight natural frequencies of a four-stiffener panel at wave number* $m = 1$ *(M. S. Cheung and Y. K. Cheung, 1971A).*

within each strip, and the idealized deck slab is shown in Fig. 9.9(b) (M. S. Cheung, Y. K. Cheung and Reddy, 1972). Since each strip has two free ends, the eigenfunctions given by Equation 4.48 were employed for the analysis.

5. Stability of Z-section under end compression

A Z-section thin-walled beam is simply supported at both ends and subjected to axial compression. The elastic critical stress was obtained using the lower order finite strips and the higher order ones with six different meshes (Khong, 1990). As the ratio of the length l to the flange width b_2 increases, the resulting critical stress σ_{cr} varies as shown in Fig. 9.10, in which k_1 is the plate buckling coefficient, defined as:

$$k_1 = \sigma_{cr} t_1 b_1^2 / \pi^2 D_1$$

and the subscript 1 represents the web of the section. Furthermore, in this figure 'Cubic' refers to the lower-order strip while 'Quintic' represents the higher-order strip with one internal nodal line (HO3). It can be seen from Fig. 9.10 that mesh A (one lower order strip for each plate component) cannot provide adequate accuracy in predicting the local mode

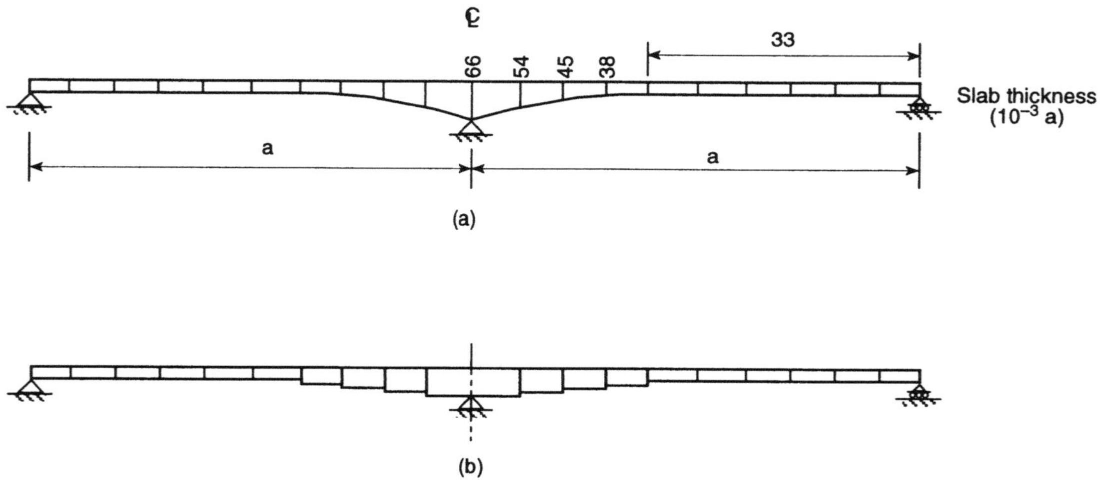

Figure 9.9 *Continuous, variable thickness deck: (a) Longitudinal section; (b) idealized section. The first five circular frequencies are 3.40, 6.76, 7.58, 9.30, 14.46. (units:* $\frac{1}{10a^2}\sqrt{\frac{D_x}{h_{\prime\prime}\rho}}$*; M. S. Cheung, Cheung and Reddy, 1972).*

of buckling. As the length increases, Euler-type buckling occurs and hence the discrepancy is reduced. Meshes C and E yield more accurate results than meshes B and D because in the local buckling mode the deformation of the web plays a more important role than those of the two narrow flanges. This reveals that more than one lower order strip is required to model each flat of a thin-walled plate structure, but that only a single higher-order strip is capable of doing so.

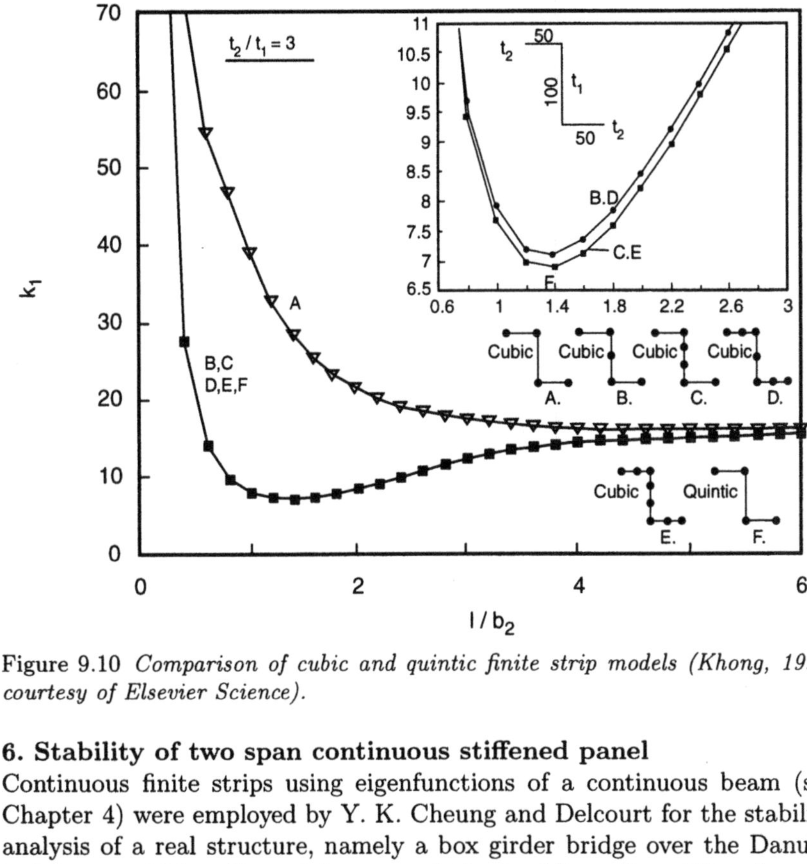

Figure 9.10 *Comparison of cubic and quintic finite strip models (Khong, 1990, courtesy of Elsevier Science).*

6. Stability of two span continuous stiffened panel

Continuous finite strips using eigenfunctions of a continuous beam (see Chapter 4) were employed by Y. K. Cheung and Delcourt for the stability analysis of a real structure, namely a box girder bridge over the Danube in Vienna which collapsed because of the buckling of a stiffened panel made up of a number of longitudinal stiffeners supported by much stiffer transverse diaphragm at midspan. The idealization of the panel is shown in Fig. 9.11(a). Because the rigidity of the longitudinal stiffeners is higher than the optimum rigidity, the buckling pattern is repetitive (local buckling) as shown in Fig. 9.11(b). Consequently, a three-strip model (Fig. 11(c)) was assumed and the conditions of zero slopes at midspans of the flanges and zero deflection at the junction of the web and the flanges were imposed. The results obtained by the finite strip method using nine series terms were compared with those due to a conventional finite element analysis. In

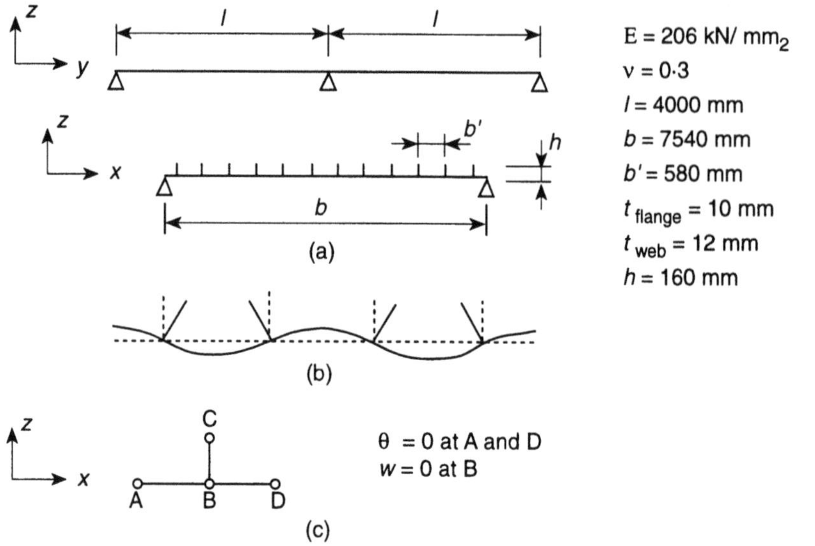

Figure 9.11 *Buckling of a continuous stiffened panel: (a) longitudinal and transverse section of the panel; (b) buckling pattern (transverse direction); (c) three strips idealization (Y. K. Cheung and Delcourt, 1977).*

Table 9.2 *Comparison between finite element method and finite strip method (Y. K. Cheung and Delcourt, 1974).*

Method	Number of degrees of freedom		
	One span	Two spans	Additional span
Finite element	223	435	212
Finite strip	60	108	24

order to keep the finite element analysis down to a manageable size, only one span was analyzed and the span length l was reduced from the original dimension of 4000 mm to 2770 mm, so that ten elements in the longitudinal direction would be adequate to simulate the buckling mode. The finite strip analysis gave a critical stress $\sigma_{ycr} = 256.14$ N mm^{-2}; the corresponding results for the finite element analysis was 267.52 N mm^{-2}. Both solutions produced a buckling mode of five halfwaves in the longitudinal direction. For an infinite long panel the analytical solution is $\sigma_{ycr} = 251.1$ N mm^{-2} (Boughan and Baab, 1944; Przemieniecki, 1973); the finite strip result for a panel of $\alpha = l/b' = 5$ (i.e. $l = 2900$ mm) is $\sigma_{ycr} = 255.75$ N mm^{-2}. Thus the error involved is only up to 1%. The number of degrees of freedom required for the finite element and finite strip analyses is given in Table 9.2.

After a series of parametric studies involving S spans with aspect ratio of flange $\alpha = l/b'$ and n terms, it was concluded that for the stability analysis

$n = (2\alpha - 1) + S - 2$ for even values of α and $S > 1$
$n = (2\alpha - 1) + 2(S - 2)$ for odd values of α and $S > 1$.
For this panel $\alpha = 5$ and $S = 2$, therefore the number of required terms is $n = (2 \times 5 - 1) + 2(2 - 2) = 9$.

7. Stability analysis of parallelogram plate by spline finite strips

The spline finite strip method, presented in Chapter 6, has been extended to the vibration and stability analyses of arbitrarily shaped plates and thin-walled structures since 1984 (Chen and Chong, 1984, 1985; Lau and Hancock, 1986; Y. K. Cheung, Tham and Li, 1988). As an example, the stability analysis of a set of parallelogram plates (Tham and Szeto, (1990)) is presented here. The skew angles varied from 45° to 75°. Each plate is simply supported (s-s-s-s) or clamped (c-c-c-c) on the four sides, and subjected to uniaxial or biaxial compression. The mesh and subparametric transformation for the analysis is shown in Fig. 9.12. For each case, the plate is first analyzed for stress distribution, then for the buckling critical load. The results are shown in Table 9.3 in terms of the buckling load factor k, defined as

$$\sigma_{cr} = k\frac{\pi^2 Dt^2}{a^2}$$

It can be seen that the results computed by this method compare fairly well with other published values.

Table 9.3 *Buckling load factors for parallelogram plates (low order 2-0-2 element; CRCJ = Column Research Committee of Japan, 1971) (Tham and Szeto, 1990, courtesy of Elsevier Science).*

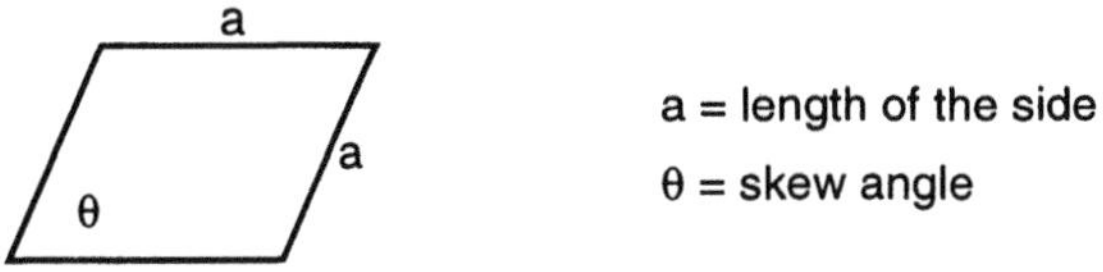

Support conditions	Analysis	Skew angle 45	60	75
A. Uniaxial load				
s-s-s-s	present	10.36	5.93	4.40
	CRCJ	9.08	5.62	4.38
c-c-c-c	present	20.76	13.60	10.84
B. Biaxial load				
s-s-s-s	present	3.71	2.55	2.12
c-c-c-c	present	9.96	6.87	5.64
	CRCJ	9.88	6.85	5.64

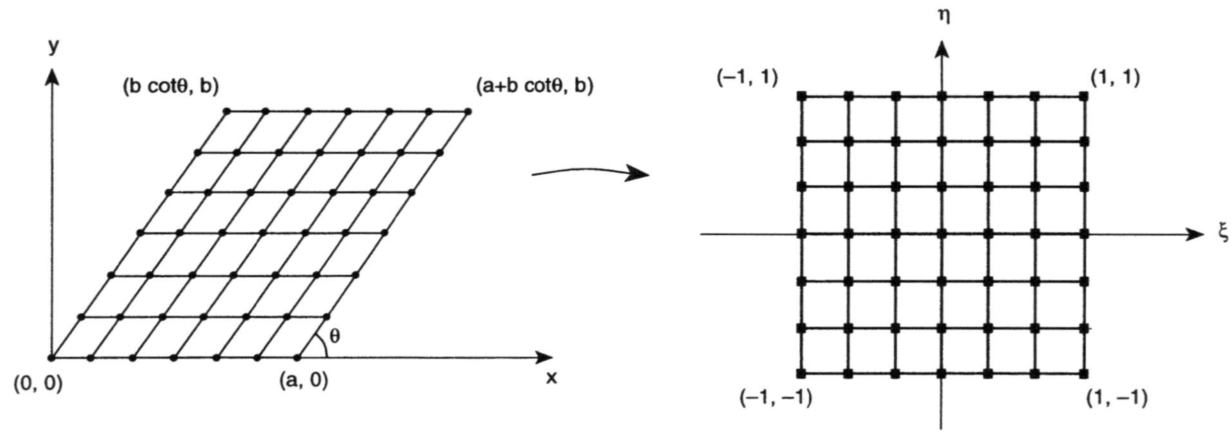

Figure 9.12 *Subparametric transformation of a skewed plate (Tham and Szeto, 1990, courtesy of Elsevier Science).*

CHAPTER 10

Nonlinear analysis

10.1 Introduction

All the topics discussed so far have been linear problems or problems that could be reasonably simplified into linear situations, each of which finally led to a system of linear algebraic equations. In elastic solid mechanics this was implied in:

1. linear forms of strain–displacement relationships;
2. linear forms of stress–strain relationships.

In practice, such linearity is not always preserved. For instance, as deformation of a structure increases beyond a certain extent, equilibrium conditions of the structure must be reconsidered according to its deformed figuration and the nonlinear items in the displacement–strain relationships must be included in the analysis. This type of nonlinearity is termed as geometrical nonlinearity. Another type of nonlinearity is referred to as material nonlinearity, which is induced by nonlinear stress–strain relationships such as plastic deformation of steel and cracking of reinforced concrete.

This chapter discusses the basic procedures applied in material and geometrical nonlinear finite strip analysis. Some numerical examples are also given to verify these procedures.

10.2 Nonlinear solution procedure

10.2.1 Introduction

In nonlinear analysis, due to nonlinear strain–displacement relationships or nonlinear stress–strain relationships, the stiffness matrix of a structure becomes a function of the displacements. Thus, the equilibrium equations of the structure have the form

$$[K(\{\delta\})]\{\delta\} - \{P\} = 0 \tag{10.1}$$

This matrix equation may be solved by iteration. In each iteration, the stiffness matrix is updated according to the displacements obtained during the previous iteration and is assumed to remain constant, then a new set of displacements can be calculated. This procedure is repeated until the error becomes smaller than a prescribed value, i.e. convergence occurs.

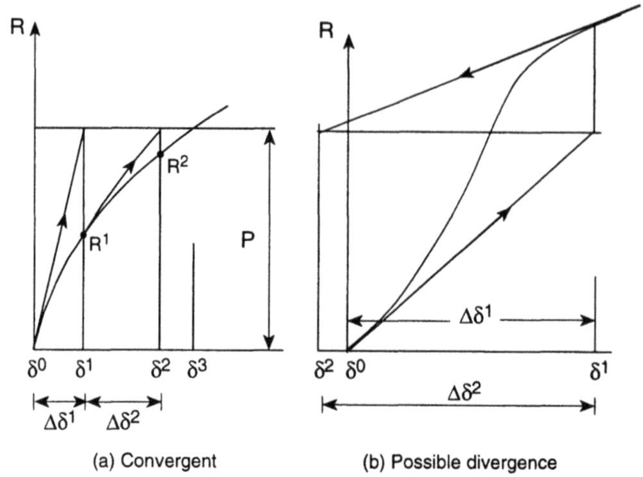

Figure 10.1 *Newton–Raphson method: (a) convergence, (b) possible divergence.*

The commonly used iteration techniques in finite strip analysis are introduced in the following subsections.

10.2.2 Newton–Raphson method

The Newton–Raphson method is also called the tangential stiffness method and is the most commonly used method in nonlinear analysis.

Initially, the displacement $\{\delta\}^1$ can be obtained as a first approximation (Fig. 10.1). The corresponding resistances $\{R\}$ of the system to these displacements are equal to

$$\{R\}^1 = \{R(\{\delta\}^1)\} = [K]\{\delta\}^1 \tag{10.2}$$

in which $[K]$ is function of $\{\delta\}$.

Because $\{\delta\}^1$ is not the true solution, the resistances $\{R\}^1$ are not equal to the external loads $\{P\}$, and the differences between them are referred to as unbalanced forces:

$$\{F\}^1 = \{P\} - \{R\}^1 \tag{10.3}$$

Therefore, the system will deform further. During this deformation, the resistances can be approximated by their truncated Taylor series expansion

$$R_i = R_i^1 + (\frac{\partial R_i}{\partial\{\delta\}})^T \Delta\{\delta\} \tag{10.4}$$

where R_i is the ith component of $\{R\}$ and the derivatives of R_i with respect to displacements are evaluated at $\{\delta\}^1$.

The above approximation means that the increments of the resistances are approximated by their differentials. Hence, the nonlinear system is replaced by the linear system

$$\{R\} = \{R\}^1 + [K_T]\Delta\{\delta\} \qquad (10.5)$$

where $[K_T]$ represents a tangential stiffness matrix, which can be derived by differentiating the resistances with respect to displacements for any particular case.

Thus, an improved approximation is obtained as follows:

$$\{\delta\}^2 = \{\delta\}^1 + [K_T]^{-1}(\{P\} - \{R\}^1) \qquad (10.6)$$

In this manner the nonlinear solution process becomes a series of linear analyses, and the solution at the kth iteration is

$$\{\delta\}^k = \{\delta\}^{k-1} + [K_T^{k-1}]^{-1}(\{P\} - \{R\}^{k-1}) \qquad (10.7)$$

This iteration is terminated when the error, i.e.

$$\Delta\{\delta\} = \{\delta\}^k - \{\delta\}^{k-1} \qquad (10.8)$$

becomes sufficiently small. Usually some 'norm' of the error is determined and iteration continues until this is sufficiently small. Various norms and convergence criteria can be used, e.g. $\mid \Delta\{\delta\} \mid = max \mid \Delta\delta_i \mid$ (maximum absolute value of component in $\Delta\{\delta\}$), or

$$\mid \Delta\{\delta\} \mid = \sqrt{\Delta\{\delta\}^T \Delta\{\delta\}} \qquad (10.9)$$

and convergence can be considered achieved under the following condition:

$$\frac{\mid \Delta\{\delta\} \mid}{\mid \{\delta\} \mid} \leq e \qquad (10.10)$$

where e is a prescribed tolerance, say 0.02–0.2%.

The process is usually convergent in the vicinity of the solution. However, if the initial 'guess' is not close, a divergence may occur (Fig. 10.1b)

10.2.3 Modified Newton–Raphson method

To overcome the difficulty of having to solve a completely new system of equations at each iteration stage, a frequently applied approximation is made by writing

$$[K_T] = [K] \qquad (10.11)$$

where $[K]$ is the elastic stiffness matrix. This modifies the algorithm in Equation 10.7 to

$$\{\delta\}^k = \{\delta\}^{k-1} + [K]^{-1}(\{P\} - \{R\}^{k-1}) \qquad (10.12)$$

and a simple resolution of the same equation system is repeatedly used, i.e. the stiffness matrix of the structure is formed and inverted only once.

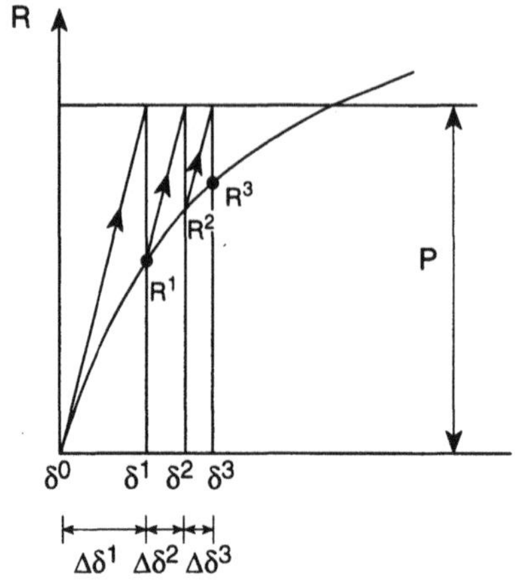

Figure 10.2 *Modified Newton–Raphson method.*

This is clearly more economical at each step but the convergence is now slower. In many cases, however, this process has an overall economy. It is illustrated in Fig. 10.2. A variation of this process, sometimes found to be useful, is the updating of $[K_T]$ after a given number of iterations, say 5 to 10 iterations.

10.2.4 Incremental method

If the loading on a structure goes far beyond the limits of linear behavior, then in order to prevent possible divergence, the loading should be applied incrementally. For example, the total load may be applied in 10 equal steps. After each incremental loading, one or more Newton–Raphson iterations are implemented. Once the total loading has been applied, Newton–Raphson iterations or modified Newton–Raphson iterations continue until convergence occurs (Fig. 10.3).

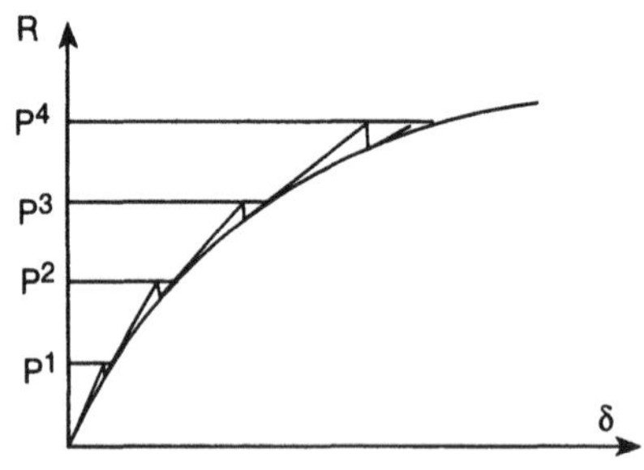

Figure 10.3 *Incremental method.*

10.2.5 Relaxation

If the results are oscillating or even diverging during iteration, the relaxation technique is often a useful approach. That is, in each iteration, the displacement increments are multiplied by a relaxation coefficient λ which isless than unity, say 0.5–0.8, before being added to the accumulated displacements. Thus, Equation 10.7 becomes

$$\{\delta\}^k = \{\delta\}^{k-1} + \lambda[K_T^{k-1}]^{-1}(\{P\} - \{R\}^{k-1}) \tag{10.13}$$

and Equation 10.12 becomes

$$\{\delta\}^k = \{\delta\}^{k-1} + \lambda[K]^{-1}(\{P\} - \{R\}^{k-1}) \tag{10.14}$$

If the convergence is too slow during modified Newton–Raphson iteration, the relaxation technique can also be used to accelerate convergence by taking a relaxation coefficient larger than unity, say 1.2–2.0.

The optimum value of λ has to be determined experimentally for a particular case.

10.3 Elastoplastic analysis

10.3.1 Introduction

In order to understand the nonlinear behavior of a steel structure during its plastic deformation and to obtain its ultimate capacity, the elasto-plastic analysis of the structure is often required. For concrete structures, the behavior of concrete in compression may also be simulated by plastic flow theory.

The finite strip method has been successfully applied to the elastoplastic analysis of steel structures (Olawale and Plank, 1988; Cheung, Ng and Zhong, 1989; Ng, Cheung and Zhong, 1991). In this analysis the tangential stiffness matrix $[K_T]$ can be derived using the virtual work principle.

It is assumed that a structure is simulated by finite strips, the stresses at any point of the structure are represented by $\{\sigma\}$, the external forces required to maintain the structure in equilibrium state are $\{R\}$, which are in magnitude equal to the resistance of the structure to its deformation $\{\delta\}$. According to the virtual work principle, during any virtual displacements $d\{\delta\}$ from the current equilibrium position the work done by the external forces should be equal to the work performed by the internal stresses. This leads to the equation

$$d\{\delta\}^T\{R\} = \int d\{\epsilon\}\{\sigma\}dv \tag{10.15}$$

Taking into consideration the following relationship

$$d\{\epsilon\} = [B]d\{\delta\} \tag{10.16}$$

Equation 10.15 can be written as

$$d\{\delta\}^T\{R\} = d\{\delta\}^T \int [B]^T\{\sigma\}dv \tag{10.17}$$

This is valid for any possible $d\{\delta\}$ only if

$$\{R\} = \int [B]^T\{\sigma\}dv \tag{10.18}$$

The differential of resistances $\{R\}$ with respect to displacements $\{\delta\}$ will produce the tangential stiffness matrix, i.e.

$$d\{R\} = [K_T]d\{\delta\} \tag{10.19}$$

This can be achieved by differentiating both sides of Equation 10.18 with respect to displacements, as follows:

$$d\{R\} = \int ([B]^T d\{\sigma\})dv = \int ([B]^T[D_T]d\{\epsilon\})dv = (\int [B]^T[D_T][B]dv)d\{\delta\} \tag{10.20}$$

where $[D_T]$ is the tangential elastoplastic matrix which can be derived from incremental stress–strain relationship as

$$d\{\sigma\} = [D_T]d\{\epsilon\} \tag{10.21}$$

Comparing Equation 10.20 and 10.19 leads to the following expression for tangential stiffness:

$$[K_T] = \int [B]^T[D_T][B]dv \tag{10.22}$$

It is seen that $[K_T]$ can be formed by following the procedures used in linear elastic analysis if $[D_T]$ is obtained.

10.3.2 Tangential elastoplastic matrix $[D_T]$

If an elastoplastic material such as steel or aluminium is elongated monotonically in the x direction, after the stress σ_x reaches a certain limit, i.e. the yield strength σ_s, a plastic deformation occurs which is irrecoverable on load removal.

Many materials show a perfect plastic behavior in which strain increases without any increment in stress (Fig. 10.4(a)). For all stresses below σ_s (in loading, unloading and reloading) a linear elasticity relationship is assumed.

A further refinement of this model is one for strain–hardening plastic materials in which further plastic deformation requires an increase in stress.

According to the Von Mises yield criterion, for a material in general three-dimensional stress state,yield occurs when the strain energy due to distortion of the material reaches a certain limit. This strain energy is equal to

$$U_{distortion} = \frac{1}{6G}\bar{\sigma}^2 = \frac{1}{2}\bar{\sigma}\bar{\epsilon} \tag{10.23}$$

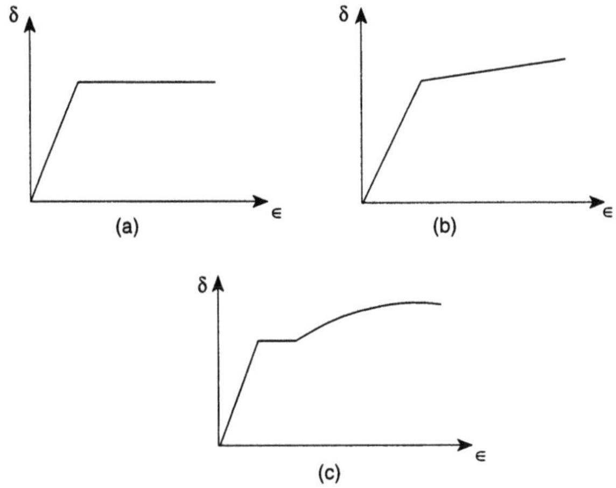

Figure 10.4 *Elastoplastic behavior.*

where $\bar{\sigma}$ is equivalent stress and is defined as

$$\bar{\sigma} = \sqrt{\frac{1}{2}[(\sigma_x - \sigma_y)^2 + (\sigma_y - \sigma_z)^2 + (\sigma_z - \sigma_x)^2 + 6(\tau_{xy}^2 + \tau_{yz}^2 + \tau_{zx}^2)]} \tag{10.24}$$

from which it can be seen that $\bar{\sigma} = \sigma_x$ for uniaxial tension in the x direction and $\bar{\sigma} = \sqrt{3}\tau_{xy}$ for pure shear in the x–y plane. The equivalent strain $\bar{\epsilon}$ is expressed as

$$\bar{\epsilon} = \frac{\bar{\sigma}}{3G} = \frac{\sqrt{2}}{3}\sqrt{(\epsilon_x - \epsilon_y)^2 + (\epsilon_y - \epsilon_z)^2 + (\epsilon_z - \epsilon_x)^2 + \frac{3}{2}(\gamma_{xy}^2 + \gamma_{yz}^2 + \gamma_{zx}^2)} \tag{10.25}$$

from which it may be noted that $\bar{\epsilon} = 2(1 + \nu)\epsilon_x/3$ for uniaxial tension in the x direction $\bar{\epsilon} = \gamma_{xy}/\sqrt{3}$ for pure shear in the x–y plane; G is the shear modulus of the material.

The critical value may be found from the uniaxial tension test in which the yield occurs when σ_x reaches the yield strength σ_s.

By substituting $\bar{\sigma} = \sigma_x = \sigma_s$ into Equation 10.23, the limit of the strain energy due to distortion is obtained as $\sigma_s^2/6G$. Thus, the yield criterion can be written as

$$U_{distortion} = \frac{1}{6G}\bar{\sigma}^2 = \frac{1}{6G}\sigma_s^2 \tag{10.26}$$

or simply

$$\bar{\sigma} = \sigma_s \tag{10.27}$$

Once strain hardening occurs, in order to maintain the plastic flow, the equivalent stress should keep increasing according to accumulated plastic deformation.

During plastic deformation, the strain component is assumed to be divisible into elastic and plastic parts as

$$d\{\epsilon\} = d\{\epsilon\}_e + d\{\epsilon\}_p \tag{10.28}$$

After unloading, the elastic part disappears but the plastic part remains. The elastic part follows elastic stress–strain relationship

$$d\{\sigma\} = [D]d\{\epsilon\}_e \tag{10.29}$$

The plastic part may be evaluated using the Prandtl–Reuss flow rule which suggests that:

1. The volume of the material does not change due to plastic strain, i.e.

$$d\epsilon_{xp} + d\epsilon_{yp} + d\epsilon_{zp} = 0 \tag{10.30}$$

   Consequently, $\nu = 0.5$ and $\bar{\epsilon_p} = \epsilon_{xp}$ for uniaxial tension in the x direction.

2. Each component of incremental plastic strain is proportional to the corresponding component of deviational stress $\{\sigma'\}$ as

$$\begin{aligned} d\epsilon_{xp} &= \lambda\sigma'_x \\ d\epsilon_{yp} &= \lambda\sigma'_y \\ &\ldots \end{aligned} \tag{10.31}$$

   in which the primes denote the deviational stress, i.e.

$$\begin{aligned} \sigma'_x = \sigma_x - \frac{\sigma_x + \sigma_y + \sigma_z}{3} \\ \ldots \end{aligned} \tag{10.32}$$

3. The material follows the Von Mises yield criterion. The equivalent stress $\bar{\sigma}$ required for a plastic flow is dependent on the accumulated equivalent plastic strain as

$$\bar{\sigma} = H(\int d\bar{\epsilon_p}) \tag{10.33}$$

   where

$$\begin{aligned} d\bar{\epsilon_p} = & \frac{\sqrt{2}}{3}\Big[(d\epsilon_{xp} - d\epsilon_{yp})^2 + (d\epsilon_{yp} - d\epsilon_{zp})^2 + (d\epsilon_{zp} - d\epsilon_{xp})^2 \\ & + \frac{3}{2}(d\gamma_{xyp}^2 + d\gamma_{yzp}^2 + d\gamma_{zxp}^2)\Big]^{1/2} \end{aligned} \tag{10.34}$$

   The function H is independent of stress state, therefore it can be obtained using the plot of uniaxial test of the material. In the plot, the uniaxial stress σ_x represents $\bar{\sigma}$, while the term ϵ_{xp} stands for $\int d\bar{\epsilon_p}$.

Based on the above theory and following the commonly used procedure in material nonlinear finite element analysis (see Zienkiewicz (1977) pp 460–464), the elastoplastic matrix $[D_T]$ can be computed as

$$[D_T] = \begin{cases} [D] & for \quad \bar{\sigma} < \sigma_s \\ [D] - [D]_p & for \quad \bar{\sigma} \geq \sigma_s \end{cases} \tag{10.35}$$

where $[D]$ is the elastic matrix of the material as defined earlier; $[D]_p$ is expanded as

$$[D]_p = \frac{[D]\frac{\partial\bar{\sigma}}{\partial\{\sigma\}}(\frac{\partial\bar{\sigma}}{\partial\{\sigma\}})^T[D]}{H' + (\frac{\partial\bar{\sigma}}{\partial\{\sigma\}})^T[D]\frac{\partial\bar{\sigma}}{\partial\{\sigma\}}} \tag{10.36}$$

where $H' = d\bar{\sigma}/d\bar{\epsilon}_p$ is the strain hardening parameter of the material.

For a perfectly elastoplastic material $H' = 0$ (Fig. 10.4(a)). If a bilinear model (Fig. 10.4(b)) is adopted, H' is constant. In general cases of strain hardening (Fig. 10.4(c)), if a plot of uniaxial test giving stress σ versus plastic uniaxial strain ϵ_p is available, H' is the slope of the plot at the particular value of $\sigma = \bar{\sigma}$.

In material nonlinear finite strip analysis of plate structures, the strips are usually divided into a number of layers in the direction of thickness, each layer being subjected to in-plane stresses according to the theory of thin plates. In this case, $\sigma_z = \tau_{yz} = \tau_{zx} = 0$, the increments of stresses and strains are

$$\begin{aligned} d\{\sigma\} &= [d\sigma_x, d\sigma_y, d\tau_{xy}]^T \\ d\{\epsilon\} &= [d\epsilon_x, d\epsilon_y, d\gamma_{xy}]^T \end{aligned} \tag{10.37}$$

The elastic matrix is

$$[D] = \frac{E}{1-\nu^2}\begin{bmatrix} 1 & \nu & 0 \\ \nu & 1 & 0 \\ 0 & 0 & \frac{1-\nu}{2} \end{bmatrix} \tag{10.38}$$

From Equation 10.24, the equivalent stress is obtained as

$$\bar{\sigma} = \sqrt{\sigma_x^2 + \sigma_y^2 - \sigma_x\sigma_y + 3\tau_{xy}^2} \tag{10.39}$$

Then, the following expressions can be derived:

$$\frac{\partial\bar{\sigma}}{\partial\{\sigma\}} = \frac{3}{2\bar{\sigma}}[\sigma_x', \sigma_y', 2\tau_{xy}]^T \tag{10.40}$$

$$[D]\frac{\partial\bar{\sigma}}{\partial\{\sigma\}} = \frac{3G}{\bar{\sigma}(1-\nu)}\begin{Bmatrix} \sigma_x' + \nu\sigma_y' \\ \sigma_y' + \nu\sigma_x' \\ (1-\nu)\tau_{xy} \end{Bmatrix} \tag{10.41}$$

in which σ_x' and σ_y' are deviational stresses, i.e.

$$\begin{aligned} \sigma_x' &= \sigma_x - \frac{\sigma_x + \sigma_y + \sigma_z}{3} = \frac{2\sigma_x - \sigma_y}{3} \\ \sigma_y' &= \sigma_y - \frac{\sigma_x + \sigma_y + \sigma_z}{3} = \frac{2\sigma_y - \sigma_x}{3} \end{aligned} \tag{10.42}$$

Substituting Equation10.40 and 10.41 into 10.36 gives

$$[D]_p = \frac{E}{Q(1-\nu^2)} \times \begin{bmatrix} (\sigma'_x+\nu\sigma'_y)^2 & & \text{symmetrical} \\ (\sigma'_x+\nu\sigma'_y)(\sigma'_y+\nu\sigma'_x)^2 & (\sigma'_y+\nu\sigma'_x)^2 & \\ (1-\nu)(\sigma'_x+\nu\sigma'_y)\tau_{xy} & (1-\nu)(\sigma'_y+\nu\sigma'_x)\tau_{xy} & (1-\nu)^2\tau_{xy}^2 \end{bmatrix} \tag{10.43}$$

in which

$$Q = \sigma'^2_x + \sigma'^2_y + 2\nu\sigma'_x\sigma'_y + 2(1-\nu)\tau_{xy}^2 + \frac{2H'(1-\nu)\bar{\sigma}^2}{9G} \tag{10.44}$$

10.3.3 Flat shell strip for elastoplastic analysis

In the material nonlinear analysis of plate structures, layered flat shell strips are employed. The displacement parameters of the strip for the mth term are taken as

$$\{\delta\}_m = [u_{im}\ v_{im}\ w_{im}\ \theta_{im}\ u_{jm}\ v_{jm}\ w_{jm}\ \theta_{jm}]^T$$

The displacements in the middle plane of the strip are

$$\begin{aligned} u &= \sum_{m=1}^{r}((1-X)u_{im}+Xu_{jm})\,Y_m(y) \\ v &= \sum_{m=1}^{r}((1-X)v_{im}+Xv_{jm})\,Y'_m(y)/\mu_m \\ w &= \sum_{m=1}^{r}((1-3X^2+2X^3)w_{im}+x(1-2X+X^2)\theta_{im} \\ &\quad +(3X^2-2X^3)w_{jm}+x(X^2-X)\theta_{jm}\,)\,Y_m(y) \end{aligned} \tag{10.45}$$

where $X = x/b, b$ is the width of the strip, r is the total number of series terms considered in the analysis, and Y_m is beam eigenfunction. If both ends of the strip are simply supported, $Y_m(y) = \sin\mu_m y$ and $\mu_m = m\pi/l$, where l is the length of the strip.

In the z direction perpendicular to the middle plane, the strip is divided into several layers of equal thickness (say 2–20 layers). In each layer, strains and stresses may be assumed to be constant in the z direction.

According to the theory of thin plates and shells, the strain at any point of each layer can be expressed in terms of nodal displacement parameters as follows:

$$\{\epsilon\} = \begin{pmatrix} \epsilon_x \\ \epsilon_y \\ \gamma_{xy} \end{pmatrix} = \begin{pmatrix} \dfrac{\partial u}{\partial x} - z\dfrac{\partial^2 w}{\partial x^2} \\ \dfrac{\partial v}{\partial y} - z\dfrac{\partial^2 w}{\partial y^2} \\ \dfrac{\partial v}{\partial x} + \dfrac{\partial u}{\partial y} - 2z\dfrac{\partial^2 w}{\partial x \partial y} \end{pmatrix} = \sum_{m=1}^{r} [B]_m \{\delta\}_m \tag{10.46}$$

where

$$[B]_m = \left[\begin{matrix} -\frac{1}{b}Y_m & 0 & (6\frac{z}{b^2} - \frac{12xz}{b^3})Y_m & (\frac{4z}{b} - \frac{6xz}{b^2})Y_m \\ 0 & (1-\frac{x}{b})Y''_m/\mu_m & -z(1 - 3\frac{x^2}{b^2} + 2\frac{x^3}{b^3})Y''_m & -z(x - 2\frac{x^2}{b} + \frac{x^3}{b^2})Y''_m \\ (1-\frac{x}{b})Y'_m & -\frac{1}{b}Y'_m/\mu_m & 12z(\frac{x}{b^2} - \frac{x^2}{b^3})Y'_m & -2z(1 - 4\frac{x}{b} + 3\frac{x^2}{b^2})Y'_m \end{matrix}\right.$$

$$\left.\begin{matrix} \frac{1}{\mathrm{b}}\mathrm{Y_m} & 0 & (\frac{-6z}{b^2} + \frac{12xz}{b^3})Y_m & (-\frac{6xz}{b^2} + \frac{2z}{b})Y_m \\ 0 & \frac{x}{b}Y''_m/\mu_m & -z(3\frac{x^2}{b^2} - 2\frac{x^3}{b^3})Y''_m & -z(\frac{x^3}{b^2} - \frac{x^2}{b})Y''_m \\ \frac{x}{b}Y'_m & \frac{1}{b}Y'_m/\mu_m & -12z(\frac{x}{b^2} - \frac{x^2}{b^3})Y'_m & -2z(3\frac{x^2}{b^2} - 2\frac{x}{b})Y'_m \end{matrix}\right] \tag{10.47}$$

in which $Y'_m = \dfrac{dY_m}{dy}$ and $Y''_m = \dfrac{d^2Y_m}{dy^2}$.

Forming the tangential stiffness matrix $[K_T]$ (using Equation 10.22) and calculating the resistances $\{R\}$ (using Equation 10.18) both require integration over entire volume of a strip. In the x and y directions, Gaussian quadrature can be utilized, while in z direction, simple algebraic summation may be carried out.

Thus, at every Gaussian point (x_G, y_G) and in every layer, the stresses $\{\sigma\}$ must be recalculated once the incremental displacements $\Delta\{\delta\}$ are obtained at each iteration step.

First, the strain increments are calculated as

$$\Delta\{\epsilon\}^k = \sum_{m=1}^{r} [B]_m \Delta\{\delta\}_m^k \tag{10.48}$$

Then, the stress increments can be computed as

$$\Delta\{\sigma\}^k = [D_T^{k-1}]\Delta\{\epsilon\}^k \tag{10.49}$$

where $[D_T^{k-1}]$ is formed, based on Equations 10.35, 10.38 and 10.43, using the stresses $\{\sigma\}^{k-1}$ obtained at the previous iteration. Thus, the stresses become

$$\{\sigma\}^k = \{\sigma\}^{k-1} + \Delta\{\sigma\}^k \tag{10.50}$$

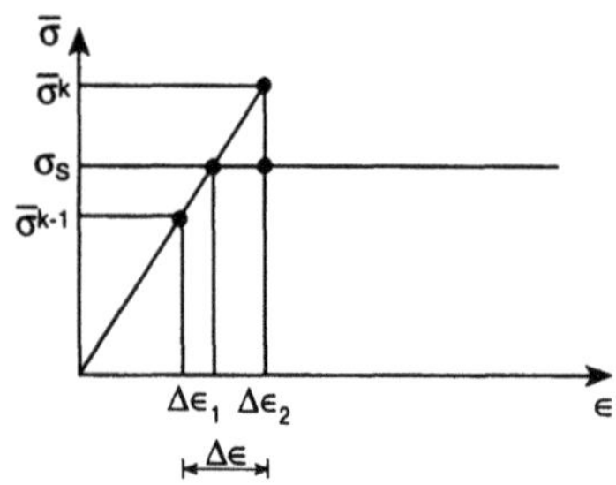

Figure 10.5 *Stress increment in the transition region.*

If material at any Gaussian point in the elastic region yields due to the stress increment, i.e. $\bar{\sigma}^{k-1} < \sigma_s$ but $\bar{\sigma}^k > \sigma_s$, then $\Delta\{\epsilon^k\}$ in Equation 10.49 should be divided into two parts, namely the part prior to yielding and the one after yielding (Fig. 10.5):

$$\begin{aligned} \Delta\{\epsilon\}_1^k &= (1-\alpha)\Delta\{\epsilon\}^k \\ \Delta\{\epsilon\}_2^k &= \alpha\Delta\{\epsilon\}^k \end{aligned} \tag{10.51}$$

Here,

$$\alpha = \frac{\bar{\sigma}^k - \sigma_s}{\bar{\sigma}^k - \bar{\sigma}^{k-1}} \tag{10.52}$$

and the stress increments due to the two parts can be evaluated as

$$\begin{aligned} \Delta\{\sigma\}_1^k &= [D]\Delta\{\epsilon\}_1^k \\ \Delta\{\sigma\}_2^k &= ([D]-[D]_p)\Delta\{\epsilon\}_2^k \end{aligned} \tag{10.53}$$

in which the $[D]_p$ matrix is formed using the stresses at yield point, i.e.

$$\{\sigma\}_1^k = \{\sigma\}^{k-1} + \Delta\{\sigma\}_1^k \tag{10.54}$$

Hence, the stresses are modified as

$$\{\sigma\}^k = \{\sigma\}^{k-1} + \Delta\{\sigma\}_1^k + \Delta\{\sigma\}_2^k \tag{10.55}$$

Thus, in all cases, the stress can be evaluated as

$$\{\sigma\}^k = \{\sigma\}^{k-1} + ([D]-\beta[D]_p)\Delta\{\epsilon\}^k \tag{10.56}$$

where

$$\beta = \begin{cases} 0 & for \quad elastic \quad region: \ \bar{\sigma}^{k-1} < \sigma_s \quad and \quad \bar{\sigma}^k < \sigma_s \\ \alpha & for \quad transition \quad region: \ \bar{\sigma}^{k-1} < \sigma_s \quad but \quad \bar{\sigma}^k \geq \sigma_s \\ 1 & for \quad plastic \quad region: \ \bar{\sigma}^{k-1} \geq \sigma_s \quad and \quad \bar{\sigma}^k \geq \sigma_s \end{cases} \tag{10.57}$$

If the plastic strain increment is concerned, it can be computed from Equations 10.28 and 10.29 as

$$\Delta\{\epsilon\}_p^k = \Delta\{\epsilon\}^k - \Delta\{\epsilon\}_e^k = \Delta\{\epsilon\}^k - [D]^{-1}\Delta\{\sigma\}^k \tag{10.58}$$

Next, the equivalent stress $\bar{\sigma}^k$ is calculated using Equation 10.39, from which the strain hardening parameter H' can be obtained from the plot of uniaxial tension test, and the tangential elastoplastic matrix $[D_T]$ is ready to be renewed using Equations 10.35 and 10.43.

Finally, the tangential stiffness matrix $[K_T]$ and the resistance $\{R\}$ can be updated using Equations 10.22 and 10.18, respectively, and the following strip matrices can be obtained:

$$[K_T]_{mn} = \int_0^b \int_0^l \sum_{l=1}^{n_l} [B]_m^T [D_T][B]_n \Delta z_l dx dy \tag{10.59}$$

$$\{R\}_m = \int_0^b \int_0^l \sum_{l=1}^{n_l} [B]_m^T \{\sigma\} \Delta z_l dx dy \tag{10.60}$$

where n_l is the number of layers in the thickness direction.

Substituting Equation 10.35 into 10.22 gives

$$\begin{aligned} [K_T] &= \int [B]^T [D_T][B] dv \\ &= \int_e [B]^T [D][B] dv + \int_p [B]^T ([D] - [D]_p)[B] dv \\ &= \int_{e+p} [B]^T [D][B] dv - \int_p [B]^T [D]_p [B] dv \\ &= [K] - \int_p [B]^T [D]_p [B] dv \\ &= [K] - [K_p] \end{aligned} \tag{10.61}$$

where e and p represent the elastic plastic regions, respectively, $[K]$ is the elastic stiffness matrix, and

$$[K_p] = \int_p [B]^T [D]_p [B] dv \tag{10.62}$$

Similarly, the following expression for $\{R\}$ can be derived:

$$\begin{aligned} \{R\}^k &= \int [B]^T\{\sigma\}^k dv \\ &= \int [B]^T(\{\sigma\}^{k-1} + ([D] - \beta[D]_p)\Delta\{\epsilon\}^k)dv \\ &= \{R\}^{k-1} + \int [B]^T([D] - \beta[D]_p)[B]\Delta\{\delta\}^k dv \\ &= \{R\}^{k-1} + ([K] - \beta[K]_p)\Delta\{\delta\}^k \end{aligned} \quad (10.63)$$

From Equations 10.61 and 10.63, it can be seen that in updating the strip matrices the volume integrations in Equations 10.59 and 10.60 need to be carried out over the plastic region only.

The coordinate transformation and assembly procedures are the same as those in the elastic analysis.

10.3.4 Numerical examples

1. Square plate

A simply supported square plate is subjected to uniform load q. The length and width of the plate are $a = 1.0$ and its thickness is $h = 0.01$. The material properties are $E = 10.92, \nu = 0.3$, σ_s=1600 and $H' = 0$. Figure 10.6 shows the results using the finite strip method (M. S. Cheung, Ng and Zhong, 1989) compared with the solution of the heterosis element method (Owen and Hinton, 1980). In this figure, $D = Eh^3/12(1-\nu^2)$ and $M_T = \sigma_s h^2/4$.

2. Box beam

A box beam is simply supported at both ends and is subjected to a uniform load p. The dimensions of cross-section are given in Fig. 10.7, and the length is 6.0 m. The material properties are E = 210000 MPa, $\nu = 0.0$, σ_s=240 MPa and $H' = 0$. The results (Ng *et al.* 1991) are also depicted in Fig. 10.7 in comparison with those from beam theory (Timoshenko and Gere, 1972) and from the finite element analysis using the package ADINA with 24 isoparametric shell elements. The difference between the finite strip analysis and the theoretical results may be attributed to shear deformation which was not included in beam theory.

In the finite strip analysis, five symmetrical series terms were used, and an 11 point Gaussian integral was employed in the longitudinal direction.

10.4 Analysis of reinforced concrete slabs

10.4.1 Introduction

The material nonlinear analysis of reinforced concrete slabs is an important technique for investigating the behavior of many civil engineering structures

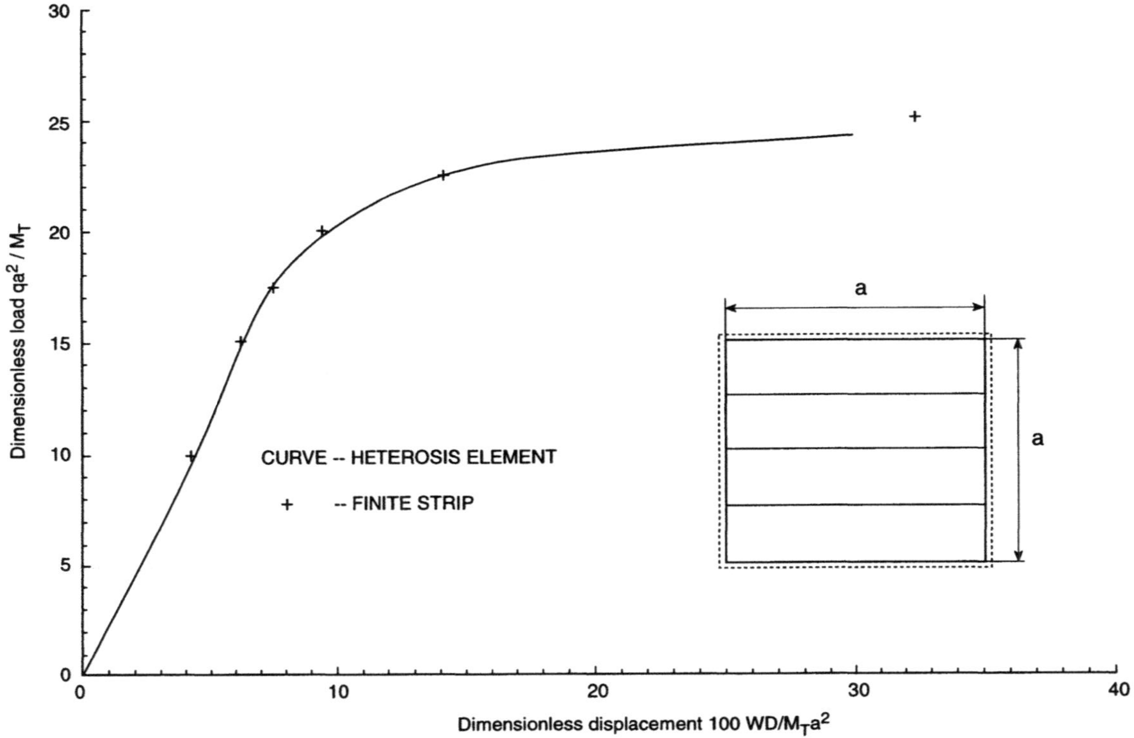

Figure 10.6 *Load–displacement curve of a simply supported plate: continuous curve shows heterosis element; and + indicates finite strip results.*

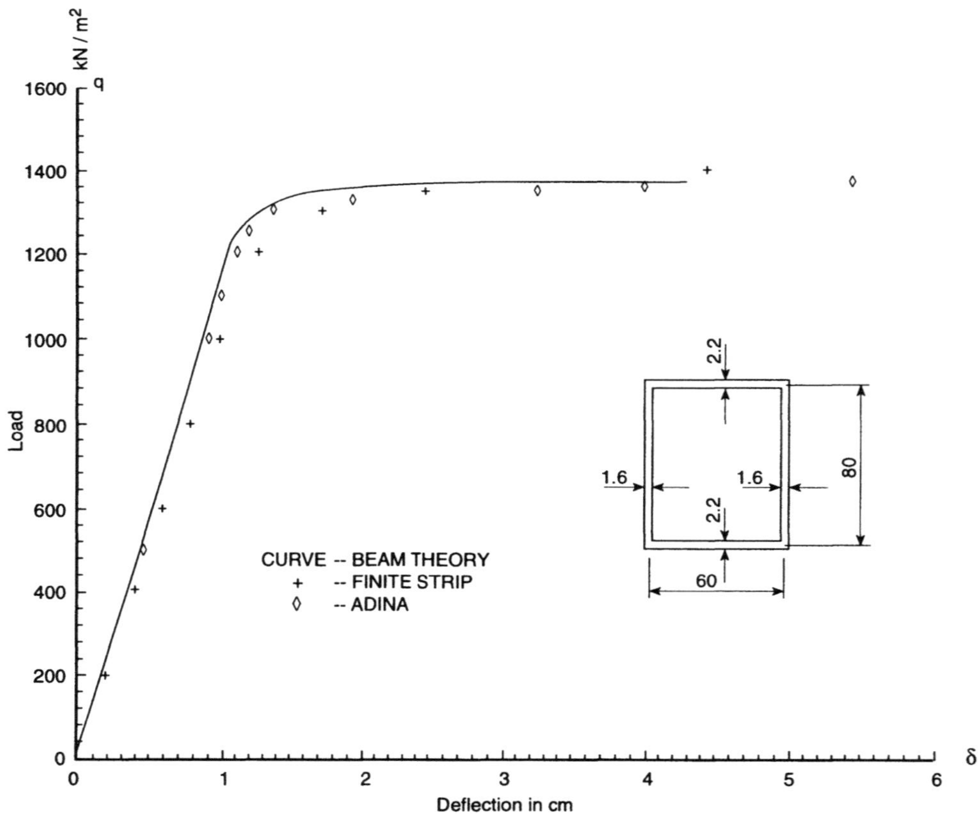

Figure 10.7 *Load-displacement curve of a simply supported box beam under uniform load: continuous curve is beam theory; + is finite strip; ◇ is ADINA.*

such as floors, roofs and bridge decks. Since the mid-1970s, finite element methods based on a number of different material models have been used successfully to simulate the material nonlinear response of RC structures. However, the finite strip method has the advantage of considerably reducing the computation, storage requirements and input data preparation. This method has been applied to such analysis by Guo *et al.* (1988), and by Cheung and Li (1990).

In this section the model based on orthotropic nonlinear elasticity is employed to simulate properties of plain concrete under biaxial stresses. This model was proposed by Darwin and Pecknold (1974), modified by Kabir (1976), and is in close agreement with experimental results (Darwin and Pecknold, 1977). For reinforcement, the bilinear elastoplastic uniaxial stress–strain relationship is employed.

The finite strip is divided into imaginary concrete layers. In addition, each layer of reinforcement is replaced by a smeared steel layer with an unchanged original reinforcement area having stiffness only in the direction of reinforcement.

In every Newton–Raphson iteration, using finite strip displacement functions and thin plate theory, the strain state in each layer can be found. After that, the stress state can be obtained by means of the above-mentioned models. Then, the tangential stiffness matrix is updated, and the unbalanced forces and corresponding displacement increments are evaluated. This procedure is repeated until convergence occurs.

In order to prevent possible oscillations and divergence during the nonlinear solution, the relaxation technique is applied, that is, in every iteration, the displacement increments are multiplied by a relaxation factor less than unity before being added to the accumulated displacements.

10.4.2 Material model of concrete

Concrete can be treated as an incrementally orthotropic linear elastic material (Darwin and Pecknold, 1977)

$$\begin{pmatrix} d\sigma_1 \\ d\sigma_2 \\ d\tau_{12} \end{pmatrix} = [D_c] \begin{pmatrix} d\epsilon_1 \\ d\epsilon_2 \\ d\gamma_{12} \end{pmatrix} \tag{10.64}$$

where

$$[D_c] = \frac{1}{1-\nu^2} \begin{pmatrix} E_1 & \nu\sqrt{E_1 E_2} & 0 \\ & E_2 & 0 \\ \text{symmetrical} & & \frac{1}{4}\beta(E_1 + E_2 - 2\nu\sqrt{E_1 E_2}) \end{pmatrix} \tag{10.65}$$

in which E_1, E_2 and $\nu = \sqrt{\nu_1 \nu_2}$ are stress-dependent material properties, $\beta = 1.0$ prior to cracking and $0 < \beta \leq 1.0$ after cracking. Material axes 1 and 2 coincide with the current principal stress axes.

The concept of equivalent uniaxial strain has been introduced in order to transform the actual biaxial stress state into two independent uniaxial stress states. It is defined as

$$\epsilon_{iu} = \sum_{all\ load\ increments} \frac{\Delta\sigma_i}{E_i} \tag{10.66}$$

where $\Delta\sigma_i$ is the incremental change in principal stress, and E_i = is the varying tangent stiffness for load increment corresponding to $\Delta\sigma_i$. The relationship between the principal stress σ_i and the equivalent uniaxial strain ϵ_{iu} is shown in Fig. 10.8.

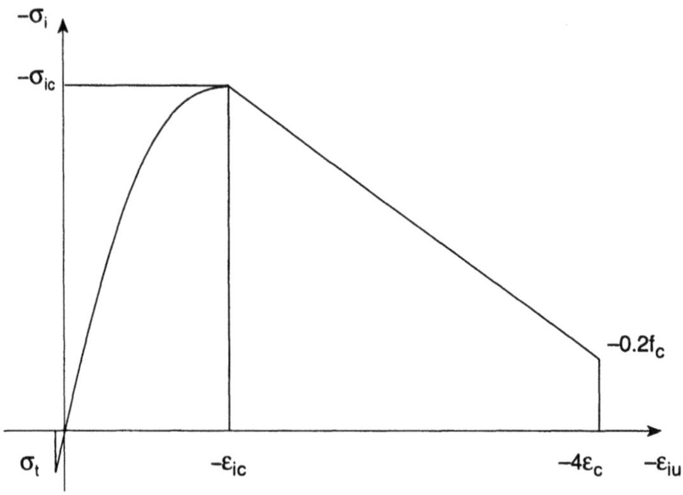

Figure 10.8 *Equivalent uniaxial stress-strain model.*

For compressive loading prior to yield, the compressive principal stress and elastic modulus can be calculated as

$$\sigma_i = \frac{\epsilon_{iu} E_0}{1 + (\frac{E_0}{E_c} - 2)q + q^2} \tag{10.67}$$

$$E_i = \frac{\partial\sigma_i}{\partial\epsilon_{iu}} = \frac{E_0(1 - q^2)}{1 + (\frac{E_0}{E_c} - 2)q + q^2} \tag{10.68}$$

$$q = \epsilon_{iu}/\epsilon_{ic}$$

where E_0 is the tangent modulus of elasticity at zero stress, $E_c = \sigma_{ic}/\epsilon_{ic}$ is the secant modulus at the point of maximum compressive stress σ_{ic}, and ϵ_{ic} is the equivalent uniaxial strain at the maximum compressive stress. Both σ_{ic} and ϵ_{ic} depend on the principal stress ratio $\alpha = \sigma_1/\sigma_2$. The analytical maximum strength envelope may be employed to determine the dependence of σ_{ic} on α (Fig. 10.9).

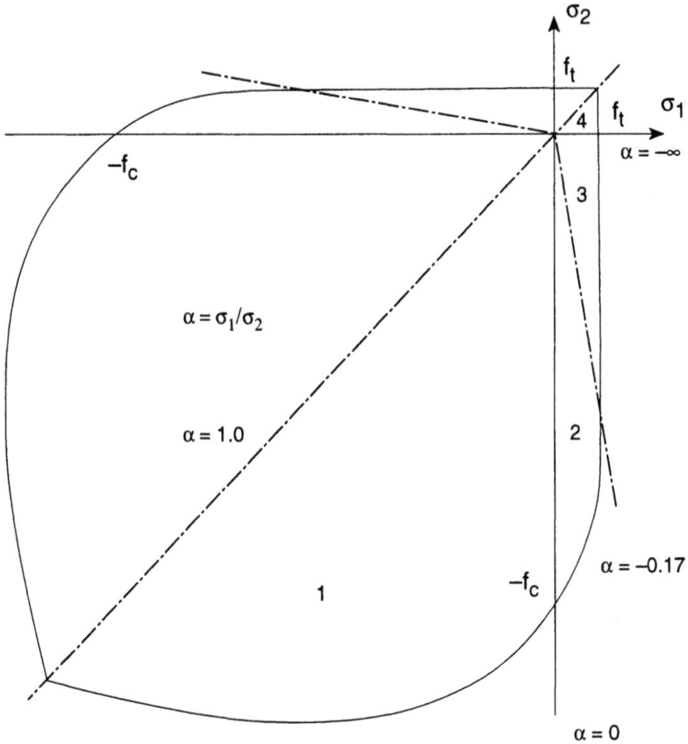

Figure 10.9 *Biaxial strength envelope.*

For tensile stress prior to cracking, the linear elastic σ_i–ϵ_{iu} relationship applied is

$$\sigma_i = E_0 \epsilon_{iu} \tag{10.69}$$

The four regions of the strength envelope with the accompanying equations for the maximum stresses σ_{ic} and corresponding equivalent strains ϵ_{ic} are summarized as follows (Greunen, 1979):

1. For σ_1 = compression, σ_2 = compression, $0 \le \alpha \le 1$

$$\begin{aligned}
\sigma_{2c} &= \frac{1 + 3.65\alpha}{(1+\alpha)^2} f'_c \\
\sigma_{1c} &= \alpha \sigma_{2c} \\
\epsilon_{2c} &= \epsilon_c (3p_2 - 2) \\
\epsilon_{1c} &= \epsilon_c (-1.6p_1^3 + 2.25p_1^2 + 0.35p_1)
\end{aligned}$$

where $p_1 = \sigma_{1c}/f'_c$, $p_2 = \sigma_{2c}/f'_c$, f'_c is the uniaxial compressive strength and ϵ_c is the uniaxial compressive strain at f'_c.

2. For σ_1 = tension, σ_2 = compression, $-0.17 \leq \alpha \leq 0$

$$\begin{aligned}\sigma_{2c} &= \frac{1+3.28\alpha}{(1+\alpha)^2} f'_c \\ \sigma_{1t} &= \alpha\sigma_{2c} \\ \epsilon_{2c} &= \epsilon_c(4.42 - 8.38p_2 + 7.54p_2^2 - 2.58p_2^3)\end{aligned}$$

where

$$\begin{aligned}p_2 &= \sigma_{2c}/f'_c \\ \epsilon_{1t} &= \sigma_{1t}/E_0\end{aligned}$$

3. For σ_1 = tension, σ_2 = compression, $-\infty \leq \alpha \leq -0.17$

$$\begin{aligned}\sigma_{2c} &= 0.65f'_c \\ \sigma_{1t} &= f'_t \\ \epsilon_{2c} &= \epsilon_c(4.42 - 8.38p_2 + 7.54p_2^2 - 2.58p_2^3) \\ p_2 &= \sigma_{2c}/f'_c \leq 0.65 \\ \epsilon_{1t} &= \sigma_{1t}/E_0\end{aligned}$$

4. For σ_1 = tension, σ_2 = tension, $1 \leq \alpha \leq \infty$

$$\begin{aligned}\sigma_{1t} &= \sigma_{2t} = f'_t \\ \epsilon_{1t} &= \epsilon_{2t} = f'_t/E_0\end{aligned}$$

in which f'_t is uniaxial tensile strength.
Cracks are assumed to open perpendicular to the highest principal tensile stress direction when the failure envelope in Fig. 10.9 has been reached, and σ_1 and E_1 are assumed to become zero as soon as the cracking occurs.

10.4.3 Material model of reinforcement

For the stress–strain relationship of reinforcement, the uniaxial bilinear elastoplastic model (Fig. 10.4(b)) is employed.

In concrete, when σ_1 reaches σ_{1t} at some point, primary cracks form at finite spacing in the surrounding region. The total tensile force is transferred across each crack by the tensile steel, but between the cracks the tensile concrete carries stress locally, mainly in the direction of the steel bars, due to the bond between the concrete and the reinforcement. This is called the tension stiffening effect. As the load increases, secondary cracks form around the reinforcement and between the primary cracks so that the load carried by the concrete and the average stress in the concrete decrease progressively. In order to incorporate this effect, after the surrounding concrete has cracked, the stress of steel σ_{st} is magnified as (Gilbert and Warners, 1978)

$$\sigma_{st} = K\sigma'_{st} \tag{10.70}$$

Table 10.1 *Coefficient K for stress in reinforcing steel after concrete cracking*

ϵ_{st}/ϵ_t	1.0–1.5	1.5–3.0	3.0–5.0	5.0–8.0	8.0–11.0	11.0–14.0	other
K	4.0	2.7	2.0	1.6	1.15	1.05	1.0

where σ'_{st} is the stress of reinforcement found by the bilinear model in Fig. 10.4(b) and K is a coefficient shown in Table 10.1, in which ϵ_{st} is the strain of steel and ϵ_t is the uniaxial strain at concrete cracking, $\epsilon_t = f'_t/E_0$. The concrete is assumed to carry no stress normal to a crack but an additional stress will be carried at the steel level. This additional stress represents the total internal tensile force in fact carried by the concrete between the cracks, conveniently lumped at the level of the tensile reinforcement and oriented in the direction of the bar.

The Young's modulus E and strain hardening parameter H' are also multiplied by the same coefficient K when formulating the strip stiffness matrices after cracking of the surrounding concrete.

10.4.4 Nonlinear solution

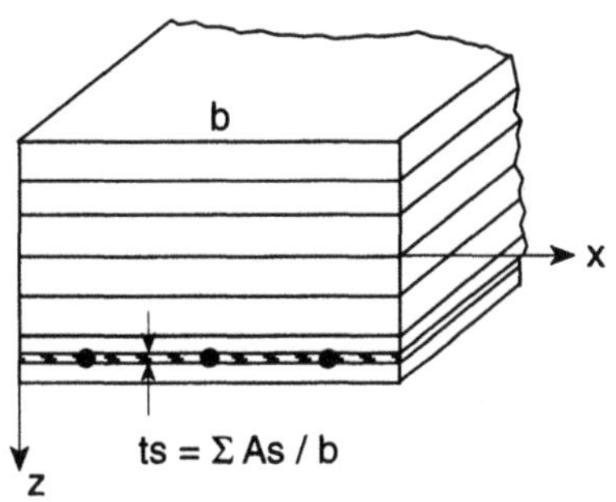

Figure 10.10 *Layered strip.*

In this analysis, a layered flat shell strip is employed, i.e. each strip is divided into a number of concrete layers, and each reinforcement layer is replaced by a smeared steel layer with unchanged original reinforcement area having stiffness only in the direction of reinforcement (Fig. 10.10). The displacement functions Equation 10.45 and the strain matrix Equation 10.47 are also applicable here.

The loads are applied in a number of steps. In each loading step, the Newton–Raphson iteration method is utilized for the nonlinear solution. For each iteration an incremental linear stress–strain relationship is assumed, the displacement increment calculated, the strain and stress state determined and the material matrix updated. The following steps are used to determine the states of strain and stress at any point in a concrete layer of a strip for the kth iteration:

1. From the displacement increment vector $\Delta\{\delta\}^k$ obtained from the previous iteration, the strain increment vector is computed as

$$\Delta\{\epsilon\}^k = (\Delta\epsilon_x^k, \Delta\epsilon_y^k, \Delta\gamma_{xy}^k)^T = \sum_{m=1}^{r} [B]_m \Delta\{\delta\}_m^k \tag{10.71}$$

where $[B]_m$ is strain matrix in Equation 10.47.

2. Using Equation 10.64 and performing coordinate transformation, the approximate stress increments can be calculated as:

$$\Delta\{\sigma\}^k = (\Delta\sigma_x^k, \Delta\sigma_y^k, \Delta\tau_{xy}^k)^T = [T_\epsilon^{k-1}]^T [D_c^{k-1}][T_\epsilon^{k-1}]\Delta\{\epsilon\}^k \tag{10.72}$$

where $[T_\epsilon]$ is the strain coordinate transformation matrix from the x–y axes to the principal axes:

$$[T_\epsilon] = \begin{bmatrix} \cos^2\theta & \sin^2\theta & \sin\theta\cos\theta \\ \sin^2\theta & \cos^2\theta & -\cos\theta\sin\theta \\ -2\sin\theta\cos\theta & 2\sin\theta\cos\theta & \cos^2\theta - \sin^2\theta \end{bmatrix} \tag{10.73}$$

in which θ is the angle between σ_1 and x, obtained from the previous iteration.

The stress increments are only approximate because of the linearized material matrix. The current total approximate stresses are then obtained as

$$\{\sigma\}^k = \{\sigma\}^{k-1} + \Delta\{\sigma\}^k \tag{10.74}$$

3. From the total approximate stresses the approximate principal stresses σ_i^k (i=1,2) and θ^k are calculated.

4. The equivalent uniaxial strains are then evaluated from Equation 10.66 as

$$\epsilon_{iu}^k = \epsilon_{iu}^{k-1} + (\sigma_i^k - \sigma_i^{k-1})/E_i^{k-1} \tag{10.75}$$

5. The biaxial stress ratio is determined from

$$\alpha^k = \sigma_1^k/\sigma_2^k \tag{10.76}$$

Then, σ_{ic}^k and ϵ_{ic}^k or σ_{it} are obtained from the biaxial strength envelope in Section 10.4.2.

6. The principal stresses σ_i^k and elastic moduli E_i^k corresponding to ϵ_i^k can then be computed from Equations 10.67–10.69 and Fig. 10.8. The

current total stress vector $\{\sigma\}^k$ is obtained by transforming the principal stresses to the x–y axes. The material matrix $[D_c^k]$ is also updated to reflect the new tangent moduli.

The contribution of concrete layers to the tangential stiffness matrix of the strip in the next iteration, corresponding to the mth and nth terms of eigenfunctions, is

$$[K_T^{k+1}]_{mn} = \int [B]_m^T [T_\epsilon^k]^T [D_c^k][T_\epsilon^k][B]_n \, dv \tag{10.77}$$

Similarly, the stress state of reinforcement and its contribution to the tangential stiffness matrix of the strip can also be computed using the corresponding material model.

According to the principle of virtual work, in each plate strip, the resistance vector corresponding to the mth term of eigenfunctions is

$$\{R\}_m^k = \int [B]_m^T \{\sigma\}^k \, dv \tag{10.78}$$

For the two integrations given above, Gaussian integration is employed in both x and y directions. In each strip, four Gauss points are used in the x direction, while in the y direction the strip is divided into a number of segments, and four Gauss points are used in each segment. The number of segments used must be sufficient to ensure satisfactory accuracy of integration for the highest order series term.

Next, the incremental displacements are evaluated as

$$[K_T^{k+1}]\Delta\{\delta\}^{k+1} = \lambda(\{F\} - \{R\}^k) \tag{10.79}$$

and the new displacements are reached:

$$\{\delta\}^{k+1} = \{\delta\}^k + \Delta\{\delta\}^{k+1} \tag{10.80}$$

where $\{F\}$ is the vector of equivalent nodal forces of external loads, and λ is the relaxation coefficient. In order to prevent possible oscillation and divergence in iterations, the value of λ must be less than unity. This procedure is repeated until convergence is achieved:

$$\frac{\sqrt{\sum(\Delta\delta_i^{k+1})^2}}{\sqrt{\sum(\delta_i^{k+1})^2}} \le e$$

in which e is a prescribed allowable error.

10.4.5 Numerical example

A simply supported two-way square reinforced concrete slab is subjected to uniform load. This slab, with a thickness of 2 in and a span of 72 in, is reinforced in both directions with uniform spacing.

Due to symmetry, only a quarter of the slab needs to be analyzed. Therefore, half of the slab is modeled by four strips, and the integration in calculating the tangential stiffness matrix and resistance to deformation is carried out over only half the span. Four symmetrical series terms are used in this analysis.

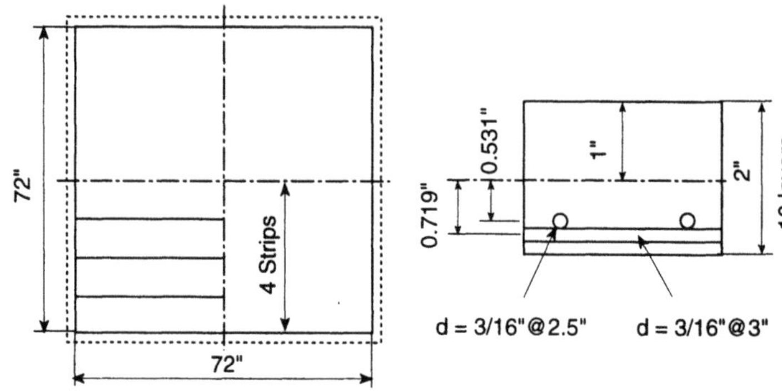

Figure 10.11 *Taylor slab (dimensions in inches, 1" = 0.0254m).*

The cross-section is divided into 10 concrete layers and two orthogonal reinforcing steel layers of 3/16 in diameter bars with the spacing shown in Fig. 10.11. The following material properties are used for the analysis: for the concrete $E_0 = 4700$ ksi, $f_c' = 5940$ psi, $f_t' = 550$ psi, $\epsilon_c = 0.0025$ and $\nu = 0.18$; for the reinforcing steel Young's modulus $E = 30000$ ksi, strain hardening parameter $H' = 2000$ ksi, yield strength $\sigma_s = 54.5$ ksi, and ultimate strain $\epsilon_u = 0.1$.

During the Newton–Raphson iteration $\lambda = 0.8$, $e = 0.002$ are adopted. The deflections at the center of the slab under increasing uniform load are shown in Fig. 10.12. For comparison, the results obtained from experiment (Taylor et al., 1966) and finite element analysis (Greunen, 1979) are also shown in the same figure.

10.5 Geometrically nonlinear analysis

10.5.1 Introduction

If the deflection of a plate is large in comparison with the plate thickness, the deflection may produce supplementary strains and stresses in the middle plane of the plate. These in-plane stresses contribute to resist the deflection of the plate. As a result, the deflections are much less than those estimated by linear elastic plate theory. Conversely, in-plane compression and shear may reduce transverse stiffness and load capability of plates significantly, and even cause buckling of the plate. Because of the interaction between deflection and in-plane deformation, the governing

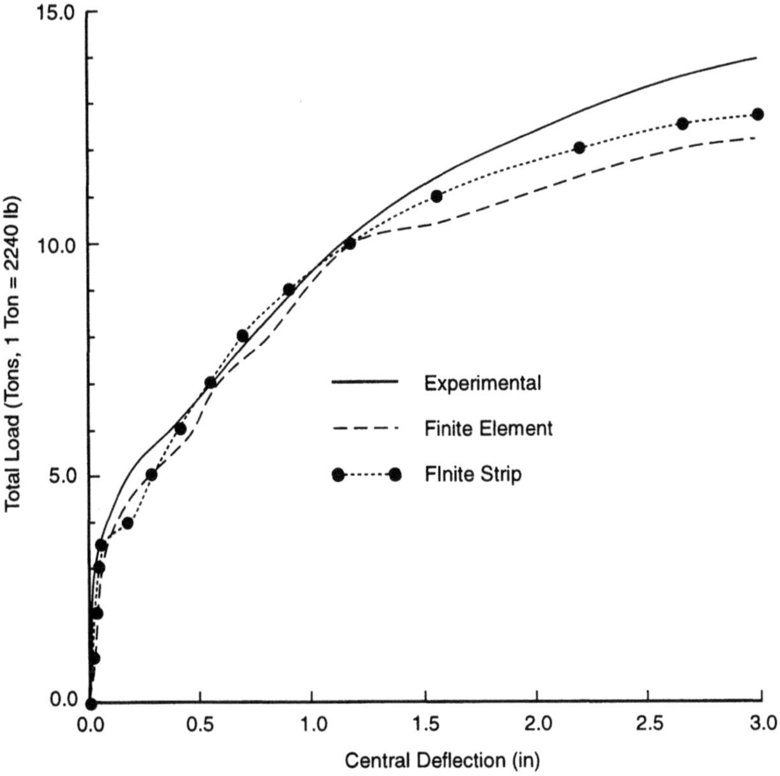

Figure 10.12 *Deflection of a Taylor slab: (—) experimental, (– –) finite element, (- - - -) finite strip. (1 in = 0.0254 m).*

differential equations of plates become nonlinear, and the solution of the problem becomes much more complicated (Timoshenko and Woinowsky–Krieger, 1959).

The finite strip method has been extended to geometrically nonlinear analysis of plate structures and has demonstrated its higher efficiency than finite element method (Graves–Smith and Sridharan, 1978; Hancock, 1981; Langyel and Cusens, 1983; Gierlinski and Graves–Smith, 1984; Azizian and Dawe, 1985; M. S. Cheung and Li, 1989B).

10.5.2 Displacement functions

Under large deflection, each strip is subjected to in-plane stresses and out-of-plane bending forces. The nodal displacement parameters of the strip corresponding to the mth term of the eigenfunction series are:

$$\{\delta\}_m = [u_{im}, v_{im}, w_{im}, \theta_{im}, u_{jm}, v_{jm}, w_{jm}, \theta_{jm}]^T$$

The displacement field within the strip of length l and width b is

$$\begin{aligned} u &= \sum_{m=1}^{r}((1-X)u_{im} + Xu_{jm})Y_m^u(y) \\ v &= \sum_{m=1}^{r}((1-X)v_{im} + Xv_{jm})Y_m^v(y) \\ w &= \sum_{m=1}^{r}(N_{1m}w_{im} + N_{2m}\theta_{im} + N_{3m}w_{jm} + N_{4m}\theta_{jm}) \end{aligned} \tag{10.81}$$

where $X = x/b$, r is the total number of series terms used in analysis, and

$$Y_m^u(y) = \sin\frac{m\pi y}{l} \tag{10.82}$$

If both ends of the strip are free to move in the longitudinal direction ($u = 0$, $\sigma_y = 0$ at $y = 0$ and $y = l$), then

$$Y_m^v(y) = \cos\frac{m\pi y}{l} \tag{10.83}$$

If both ends are immovable ($u = 0$, $v = 0$ at $y = 0$ and $y = l$), then

$$Y_m^v(y) = \sin\frac{(m+1)\pi y}{l} \tag{10.84}$$

In the expression for w,

$$\begin{aligned} N_{1m} &= (1-3X^2+2X^3)Y_m^w(y) \\ N_{2m} &= x(1-2X+X^2)Y_m^w(y) \\ N_{3m} &= (3X^2-2X^3)Y_m^w(y) \\ N_{4m} &= x(X^2-X)Y_m^w(y) \end{aligned} \tag{10.85}$$

If both ends are simply supported in bending ($w = 0$, $M_y = 0$ at $y = 0$ and $y = l$), then

$$Y_m^w(y) = \sin\frac{m\pi y}{l} \tag{10.86}$$

If both ends are clamped ($w = 0$, $\frac{\partial w}{\partial y} = 0$ at $y = 0$ and $y = l$), then

$$Y_m^w(y) = \sin\frac{\mu_m y}{l} - \sinh\frac{\mu_m y}{l} - \alpha_m[\cos\frac{\mu_m y}{l} - \cosh\frac{\mu_m y}{l}] \tag{10.87}$$

where

$$\alpha_m = \frac{\sin\mu_m - \sinh\mu_m}{\cos\mu_m - \cosh\mu_m}$$

μ_m is the mth solution of the equation $1 - \cos\mu\cosh\mu = 0$; its values are listed in Section 4.3.5.

In order to obtain the initial stiffness matrix, the following linear strain–displacement and stress–strain relationships are applicable:

$$\epsilon_x = \frac{\partial u}{\partial x}$$

$$\begin{aligned}
\epsilon_y &= \frac{\partial v}{\partial y} \\
\gamma_{xy} &= \frac{\partial u}{\partial y} + \frac{\partial v}{\partial x} \\
\chi_x &= -\frac{\partial^2 w}{\partial x^2} \\
\chi_y &= -\frac{\partial^2 w}{\partial y^2} \\
\chi_{xy} &= 2\frac{\partial^2 w}{\partial x \partial y}
\end{aligned} \tag{10.88}$$

$$\{\sigma\} = \begin{Bmatrix} N_x \\ N_y \\ N_{xy} \\ M_x \\ M_y \\ M_{xy} \end{Bmatrix} = \begin{bmatrix} K_x & K_1 & & & & \\ K_1 & K_y & & & & \\ & & K_{xy} & & & \\ & & & D_x & D_1 & \\ & & & D_1 & D_y & \\ & & & & & D_{xy} \end{bmatrix} \begin{Bmatrix} \epsilon_x \\ \epsilon_y \\ \gamma_{xy} \\ \chi_x \\ \chi_y \\ \chi_{xy} \end{Bmatrix} = [D]\{\epsilon\} \tag{10.89}$$

where

$$\begin{array}{ll}
K_x = E_x h/(1-\nu_x\nu_y) & K_y = E_y h/(1-\nu_x\nu_y) \\
K_{xy} = E_{xy} h & K_1 = \nu_x K_x = \nu_y K_y \\
D_x = E_x h^3/12(1-\nu_x\nu_y) & D_y = E_y h^3/12(1-\nu_x\nu_y) \\
D_{xy} = E_{xy} h^3/12 & D_1 = \nu_x D_x = \nu_y D_y
\end{array} \tag{10.90}$$

where h is the thickness of the plate.

By substituting Equation 10.81 into 10.88, the strains can be expressed in terms of displacement parameters of the form

$$\{\epsilon\} = [\epsilon_x, \epsilon_y, \gamma_{xy}, \chi_x, \chi_y, \chi_{xy}]^T = \sum_{m=1}^{r} [B]_m \{\delta\}_m \tag{10.91}$$

where $[B]_m$ is the linear strain matrix

$$[B]_m = \left[\begin{matrix} -\frac{1}{b}Y_m^u & 0 & 0 & 0 \\ 0 & (1-X)Y_{m,y}^v & 0 & 0 \\ (1-X)Y_{m,y}^u & -\frac{1}{b}Y_m^v & 0 & 0 \\ 0 & 0 & -N_{1m,xx} & -N_{2m,xx} \\ 0 & 0 & -N_{1m,yy} & -N_{2m,yy} \\ 0 & 0 & 2N_{1m,xy} & 2N_{2m,xy} \end{matrix}\right. \\
\left.\begin{matrix} \frac{1}{b}Y_m^u & 0 & 0 & 0 \\ 0 & XY_{m,y}^v & 0 & 0 \\ XY_{m,y}^u & \frac{1}{b}Y_m^v & 0 & 0 \\ 0 & 0 & -N_{3m,xx} & -N_{4m,xx} \\ 0 & 0 & -N_{3m,yy} & -N_{4m,yy} \\ 0 & 0 & 2N_{3m,xy} & 2N_{4m,xy} \end{matrix}\right] \tag{10.92}$$

in which a comma denotes differentiation for example, $Y^u_{m,y} = dY^u_m/dy$ and $N_{1m,xx} = \partial^2 N_{1m}/\partial x^2$.

Then, the linear stiffness matrix of the strip is obtained as

$$[K_{mn}] = \int [B]^T_m [D][B]_n dxdy \tag{10.93}$$

10.5.3 General solution procedure

Taking into consideration the effect of large deflections on in-plane strains, the strain–displacement relationships become (see Chapter 9)

$$\begin{aligned}
\epsilon_x &= \frac{\partial u}{\partial x} + \frac{1}{2}(\frac{\partial w}{\partial x})^2 \\
\epsilon_y &= \frac{\partial v}{\partial y} + \frac{1}{2}(\frac{\partial w}{\partial y})^2 \\
\gamma_{xy} &= \frac{\partial u}{\partial y} + \frac{\partial v}{\partial x} + \frac{\partial w}{\partial x}\frac{\partial w}{\partial y} \\
\chi_x &= -\frac{\partial^2 w}{\partial x^2} \\
\chi_y &= -\frac{\partial^2 w}{\partial y^2} \\
\chi_{xy} &= 2\frac{\partial^2 w}{\partial x \partial y}
\end{aligned} \tag{10.94}$$

In order to form the tangential stiffness matrix $[K_T]$ and resistance vector $\{R\}$, the incremental strain–displacement relationships should be derived first. For this purpose, by substituting displacement functions Equation 10.81 into the expression for ϵ_x in Equation 10.94 and giving the displacement parameters infinitesimal increments $d\{\delta\}$, the increment of strain ϵ_x can then be obtained as

$$\begin{aligned}
d\epsilon_x &= d\left(\frac{\partial u}{\partial x} + \frac{1}{2}(\frac{\partial w}{\partial x})^2\right) \\
&= d\frac{\partial u}{\partial x} + \frac{\partial w}{\partial x} d\frac{\partial w}{\partial x} \\
&= \sum_{m=1}^{r}(-\frac{1}{b}du_{im} + \frac{1}{b}du_{jm})Y^u_m(y) + \frac{\partial w}{\partial x}\sum_{m=1}^{r}(\frac{\partial N_{1m}}{\partial x}dw_{im} + \\
&\quad \frac{\partial N_{2m}}{\partial x}d\theta_{im} + \frac{\partial N_{3m}}{\partial x}dw_{jm} + \frac{\partial N_{4m}}{\partial x}d\theta_{jm})
\end{aligned} \tag{10.95}$$

in which $\frac{\partial w}{\partial x}$ is the current value of slope and is calculated from current displacement parameters $\{\delta\}$ as

$$\frac{\partial w}{\partial x} = \sum_{m=1}^{r} (\frac{\partial N_{1m}}{\partial x} w_{im} + \frac{\partial N_{2m}}{\partial x} \theta_{im} + \frac{\partial N_{3m}}{\partial x} w_{jm} + \frac{\partial N_{4m}}{\partial x} \theta_{jm}) \quad (10.96)$$

Similarly, other incremental strain components due to $d\{\delta\}$ can also be obtained. Thus, the incremental strain–displacement relationships are derived in matrix form as

$$d\{\epsilon\} = d[\epsilon_x, \epsilon_y, \gamma_{xy}, \chi_x, \chi_y, \chi_{xy}]^T = \sum_{m=1}^{r} [\bar{B}]_m d\{\delta\}_m \quad (10.97)$$

and

$$[\bar{B}]_m = [B]_m + [B_L]_m \quad (10.98)$$

in which $[B]_m$ is the linear strain matrix of Equation 10.92, and $[B_L]_m$ is a function of $\{\delta\}$ which reflects the nonlinear effect of a large deflection:

$$[B_L]_m = \begin{bmatrix} 0 & 0 & w_{,x}N_{1m,x} & w_{,x}N_{2m,x} & 0 & 0 & w_{,x}N_{3m,x} & w_{,x}N_{4m,x} \\ 0 & 0 & w_{,y}N_{1m,y} & w_{,y}N_{2m,y} & 0 & 0 & w_{,y}N_{3m,y} & w_{,y}N_{4m,y} \\ 0 & 0 & w_{,x}N_{1m,y} + w_{,y}N_{1m,x} & w_{,x}N_{2m,y} + w_{,y}N_{2m,x} & 0 & 0 & w_{,x}N_{3m,y} + w_{,y}N_{3m,x} & w_{,x}N_{4m,y} + w_{,y}N_{4m,x} \\ 0 & 0 & 0 & 0 & 0 & 0 & 0 & 0 \\ 0 & 0 & 0 & 0 & 0 & 0 & 0 & 0 \\ 0 & 0 & 0 & 0 & 0 & 0 & 0 & 0 \end{bmatrix} \quad (10.99)$$

Here the subscripts ', x' and ', y' denote $\frac{\partial}{\partial x}$ and $\frac{\partial}{\partial y}$, respectively, and

$$w_{,x} = \sum_{m=1}^{r} (N_{1m,x} w_{im} + N_{2m,x} \theta_{im} + N_{3m,x} w_{jm} + N_{4m,x} \theta_{jm})$$

$$w_{,y} = \sum_{m=1}^{r} (N_{1m,y} w_{im} + N_{2m,y} \theta_{im} + N_{3m,y} w_{jm} + N_{4m,y} \theta_{jm}) \quad (10.100)$$

or, in matrix form,

$$\begin{Bmatrix} w_{,x} \\ w_{,y} \end{Bmatrix} = \sum_{m=1}^{r} \begin{bmatrix} 0 & 0 & N_{1m,x} & N_{2m,x} & 0 & 0 & N_{3m,x} & N_{4m,x} \\ 0 & 0 & N_{1m,y} & N_{2m,y} & 0 & 0 & N_{3m,y} & N_{4m,y} \end{bmatrix} \{\delta\}_m$$

$$= \sum_{m=1}^{r} [G]_m \{\delta\}_m \quad (10.101)$$

If the strains are not large and material remains in the linear elastic state, the stresses can be evaluated from current strains using the linear elastic stress–strain relationships given in Equation 10.89.

The resistances $\{R\}$ to the large deflections of the plate can then be calculated by means of the principle of virtual work. The so-called resistances $\{R\}$ are referred to the external forces required to maintain a given deformation. If any virtual displacement $d\{\delta\}_m$ occurs from the current equilibrium position of a strip, the work done by the external forces during the virtual displacement is

$$dW_{ex} = d\{\delta\}_m^T\{R\}_m$$

where $\{R\}_m$ is the resistance vector of a strip corresponding to mth term of eigenfunctions.

During this virtual displacement, the work performed by internal forces is

$$\begin{aligned} dW_{in} &= \int d\{\epsilon\}^T\{\sigma\}dxdy = \int ([\bar{B}]_m d\{\delta\}_m)^T\{\sigma\}dxdy \\ &= d\{\delta\}_m^T \int [\bar{B}]_m^T\{\sigma\}dxdy \end{aligned}$$

in which $\{\sigma\}$ is the stresses in the strip,

$$\{\sigma\} = [N_x, N_y, N_{xy}, M_x, M_y, M_{xy}]^T.$$

According to the principle of virtual work, dW_{ex} must be equal to dW_{in} for any possible $d\{\delta\}_m$, which leads to

$$\{R\}_m = \int [\bar{B}]_m^T\{\sigma\}\, dx\, dy \tag{10.102}$$

The integration in Equation 10.102 can be implemented by Gaussian quadrature.

Thus, at the kth iteration, the resistance $\{R\}^k$ can be computed once the stresses $\{\sigma\}^k$ are obtained. If the current resistances $\{R\}^k$ are not equal to the external loads $\{P\}$, under so-called 'unbalanced forces', i.e. the difference between $\{P\}$ and $\{R\}^k$, the structure deforms further to a new position $\{\delta\}^{k+1}$ as

$$\{\delta\}^{k+1} = \{\delta\}^k + \Delta\{\delta\}^{k+1} \tag{10.103}$$

where

$$[K]\Delta\{\delta\}^{k+1} = \{P\} - \{R\}^k \tag{10.104}$$

in which $[K]$ is the initial linear elastic stiffness matrix if the modified Newton–Raphson method is used, or $[K]$ is the tangential stiffness matrix $[K] = [K_T^k]$ if the Newton–Raphson method is applied.

Next, the slopes $\frac{\partial w}{\partial x}$ and $\frac{\partial w}{\partial y}$ may be obtained using Equation 10.100.

Substituting the results into Equation 10.94 gives updated strains as

$$\{\epsilon\}^{k+1} = \sum_{m=1}^{r} [B]_m \{\delta\}^{k+1} + [\frac{1}{2}(\frac{\partial w}{\partial x})^2, \frac{1}{2}(\frac{\partial w}{\partial y})^2, \frac{\partial w}{\partial x}\frac{\partial w}{\partial y}, 0, 0, 0]^T_{k+1} \tag{10.105}$$

Then, new stresses can be evaluated from Equation 10.89:

$$\{\sigma\}^{k+1} = [D]\{\epsilon\}^{k+1} \tag{10.106}$$

Further, substituting the slopes into Equation 10.99 gives the new $[B_L^{k+1}]_m$, and $[\bar{B}^{k+1}]_m$ is readily formed.

Thus, the resistances can be renewed by virtue of Equation 10.102 as

$$\{R\}^{k+1}_m = \int [\bar{B}^{k+1}]^T_m \{\sigma\}^{k+1} dxdy \tag{10.107}$$

This procedure is repeated until convergence is achieved, i.e. until

$$\frac{\sqrt{\sum(\Delta\delta_i^{k+1})^2}}{\sqrt{\sum(\delta_i^{k+1})^2}} \leq e \tag{10.108}$$

where e is a prescribed tolerance.

10.5.4 Modified Newton–Raphson method

If the modified Newton–Raphson method is applied to geometrically nonlinear finite strip analysis, the initial linear elastic stiffness matrix of the plate is employed during all the iterations. Consequently, the uncoupled problem in linear-elastic analysis will remain uncoupled in the nonlinear analysis, the half-bandwidth of the stiffness matrix also remaining unchanged. Thus, the high efficiency of the finite strip method will not be lost in nonlinear analysis. Also, the stiffness matrix need not be reformed and inverted in each new iteration. Although the number of iterations will be increased, an overall economy may be achieved in many cases, especially for problems with only slight nonlinearities.

However, because the true flexural stiffness of a plate under large deflection is much greater than its linear elastic stiffness, divergence may possibly occur if the initial stiffness matrix remains unchanged through all iterations (Fig. 10.13). In order to prevent this situation, the flexural stiffness matrix may be modified by a simple method, i.e. by multiplying all the items of this matrix by a flexural stiffness amplification factor, α, which is a function of the ratio of current maximum deflection to thickness of the plate. In practice, this method may be replaced by a more efficient equivalent approach, i.e. by dividing the 'unbalanced force' corresponding to every flexural degree of freedom on the right-hand side of Equation 10.104 by the same factor at each iteration. It has been shown that this

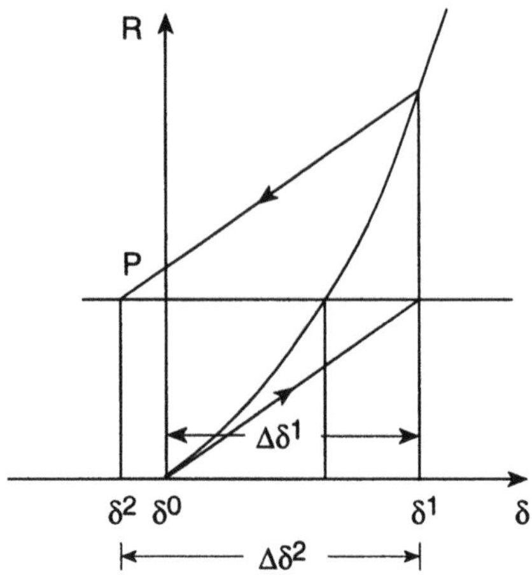

Figure 10.13 *Possible divergence.*

method can ensure convergence effectively if α is reasonably chosen (see M. S. Cheung and Li, 1989B). In the meantime, the in-plane nodal forces should not be changed in order to accelerate convergence.

After a number of trials, appropriate values for α have been found as follows:

1. for plates with two opposite clamped edges

$$\alpha = 1 + 5.5(\frac{\delta_{max}}{h})^2 \tag{10.109}$$

2. for plates with two opposite simply supported edges

$$\alpha = 1 + 10.0(\frac{\delta_{max}}{h})^2 \tag{10.110}$$

In these expressions, δ_{max} is the maximum displacement parameter found in the current iteration.

10.5.5 Newton–Raphson method

If the Newton–Raphson method is adopted for geometrically nonlinear analysis, the tangential stiffness matrix may be derived by taking appropriate variations of resistances in Equation 10.102 with respect to $d\{\delta\}_n$ as

$$d\{R\}_m = \int d[\bar{B}]_m\{\sigma\}dxdy + \int [\bar{B}]_m d\{\sigma\}dxdy = [K_T]_{mn}d\{\delta\}_n \tag{10.111}$$

Using Equations 10.89 and 10.97 we have

$$d\{\sigma\} = [D]d\{\epsilon\} = [D][\bar{B}]_n d\{\delta\}_n \tag{10.112}$$

Because the linear strain matrix $[B]_m$ is not related to displacements, according to Equation 10.98, the following relationship is obtained:

$$d[\bar{B}]_m = d([B]_m + [B_L]_m) = d[B_L]_m \tag{10.113}$$

Equation 10.111 then becomes

$$d\{R\}_m = \int d[B_L]_m\{\sigma\}dxdy + [\bar{K}_T]_{mn}d\{\delta\}_n \tag{10.114}$$

where

$$[\bar{K}_T]_{mn} = \int [\bar{B}]_m^T[D][\bar{B}]_n dxdy \tag{10.115}$$

The first term of Equation 10.114 represents the contribution of the in-plane forces to the flexural stiffness if the in-plane forces in $\{\sigma\}$ are assumed to be constant with respect to displacements. This is the assumption adopted in the stability analysis (Chapter 9). Therefore, it is conceivable that after some mathematical process (see Zienkiewicz, 1977, pp 509–511) this term can be converted to the form

$$\int d[\bar{B}]_m^T\{\sigma\} = [K_G]_{mn}d\{\delta\}_n \tag{10.116}$$

where $[K_G]_{mn}$ is the geometrical stiffness matrix which is the same as for the stability analysis:

$$[K_G]_{mn} = \int [G]_m \begin{bmatrix} N_x & N_{xy} \\ N_{xy} & N_y \end{bmatrix} [G]_n dxdy \tag{10.117}$$

in which $[G]$ is the slope matrix defined in Equation 10.101, and N_x, N_y and N_{xy} represent the current values of in-plane forces.

Thus, the tangential stiffness matrix $[K_T]_{mn}$ can be expressed as

$$[K_T]_{mn} = [\bar{K}_T]_{mn} + [K_G]_{mn} \tag{10.118}$$

10.5.6 Numerical examples

The following two examples are given to illustrate efficiency and accuracy of the geometrically nonlinear finite strip analysis. In both examples the modified Newton–Raphson method was employed (M. S. Cheung and Li, 1989B).

1. Cylindrical bending of a uniformly loaded rectangular plate with simply supported immovable edges

The plate is very long in the x direction, so that, the deflected surface of a portion of such a plate at a considerable distance from the ends can be assumed cylindrical. Therefore, the analysis needs only one strip with boundary conditions $u_1 = u_2 = 0$ and $\theta_1 = \theta_2 = 0$ along two nodal lines. The length of the strip is $l = 50$ in. The thickness is $h = 0.5$ in. The elastic constants of the material are $E = 30000000$ psi and $\nu = 0.3$. The intensity

Table 10.2 *Deflection and stresses in cylindrical plate bending*

m_{max}	w_{max}	σ_y	M_y	N_{iter}
1	0.6972	15.820	945.20	17
3	0.6827	15.770	778.70	47
5	0.6841	15.840	825.60	48
Theory	0.688	15.830	830.14	–

Table 10.3 *Deflection and stresses in a clamped square plate*

qa^4/Dh	w_0/h	w_{max}/h	$\sigma_n a^2(1-\nu^2)/Eh^2$	N_{iter}	w_{max}/h^*
50	1.008	0.7666	4.22	7	0.758
100	2.016	1.1578	7.09	13	1.149
150	3.024	1.4182	9.37	17	1.410
200	4.032	1.6178	11.33	21	1.600
250	5.040	1.7821	13.05	25	1.777

$*$ Timoshenko and Woinowsky–Krieger, 1959

of the uniformly distributed load is $q = 20$ psi. The maximum linear elastic deflection is $w_0 = 4.740$, $w_0/h = 9.48$. Because of the symmetry of deformation, only the eigenfunctions symmetrical to the midspan are used, i.e. $m = 1, 3, 5, \ldots$. The numerical results of the deflection w_{max} (in inches), the longitudinal stress of the middle plane σ_y in psi, and the longitudinal bending moment M_y in lb in, at midspan are listed in Table 10.2. For comparison, the theoretical answers (Timoshenko and Woinowsky–Krieger, 1959) are also given in this table, in which m_{max} is the highest order of eigenfunctions used in these analysis and N_{iter} is the number of iterations up to convergence.

2. Large deflections of a uniformly loaded square plate with clamped edges

A square plate of side $2a$ and thickness h, clamped along all the immovable edges, is subjected to a uniformly distributed load q. Using symmetry, only half the plate is analyzed; it is divided into five equal width strips and analyzed by means of five eigenfunctions symmetrical to the midspan ($m_{max} = 9$). The numerical results are listed in Table 10.3 in which D, the flexural rigidity of the plate, is $Eh^3/12(1-\nu^2)$, $\nu = 0.3$, w_0 is the linear elastic deflection at the center of plate, w_{max} is the result obtained from nonlinear analysis, and σ_n is the maximum normal bending stress at the center of each edge. For comparison, the answers obtained by an energy method with 11 parameters (Timoshenko and Woinowsky–Krieger, 1959, pp 421–433) are also given in the same table. In every case, the whole load is imposed within one step. The number of iterations up to convergence N_{iter} is 7 to 25, depending on the ratio of w_0/h. In ordinary structures, this ratio will not be too large; it is concluded that the analysis does not need too many iterations.

CHAPTER 11

Combined analysis

11.1 Introduction

If a plate structure has constant cross-section and its end support condition does not change transversely, the finite strip method has proved to be a very efficient numerical structural analysis method. However, if the structure has irregularities, e.g. a rectangular plate with openings, the finite strip method is no longer applicable on its own and the finite element method or the boundary element method has to be used. In this case, however, if these methods can be combined together, with the finite strips being used for the regular part of the plate and the finite elements or boundary elements modelling the irregular part, then the efficiency of the finite strip method and the universality of the latter methods are both utilized to their full advantage. Another case favorable for combined analysis is a slab-on-girder bridge or box girder bridge under moving-wheel loading. If the local effect of the load is concerned, the semianalytical finite strip method becomes uneconomical because too many series terms must be used (see Chapter 14). In this situation, the spline finite strip method may give better efficiency. However, in order to obtain accurate results for the maximum bending moments, the load application point must be taken as a node and a dense mesh is required around this point. When the load is moving along the bridge, the mesh must be modified for each new position of the load. This problem can be solved by using the boundary element method to analyze the slab, because only the boundary of the slab panel is divided into elements and there is no mesh inside the panel. Consequently, no mesh modification is necessary while the load is moving. In this case, the girders are still analyzed by the finite strip method since only their deformation, not stresses, have significant influence on the local bending moments in the slab.

In the present study, special transition elements are used to connect the two different regions. One side of such a transition element coincides with the nodal line of adjacent finite strip and has the same degrees of freedom; on the opposite side there are a number of nodes which are connected with the finite element or boundary element. Inside the transition element the deflection is expressed in terms of the degrees of freedom of the finite strip nodal line and the nodes of the finite element or boundary element by their corresponding shape functions. Using the principle of minimum total potential energy, the stiffness matrix and load vector of the transition element can be obtained.

This chapter presents the basic procedures for:

1. combined finite strip/finite element analysis of irregular plates;
2. combined finite strip/boundary element analysis of irregular plates;
3. combined finite strip/boundary element analysis of slab-on-girder bridges subjected to moving wheel loads.

Numerical examples are given to show the applicability of above approaches.

11.2 Combined FS/FE analysis of irregular plates

11.2.1 Finite strips for regular region

The regular region of a rectangular plate is subdivided into a number of finite strips. Within each strip, the deflection is expressed as

$$w = \sum_{m=1}^{r} (f_1(\xi)w_{1m} + bg_1(\xi)\theta_{1m} + f_2(\xi)w_{2m} + bg_2(\xi)\theta_{2m})\,Y_m(y) \quad (11.1)$$

where l and b are the length and width of the strip respectively, $f_1(\xi)$, $f_2(\xi)$, $g_1(\xi)$ and $g_2(\xi)$ are transverse shape functions which are the Hermitian cubic polynomials

$$f_1(\xi) = 1 - 3\xi^2 + 2\xi^3, \quad f_2(\xi) = 3\xi^2 - 2\xi^3,$$
$$g_1(\xi) = \xi - 2\xi^2 + \xi^3, \quad g_2(\xi) = \xi^3 - \xi^2, \quad (\xi = x/b) \quad (11.2)$$

and $Y_m(y)$ is the mth series term of beam vibration eigenfunctions, as given in Section 4.3.5 according to the end conditions of the plate strip.

The stiffness matrix and load vector of this strip have been given in Chapter 4.

11.2.2 Finite elements for irregular regions

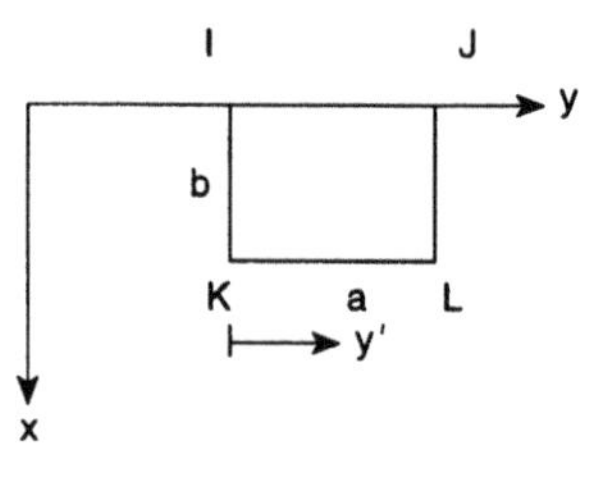

Figure 11.1 *Rectangular finite element.*

The irregular region of a plate can be simulated by a number of conforming rectangular plate elements (Fig. 11.1) (Brebbia and Connor, 1973) with four degrees of freedom at each node which are

$$\{\delta\} = (w, \frac{\partial w}{\partial x}, \frac{\partial w}{\partial y}, \frac{\partial^2 w}{\partial x \partial y})^T$$

The shape function of this element is

$$\begin{aligned} w &= f_1(\xi)f_1(\eta)w_i + bg_1(\xi)f_1(\eta)\frac{\partial w_i}{\partial x} + af_1(\xi)g_1(\eta)\frac{\partial w_i}{\partial y} \\ &+abg_1(\xi)g_1(\eta)\frac{\partial^2 w_i}{\partial x \partial y} \\ &+f_1(\xi)f_2(\eta)w_j + bg_1(\xi)f_2(\eta)\frac{\partial w_j}{\partial x} + af_1(\xi)g_2(\eta)\frac{\partial w_j}{\partial y} \\ &+abg_1(\xi)g_2(\eta)\frac{\partial^2 w_j}{\partial x \partial y} \\ &+f_2(\xi)f_1(\eta)w_k + bg_2(\xi)f_1(\eta)\frac{\partial w_k}{\partial x} + af_2(\xi)g_1(\eta)\frac{\partial w_k}{\partial y} \\ &+abg_2(\xi)g_1(\eta)\frac{\partial^2 w_k}{\partial x \partial y} \\ &+f_2(\xi)f_2(\eta)w_l + bg_2(\xi)f_2(\eta)\frac{\partial w_l}{\partial x} + af_2(\xi)g_2(\eta)\frac{\partial w_l}{\partial y} \\ &+abg_2(\xi)g_2(\eta)\frac{\partial^2 w_l}{\partial x \partial y} \end{aligned} \tag{11.3}$$

where $\xi = x/b$ and $\eta = y'/a$.

This type of element meets the continuity requirement between any adjacent elements not only for the deflection w and the tangential derivative $\frac{\partial w}{\partial s}$ but also for the normal derivative $\frac{\partial w}{\partial n}$. Therefore, this element is a conforming plate bending element, and it achieves considerable improvement in accuracy in comparison with rectangular elements having only three degrees of freedom at each node.

11.2.3 Transition element

A row of rectangular transition elements is used to connect the finite element region with the finite strip region. One side of each transition element coincides with the nodal line of the adjacent strip and has the same degrees of freedom. On the opposite side there are two corner nodes which are attached to the nodes of adjacent finite elements and which have the same degrees of freedom as these finite element nodes (Fig. 11.2). Inside the transition element, the deflection w is expressed in terms of degrees of freedom of the nodal line and the two nodes by their corresponding shape

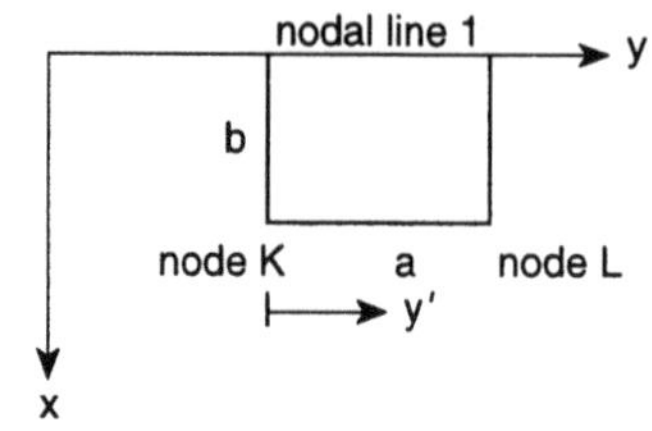

Figure 11.2 *Transition element.*

functions in the form

$$
\begin{aligned}
w \quad = \quad & \sum_{m=1}^{r}(f_1(\xi)w_{1m} + bg_1(\xi)\theta_{1m})Y_m(y) \\
& + f_2(\xi)f_1(\eta)w_k + bg_2(\xi)f_1(\eta)\frac{\partial w_k}{\partial x} + af_2(\xi)g_1(\eta)\frac{\partial w_k}{\partial y} \\
& + abg_2(\xi)g_1(\eta)\frac{\partial^2 w_k}{\partial x \partial y} \\
& + f_2(\xi)f_2(\eta)w_l + bg_2(\xi)f_2(\eta)\frac{\partial w_l}{\partial x} + af_2(\xi)g_2(\eta)\frac{\partial w_l}{\partial y} \\
& + abg_2(\xi)g_2(\eta)\frac{\partial^2 w_l}{\partial x \partial y} \qquad (11.4)
\end{aligned}
$$

where $\xi = x/b$ and $\eta = y'/a$.

It should be noted that the nodal line 1 and nodes k and l have the same type of shape function in the direction x, which are Hermitian cubic polynomials, in order to simulate some basic deformation patterns more accurately, such as one-dimensional bending in the direction y or torsion about the axis y.

The stiffness matrix and the load vector can be obtained in accordance with standard finite strip formulation. Simply assembling the stiffness matrices and load arrays of these transition elements with those of finite strips and finite elements will lead immediately to the formulation for the whole plate structure.

11.2.4 Numerical examples

1. Clamped square plate under uniform load

This example is chosen to verify the present method. The square plate is clamped along four edges and is subjected to a uniform load. The length of each side is 10.0 m, the thickness is 0.5 m. The material properties are $E = 100000$ MPa and $\nu = 0.3$. The intensity of the load is $q = 1.0$ MN m^{-2}. Half the plate is divided into finite strips, the other half into finite elements. The mesh is shown in Fig. 11.3 and the results are listed in Table 11.1. The

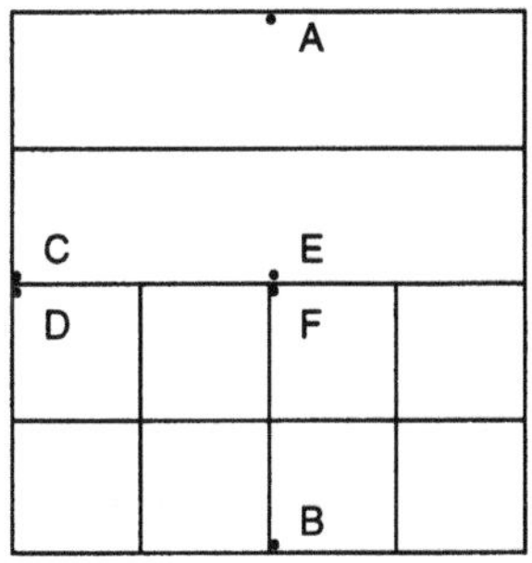

Figure 11.3 *Square plate.*

Table 11.1 *Deflection and moments in clamped square plate*

No.of strips	2	3	4	
No.of sym.terms	3	3	5	Theory
Mesh of elements	2 by 4	3 by 6	4 by 8	
w_{max}	0.0113105	0.0111349	0.0110848	0.0110074
M_x at A	−4.390	−4.747	−4.896	
M_x at B	−4.719	−4.916	−5.001	
M_y at C	−5.111	−4.915	−5.071	−5.130
M_x at E	2.496	2.380	2.331	
M_y at E	2.445	2.376	2.322	2.375

analytical results are also given in this table for comparison. From this table it can been seen that the method achieves satisfactory accuracy in evaluating the maximum deflection and bending moments.

2. Square plate supported by walls and columns

A square plate with a rectangular opening is supported by walls and columns (Fig. 11.4). The length of each side is 9.0 m and the thickness is 0.2 m. The material properties are $E = 25000$ MPa and $\nu = 0.15$. The plate is subjected to a uniform load of intensity $q = 10$ kN m^{-2}.

The part supported by columns is divided into 6 by 18 conforming rectangular plate elements with four degrees of freedom at each node; the rest of the plate is divided into 11 strips and one row of transition elements that connect the strips to the finite elements. Five symmetrical series terms are used. For simplicity, the walls are regarded as simple supports, and the columns as point supports.

The resulting deflection along line C–C and the bending moment M_x along the lines B–B and C–C are shown in Fig. 11.5. The CPU time spent for the analysis is 4.2s on an AMDAHL-5860 mainframe.

The structure was also analyzed using an 18 × 18 array of the above-

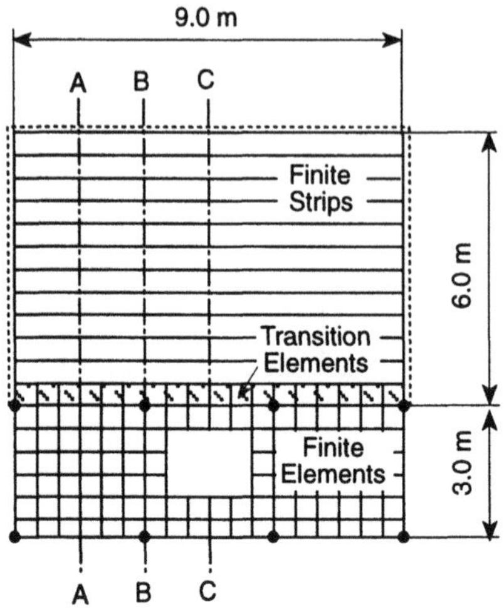

Figure 11.4 *Plate supported by walls and columns.*

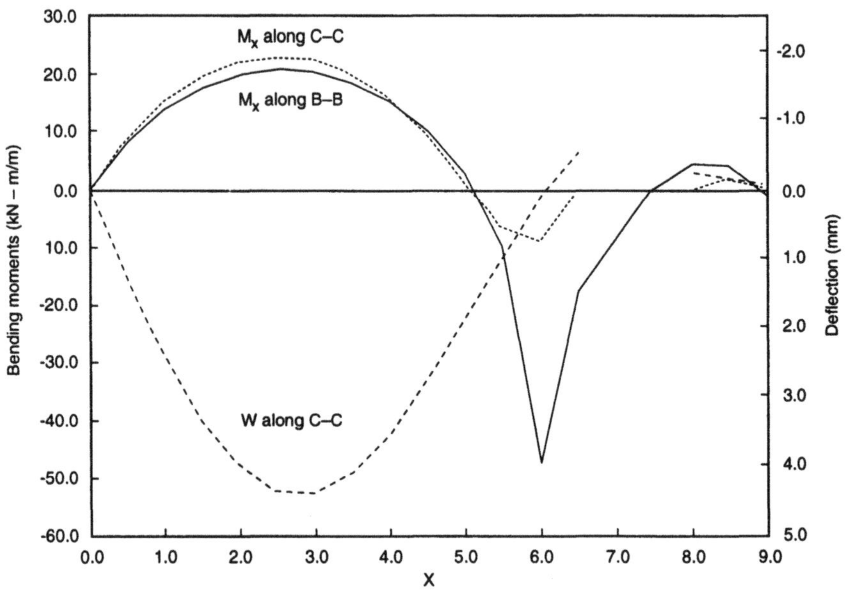

Figure 11.5 *Deflection and bending moments of the plate in Fig. 11.4.*

mentioned elements at a cost of 12.1s of CPU time. However, the results show no discernible difference from the previous solution when the results of both methods are drawn on the same figure.

11.3 Combined FS/BE analysis of irregular plates

11.3.1 Boundary elements for irregular regions

In this approach, the boundary of the irregular region of the plate is divided into a number of boundary elements. Each element has between 2 and 11 nodes, and all the boundary functions along the element are expressed in terms of their nodal values by Lagrange interpolation (Abdel–Akher and Hartley, 1989). A source point is located at a little distance from the boundary on the outside normal line passing through each node. Applying a unit force and a unit normal moment at each source point respectively and implementing the Maxwell–Betti theorem provide the two integrate equations

$$\int_s (WV^f + \Theta M^f)\,ds + \sum_{i=1}^{N_c} W_i C_i^f$$
$$= \int_A pw^f\,dA + \int_s (W^f V + \Theta^f M)\,ds + \sum_{i=1}^{N_c} W_i^f C_i \tag{11.5}$$

$$\int_s (WV^m + \Theta M^m)\,ds + \sum_{i=1}^{N_c} W_i C_i^m$$
$$= \int_A pw^m\,dA + \int_s (W^m V + \Theta^m M)\,ds + \sum_{i=1}^{N_c} W_i^m C_i \tag{11.6}$$

where A and s denote the plate area and the boundary coordinate, w is the deflection in the domain, W and Θ are the boundary deflection and normal rotation, M, V and C are boundary moment, equivalent shear and corner force, f and m are superscripts identifying the fundamental solutions for the unit force and the unit moment respectively, and N_c is the number of corners.

All four boundary functions W, Θ, M and V are unknowns on the interface between the transition strip and the neighboring boundary element, but among them only two are unknown on the other boundaries after imposing displacement and force boundary conditions.

At the ends of the interface, double nodes are used (Brebbia *et al.*, 1984); Fig. 11.6). Both nodes A and B have the same coordinates, but may have different boundary conditions. In order to make the number of unknowns at nodes A and B match the number of the boundary integration equations, the corner forces at these two nodes may be expressed in terms of the normal rotations of their respective elements in the form

$$C_A = D(1-\nu)\left(\frac{\partial^2 w}{\partial s \partial n}\right)^{(1)} = D(1-\nu)\sum_{j=1}^{N_1} \frac{\partial L_j^{(1)}}{\partial s}\Theta_j \tag{11.7}$$

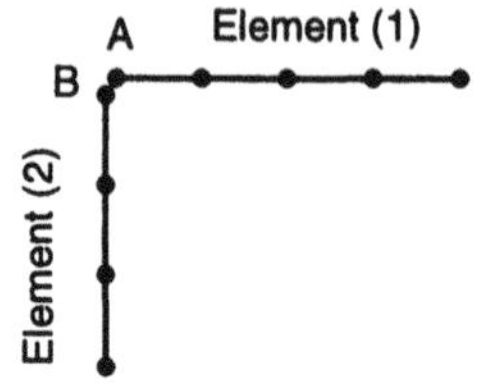

Figure 11.6 *Double nodes.*

$$C_B = -D(1-\nu)(\frac{\partial^2 w}{\partial s \partial n})^{(2)} = -D(1-\nu)\sum_{j=1}^{N_2} \frac{\partial L_j^{(2)}}{\partial s}\Theta_j \quad (11.8)$$

where the superscripts (1) and (2) identify the elements ending at node A and starting at node B, respectively, N_1 and N_2 are the numbers of nodes on boundary elements 1 and 2, respectively, and L_j is the Lagrange shape function of node j and is expressed as

$$L_j(s) = \prod_{\substack{i=1 \\ i \neq j}}^{N} \frac{s - s_i}{s_j - s_i} \quad (11.9)$$

Substituting the expressions for fundamental solutions into the boundary integral equations and implementing the boundary element discretization technique (Abdel–Akher and Hartley, 1989) leads to the boundary element matrices

$$[H^1, H_I^1]\left\{ \begin{array}{c} U^1 \\ U_I^1 \end{array} \right\} = [G^1, G_I^1]\left\{ \begin{array}{c} P^1 \\ P_I^1 \end{array} \right\} + \{B^1\} \quad (11.10)$$

where U denotes the values of deflection and normal rotation at all the nodes, P represents the values of equivalent shear and normal moment at all the nodes, B is the effect of the loads in domain, 1 is the superscript identifying the boundary element region, and I is the subscript defining the interface between the transition strip and the boundary elements.

For each node on the interface there are four unknowns (W, Θ, V and M) versus two equations. Therefore, two more equations are required for each of these nodes.

11.3.2 Transition strip and combined solution

A transition strip is inserted between the finite strip region and the boundary element region and connects the two regions together. One side of the transition strip coincides with the nodal line of the adjacent strip and has the same degrees of freedom as this nodal line. On the opposite side there are a number of nodes which are attached to the nodes

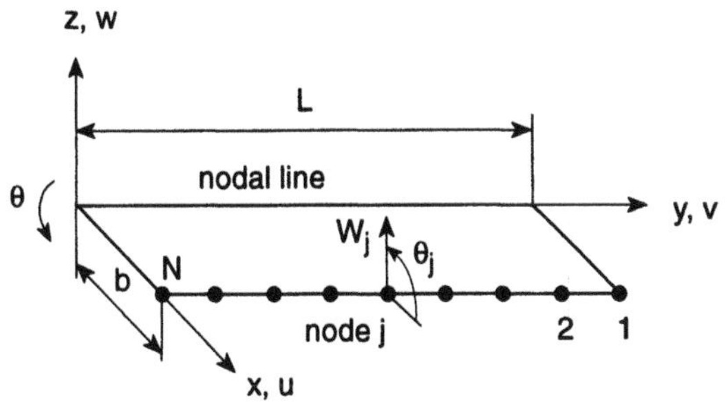

Figure 11.7 *Transition strip.*

of the neighboring boundary elements and have the same displacement parameters (deflection W and normal rotation Θ) (Fig. 11.7). Inside the transition strip the deflection w is expressed in terms of the degrees of freedom of the nodal line and the nodes by their corresponding shape functions in the form:

$$w = \sum_{m=1}^{r} \left(f_1(\xi) w_{1m} + b g_1(\xi)\theta_{1m}\right) Y_m(y) + \sum_{j=1}^{N} L_j \left(f_2(\xi) w_j + a g_2(\xi)\theta_j\right) \tag{11.11}$$

where N is the number of nodes of this strip along the interface, and L_j is the Lagrange shape function of node j.

It should be noted that the nodal line and nodes 1 to N have the same type of shape functions in the direction x, which are Hermitian cubic polynomials, in order to simulate some basic deformation patterns more accurately as noted earlier.

The stiffness matrix and the load vector of the transition strip are generated using the principle of minimum total potential energy.

Assembling the stiffness matrix and load vector of the transition strip with those of the finite strips and applying the condensation technique (Cook *et al.*, 1989) will transform the finite strips and transition strip into a substructure having only the degrees of freedom of the nodes on the interface. The matrix equation of this substructure is of the form:

$$[K]\{U_I^2\} = \{F_I^2\} + \{B^2\} \tag{11.12}$$

where F_I^2 is the vector of unknown interaction nodal forces, B represents the effects of external load, and 2 is the superscript identifying the finite strip and transition strip region.

At all nodes on the interface, the displacement on both sides must be the same, and the interaction force must be the same in magnitude and opposite in direction. These requirements yield the relationships

$$\{U_I^2\} \quad = \quad \{U_I^1\} \tag{11.13}$$

$$\{F_I^2\} = -[M]\{P_I^1\} \tag{11.14}$$

where $[M]$ is the transformation matrix from the distributed boundary forces on the interface boundary element to the nodal forces at the nodes of the transition strip, and its elements M_{ij} are given by

$$M_{ij} = \int_e L_i(y)L_j(y)\,dy$$

Substituting Equations 11.13 and 11.14 into 11.12 gives the set of equations

$$[K]\{U_I^1\} = -[M]\{P_I^1\} + \{B^2\} \tag{11.15}$$

Combining matrix equations 11.10 and 11.15 leads to a set of equations that are just sufficient in number to solve for all the unknowns in the boundary element region, including the unknowns at the nodes on the interface. These latter can be substituted back into the matrix of the finite strip and transition strip region, and all the unknown nodal parameters in this region can then be readily obtained. Thereafter it is a straightforward procedure to compute the displacement and stress components at any point of interest.

11.3.3 Numerical examples

1. Simply supported square plate under uniform load

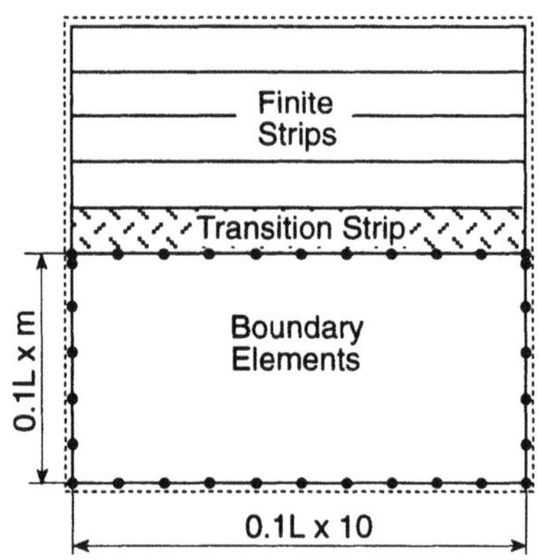

Figure 11.8 *Simply supported square plate under uniform load.*

This example is chosen to verify the present method. A square plate is hinged along four edges and subjected to uniform load. The length of each side is L, Poisson's ratio is $\nu = 0.3$, the flexural rigidity is D, and the intensity of the load is q. Three different meshes are used to analyze the plate (Fig. 11.8). The upper part of each mesh is divided into finite strips with equal width 0.1 L. Five symmetrical terms are used in the series of

Table 11.2 *Deflection and stresses in simply supported square plate*

Mesh	$w_{max} = \alpha qL^4/D$ α	$(M_x)_{max} = \beta qL^2$ β	$V_{max} = \delta qL$ δ	R nqL^2 n
$m = 3$	0.004063	0.04804	0.4205	0.0618
$m = 5$	0.004062	0.04790	0.4205	0.0646
$m = 7$	0.004063	0.04789	0.4205	0.0654
Theory	0.00406	0.0479	0.420	0.065

longitudinal shape functions. The boundary of the lower part is simulated by four boundary elements with equal nodal spacing 0.1 L. Both parts are connected by a transition strip with a width of 0.1 L. The numerical results are listed in Table 11.2. In this table, W_{max} and $(M_x)_{max}$ are the deflection and bending moment at the plate center respectively, V_{max} is the equivalent shear force at the center of the bottom edge, and R is the corner force at the lower corners. In comparison with plate theory (Timoshenko and Woinowsky–Krieger, 1959), it can be seen that the accuracy of the present method is satisfactory.

2. Simply supported square plate with an opening and a skew corner

A square plate is simply supported along all four sides and subjected to uniform load $q = 10$ kN m^{-2}. The length of each side is 10.0 m, and the thickness 0.2 m. The material properties are $E = 25000$ Mpa and $\nu = 0.15$. One corner is cut off at 45°, and a square opening is located near another corner, as shown in Fig. 11.9.

The part including the opening and the skew corner is analyzed by 9 boundary elements with 50 nodes, while the rest of the plate is analyzed by five finite strips and one transition strip with 10 series terms.

The resulting bending moments M_x and M_y along line A–B–C are depicted in Fig. 11.10.

The plate was also analyzed by the boundary element method alone, with 9 elements and 58 nodes. The results of this analysis, if plotted on Fig. 11.10, would be indistinguishable from those already shown.

11.4 Combined BE/FS analysis of slab girder bridges

11.4.1 Finite strips for girders

As mentioned earlier, if a slab girder bridge is subjected to moving wheel loads and the local bending moments are concerned, the slab can be analyzed by the boundary element method and the girders by

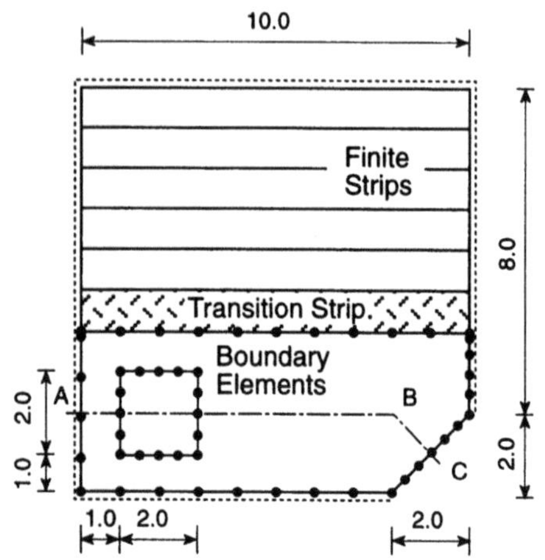

Figure 11.9 *Plate with opening and skew corner.*

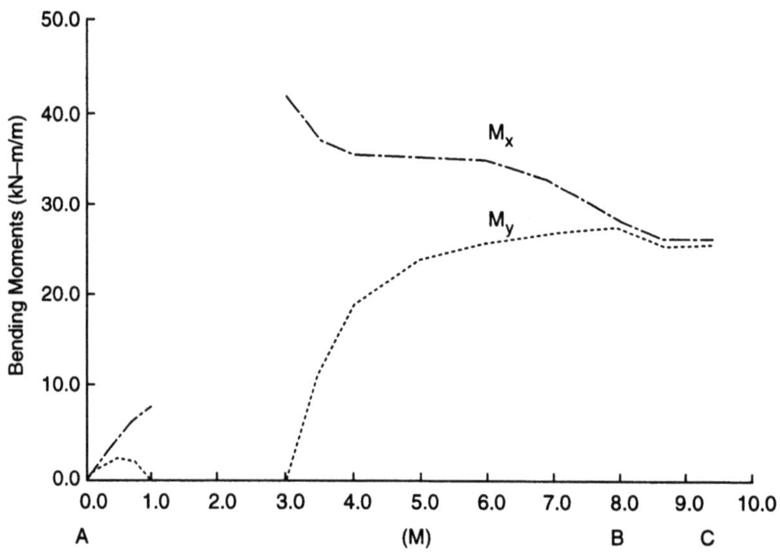

Figure 11.10 *Bending moments along A–B–C.*

finite strip method (Fig. 11.11); (M. S. Cheung, Akhras and Li, 1994). In this application, the flat shell strips (Chapter 4) are employed. The displacement components in the local system of the strip are expressed in

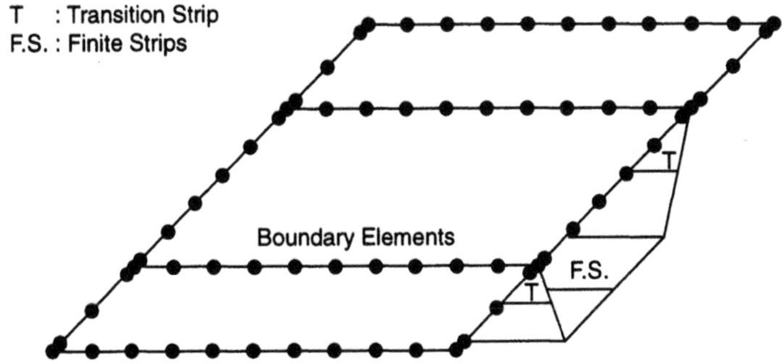

Figure 11.11 *Combined analysis of a bridge; T = transition strip, F.S. = finite strip.*

terms of the displacement parameters as (Fig. 11.12)

$$
\begin{aligned}
u &= \sum_{m=1}^{r} [(1-\xi)u_{1m} + \xi u_{2m}] Y_m(y) \\
v &= \sum_{m=1}^{r} [(1-\xi)v_{1m} + \xi v_{2m}] Y'_m(y)/\mu_m \\
w &= \sum_{m=1}^{r} \left(f_1(\xi)w_{1m} + bg_1(\xi)\theta_{1m} + f_2(\xi)w_{2m} + bg_2(\xi)\theta_{2m}\right) Y_m(y)
\end{aligned}
\tag{11.16}
$$

where $f_1(\xi)$, $g_1(\xi), \ldots$ are the transverse shape functions defined in Equation 11.2.

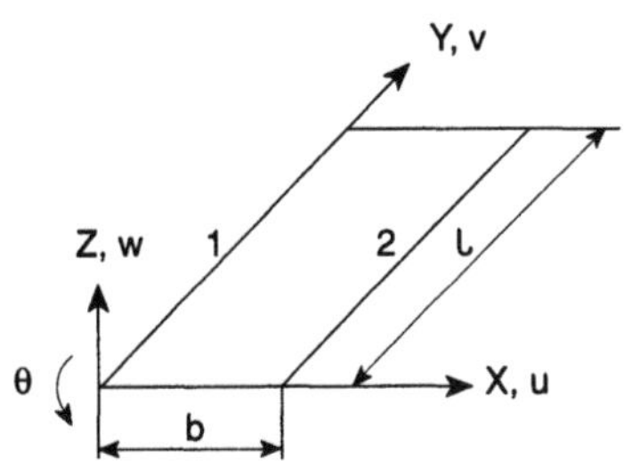

Figure 11.12 *Finite flat shell strip.*

The stiffness matrix and the load vector of the strip have been explicitly given in Chapter 4.

11.4.2 Boundary elements for slab panels

In this approach, the boundary of each slab panel is divided into a number of boundary elements which include actions of both in-plane stress and plate bending. Each element has 2 to 11 nodes, and all the boundary functions along the element are expressed in terms of their nodal values by Lagrange interpolation as follows:

$$[u, v, w, \theta, P_x, P_y, V, M]^T = \sum_{j=1}^{N} L_j(s)[u_j, v_j, w_j, \theta_j, P_{xj}, P_{yj}, V_j, M_j]^T \tag{11.17}$$

where $L_j(s)$ is the Lagrange shape function given in Equation 11.9, N is the number of nodes in the element, s is the boundary coordinate of the element, u, v and w are the boundary displacements and θ is the normal rotation, P_x and P_y are the boundary in-plane forces, M is the boundary normal bending moment, and V is the equivalent transverse shear which also includes the effects of corner forces.

At the ends of each girder, double nodes (Fig. 11.6) are used: both have the same coordinates but different boundary conditions.

For the in-plane deformation, each boundary node is selected as a source point. For the plate bending, a different source point, located at a short distance from the boundary on the outside normal line passing through each node, is selected. Applying unit forces and unit moment at the source points of each node, implementation of the Maxwell–Betti theorem provides four integration equations for every node:

$$\begin{aligned}
c_x u + \int_s (uP_x^{f_x} + vP_y^{f_x})ds &= \int_s (u^{f_x}P_x + v^{f_x}P_y)ds \\
c_y v + \int_s (uP_x^{f_y} + vP_y^{f_y})ds &= \int_s (u^{f_y}P_x + v^{f_y}P_y)ds \\
\int_s (wV^{f_z} + \theta M^{f_z})ds + \sum_{i=1}^{N_c} w_i C_i^{f_z} &= \int_a p_z w^{f_z} dA + \int_s (w^{f_z}V + \theta^{f_z}M)ds \\
\int_s (wV^m + \theta M^m)ds + \sum_{i=1}^{N_c} w_i C_i^m &= \int_a p_z w^m dA + \int_s (w^m V + \theta^m M)ds
\end{aligned} \tag{11.18}$$

where c_x and c_y are coefficients which may be determined by the conditions of the rigid body motion (Brebbia et al., 1984), N_c is the number of corners, C is the corner force of plate bending, p_z is the vertical distributed load on the slab, A denotes the area of plate midplane, f_x, f_y, f_z and m are the superscripts identifying the fundamental solutions of the unit forces in x, y and z directions and the unit normal moment, respectively.

By substituting Equation 11.17 and the expressions for the fundamental solutions into the boundary integration 11.18 for all the boundary nodes, and performing the boundary and area integrations analytically (Abdel–Akher and Hartley, 1989), the following boundary element matrix equation is obtained for each panel:

$$[H^1, H_I^1]\left\{ \begin{array}{c} U^1 \\ U_I^1 \end{array} \right\} = [G^1, G_I^1]\left\{ \begin{array}{c} P^1 \\ P_I^1 \end{array} \right\} + \{B^1\} \qquad (11.19)$$

where U denotes the values of the displacements and normal rotation at all the boundary nodes, P represents the values of the boundary forces and normal moment at all the nodes, B is the effect of the loads in the domain, I is the subscript defining the interfaces between the transition strips and the boundary elements, and 1 is the superscript identifying the boundary element analysis.

For each panel, Equation 11.19 includes four equations for each node. For every node beyond the interfaces, only four variables out of the eight (u_j, v_j, w_j, θ_j, P_{xj}, P_{yj}, V_j and M_j) are unknown after imposing the displacement and force boundary conditions. However, for every node at the interfaces, these eight parameters are all unknown. Therefore, more equations are required for interface nodes. These equations can be obtained by taking into consideration the compatibility and equilibrium conditions across the interfaces as introduced in the following section.

11.4.3 Transition strips and combined solution

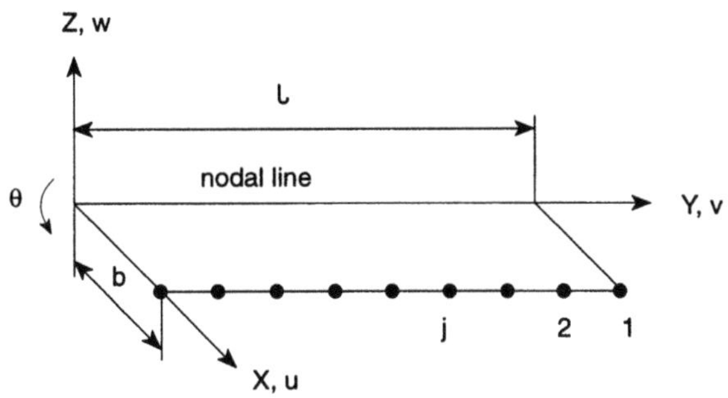

Figure 11.13 *Transition strip.*

Transition strips are inserted between the finite strips and the boundary elements to connect them together. One side of a transition strip coincides with the nodal line of the adjacent finite strip and has the same degrees of freedom of this nodal line. On the opposite side there are a number of nodes which are attached to the nodes of neighboring boundary elements and have the same displacement parameters (displacements u, v, w and normal rotation θ) as those of the nodes of the boundary elements (Fig.

11.13). Within each transition strip, the displacements are expressed in terms of the degrees of freedom of the nodal line, and of the nodes. The corresponding shape functions are as follows:

$$\begin{aligned}
u &= \sum_{m=1}^{r}[(1-\xi)u_{1m}]Y_m(y) + \sum_{j=1}^{N}\xi L_j(y)u_j \\
v &= \sum_{m=1}^{r}[(1-\xi)v_{1m}]Y'_m(y)/\mu_m + \sum_{j=1}^{N}\xi L_j(y)v_j \\
w &= \sum_{m=1}^{r}(f_1(\xi)w_{1m} + bg_1(\xi)\theta_{1m})Y_m(y) + \\
&\quad + \sum_{j=1}^{N} L_j(f_2(\xi)w_j + bg_2(\xi)\theta_j)
\end{aligned} \tag{11.20}$$

The stiffness matrix of the transition strip can be obtained using the principle of minimum total potential energy. After assembling the stiffness matrices of the transition strips with those of the finite strips, the combined finite strips and transition strips are transformed into a substructure by applying the condensation technique (Cook et al. 1989) so that the resulting substructure only has degrees of freedom at the nodes on the interfaces. The matrix equation of this substructure has the form

$$[K]\{U_I^2\} = \{F_I^2\} + \{B^2\} \tag{11.21}$$

where F_I^2 is the vector of the unknown interaction nodal forces, B represents the effect of the external load on the substructure, and 2 is the superscript identifying the finite strip and transition strip substructure.

In general, two panels and one girder meet on each interface, as shown in Fig. 11.11. At all the nodes on the interface, the displacements of the panels and the girder should be the same, and the interaction forces should satisfy the equilibrium conditions. This yields the relationships

$$\{U_I^2\} = \{U_I^{1l}\} = \{U_I^{1r}\} = \{U_I^1\} \tag{11.22}$$

and

$$\{F_I^2\} + [M](\{P_I^{1l}\} + \{P_I^{1r}\}) = \{0\} \tag{11.23}$$

where the superscripts $1l$ and $1r$ refer, respectively, to the left and the right panels of the interface, and $[M]$ is the transformation matrix from the distributed boundary forces on the interface boundary elements to the nodal forces at the nodes of the transition strips. Its entries have the following expression:

$$M_{ij} = \int_e L_i(y)L_j(y)\,dy$$

Substituting Equation 11.22 and 11.23 into 11.21 gives a set of equations:

$$[K]\{U_I^1\} = -[M](\{P_I^{1l}\} + \{P_I^{1r}\}) + \{B^2\} \qquad (11.24)$$

Considering all the panels and girders, and using Equations 11.19 and 11.24 together, will yield enough equations for solving all the unknowns of the boundary elements, including those at the nodes on the interfaces. By backsubstituting into the matrix of the finite strip and transition strip substructure, all the unknown nodal parameters are readily obtained. Then, the displacement and the stress components at any points of interest are computed in a straightforward procedure.

11.4.4 Numerical examples

In order to illustrate the proposed methodology, a twin-girder rectangular bridge and a twin-girder skewed bridge are analyzed in the following examples, although the method is general enough to be applicable to multipanel slab-on-girder or box girder bridges.

1. Rectangular slab-on-girder bridge under a wheel load

This example is chosen to verify the present approach. A simply supported rectangular slab-on-girder bridge is shown in Fig. 11.14. The material properties are E =25 000 MPa and ν=0.15. A vertical wheel load, P = 11.25 × 9.81 kN, distributed uniformly on a small patch (Rowe, 1976) is applied at the center of the slab as illustrated in Fig. 11.14.

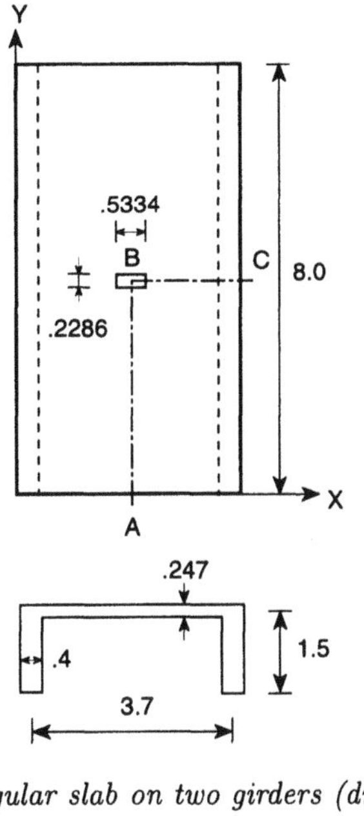

Figure 11.14 *Rectangular slab on two girders (dimensions in meters).*

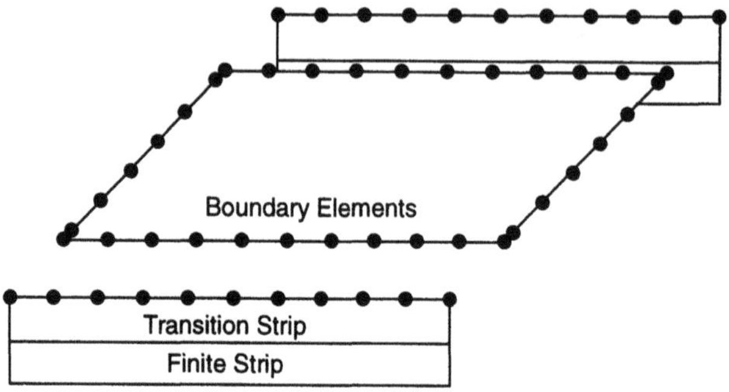

Figure 11.15 *Mesh of combined analysis.*

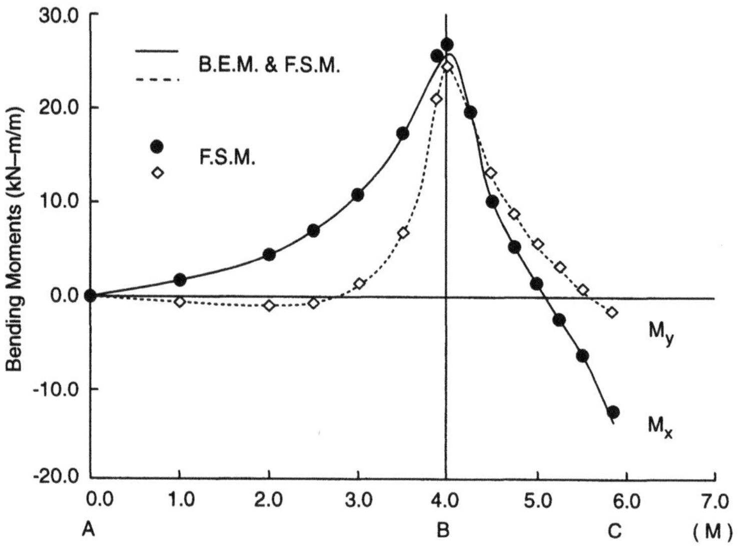

Figure 11.16 *Bending moments along A–B–C.*

The bridge is first analyzed by the finite strip method. The entire slab is divided into 14 third-order flat shell strips, and each girder into two strips. Fifty symmetrical series terms are taken in the analysis. Little improvement can be obtained if more strips or series terms are employed.

Then, the bridge is analyzed by the present method with the mesh shown in Fig. 11.15.

The resulting bending moments along line A–B–C by both methods are depicted in Fig. 11.16. The maximum bending moments occurring at the center of the wheel loading area are $M_x = 25.81$ kN-m/m and $M_y = 24.39$ kN-m/m by the present method comparatively to $M_x = 26.82$ kN-m/m and $M_y = 24.69$ kN-m/m by the finite strip analysis.

Table 11.3 *Effect of simplification of girders*

Simplification of two girders	M_x (kN-m/m)	M_y (kN-m/m)	M_{xy} k(N-m/m)
Simple supports	30.79	27.11	0.00
Clamped supports	23.40	22.68	0.00
Two girders	25.81	24.39	0.00

It can be seen that the agreement between the two methods is satisfactory. However, in order to achieve the same accuracy, the finite strip method needed 3800 degrees of freedom, whereas the present method employed only 224 degrees of freedom for its final solution.

If the two girders are replaced by two simply supported edges or two clamped edges, the resulting maximum bending moments are listed in Table 11.3. The results show that the replacements either overestimate or underestimate the maximum bending moments with errors of 10–20%. This suggests that the flexibility of the girder should not be neglected in the local effect analysis of the wheel load.

2. Skewed slab-on-girder bridge under a wheel load

In this example, a number of skewed slab-on-girder bridges is analyzed to investigate the effect of the skew angle on the local bending moments due to a wheel loading.

All the parameters used for this example, including the dimensions, material properties, loading and the mesh are the same as for the previous example, except the ends of the slab which are skewed at an angle θ as shown in Fig. 11.17.

The resulting bending moments along line D–E–F are drawn in Fig. 11.18. The maximum bending moments occur at the center of the loading area. The values of the bending moments for different skew angles are listed in Table 11.4. It can be seen that the skew angle of the slab ends has little influence on the local wheel effect.

Table 11.4 *Effects of end slew angles*

Skew angle	M_x (kN-m/m)	M_y (kN-m/m)	M_{xy} (kN-m/m)
0°	25.81	24.39	0.00
10°	25.77	24.36	−0.18
20°	25.63	24.25	−0.36
30°	25.40	24.03	−0.51
40°	25.09	23.72	−0.60

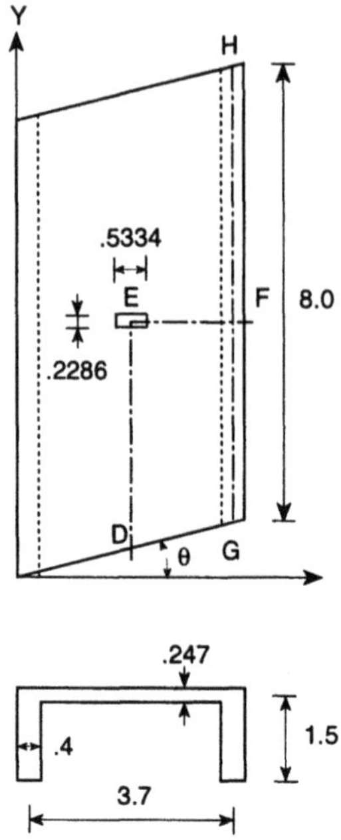

Figure 11.17 *Skew slab on two girders (dimensions in meters).*

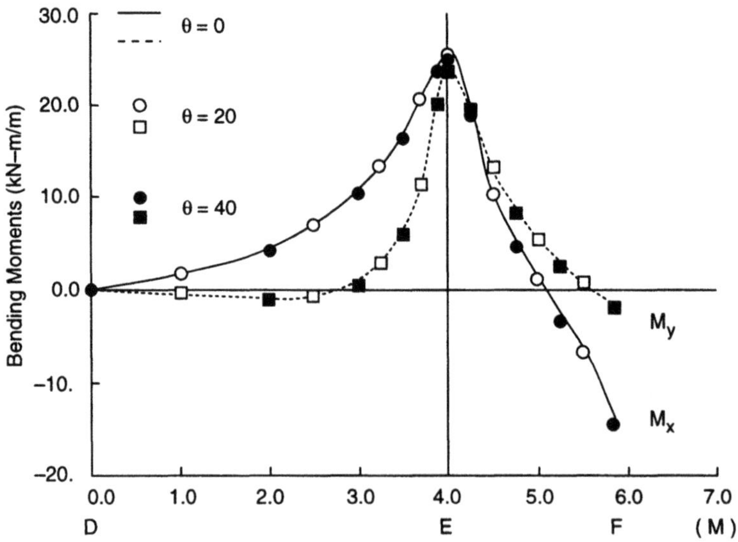

Figure 11.18 *Bending moments along D–E–F.*

PART III

Finite Strip Analysis of Bridges

CHAPTER 12

Slab and slab-on-girder bridges

12.1 Introduction

Many short or medium span highway bridges are slab bridges or slab-on-girder bridges. If a straight or circularly curved bridge has regular geometry, i.e. the cross-section remains constant longitudinally and the boundary conditions at both ends do not change transversely, the semianalytical finite strip method is a very efficient numerical method for the structural analysis of the bridges, especially for simply supported rectangular bridges under distributed loads.

Some bridge decks are voided or stiffened slabs. For acquaintance of the overall behavior of such bridges, the decks can be idealized as orthotropic plates with equivalent elasticity properties, and the plate strip (see Section 4.3) can be employed.

If the longitudinal stiffeners of a slab deck are unequally spaced or far apart from each other, the actual stress distribution in the slab becomes more complex. Thus, it is necessary to divide each slab panel into a number of plate strips, and to model each stiffener by a beam, whose stiffness matrix is then assembled with those of the plate strips (see next section).

If the stiffener is relatively deep, its eccentricity must be taken into consideration and the in-plane deformation and membrane forces in the slab must be included. Then, the slab should be simulated by flat shell strips (see Section 4.5) and the beam strip should include the effect of eccentricity. In some slab-on-girder bridges, the girders are composed of steel plates. The stress distribution in such girders is not comprehensible by beam theory, and consequently, the beam strips are no longer applicable. However, each girder can be regarded as a shell structure, and its each component (web and bottom flange) can be modeled by the flat shell strips.

If a bridge is continuous over intermediate supports, the flexibility approach (see Section 4.7) can be taken, and the deformation of the supports and the settlement of foundations can also be included by some minor modifications in this approach.

This chapter describes the above-mentioned procedures and some numerical examples in the semianalytical finite strip analysis of rectangular slab and slab-on-girder bridges.

The spline finite strip method and compound strip method are also applicable to the analysis of the slab and slab girder bridges. They have been respectively introduced in Chapter 6 and Chapter 7 already. However, some numerical examples appear in this chapter. The analysis of curved or skewed bridges is presented in the next chapter.

12.2 Stiffness matrix of a longitudinal beam

12.2.1 Beam without eccentricity

If the eccentricity of a longitudinal beam is negligible, the centerline of the beam can be assumed to coincide with one nodal line of the plate strip and have the same deflection and rotation. Thus, the displacements of the beam can be expressed as

$$\{f_b\} = \left\{ \begin{array}{c} w_b \\ \theta_b \end{array} \right\} = \sum_{m=1}^{r} Y_m(y) \left\{ \begin{array}{c} w_{im} \\ \theta_{im} \end{array} \right\} \tag{12.1}$$

The strain energy of the beam is

$$U_b = \frac{1}{2}\int_0^l [EI_x(\frac{d^2w_b}{dy^2})^2 + GJ(\frac{d\theta_b}{dy})^2]dy \tag{12.2}$$

where EI_x and GJ are the flexural rigidity about the x axis and the torsional rigidity, respectively. Substituting Equation 12.1 into 12.2, the strain energy of the beam can be expressed in terms of the displacement parameters of the nodal line as

$$U_b = \frac{1}{2}\sum_{m=1}^{r}\sum_{n=1}^{r}\{\delta\}_{in}^T[k]_{mn}^b\{\delta\}_{im} \tag{12.3}$$

in which, $[k]_{mn}^b$ is the beam stiffness matrix defined as

$$[k]_{mn}^b = \left[\begin{array}{cc} k_{mn}^f & 0 \\ 0 & k_{mn}^t \end{array} \right] \tag{12.4}$$

and

$$\begin{aligned} k_{mn}^f &= \int_0^l EI_x Y_m'' Y_n'' dy \\ k_{mn}^t &= \int_0^l GJ Y_m' Y_n' dy \end{aligned} \tag{12.5}$$

For a simply supported beam, the series decouples when EI and GJ are kept constant, and the stiffness coefficients become simply

$$\begin{aligned} k_m^f &= \frac{(m\pi)^4}{2}\frac{EI}{l^3} \\ k_m^t &= \frac{(m\pi)^2}{2}\frac{GJ}{l} \end{aligned} \tag{12.6}$$

12.2.2 Beam with eccentricity

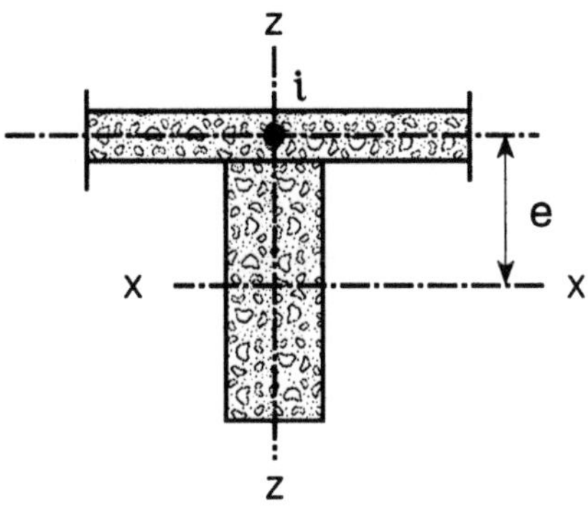

Figure 12.1 *Slab and beam.*

For a deep longitudinal beam attached to nodal line i (Fig. 12.1) the eccentricity e must be taken into consideration in the analysis, and the flat shell strips should be used to model the slab in order to include the in-plane stresses. According to the compatibility conditions, the deformation of the beam centerline can be expressed in terms of the displacements of nodal line as

$$\begin{aligned} u_b &= u_i - e\theta_i \\ v_b &= v_i - e\frac{dw_i}{dy} \\ w_b &= w_i \\ \theta_b &= \theta_i \end{aligned} \tag{12.7}$$

The strain energy of the beam is

$$U_b = \frac{1}{2}\int_0^l [EA(\frac{dv_b}{dy})^2 + EI_z(\frac{d^2u_b}{dy^2})^2 + EI_x(\frac{d^2w_b}{dy^2})^2 + GJ(\frac{d\theta_b}{dy})^2]dy \tag{12.8}$$

where A is the area of cross-section EI_z is the flexural rigidity about the z axis.

Substituting Equation 12.7 into 12.8 and using the displacement functions for the flat shell strip, the strain energy of the beam can be expressed in terms of the displacement parameters of the attached nodal line in the form

$$U_b = \frac{1}{2}\sum_{m=1}^{r}\sum_{n=1}^{r}\{\delta\}_{in}^T[k]_{mn}^b\{\delta\}_{im} \tag{12.9}$$

and the beam stiffness matrix $[k]_{mn}^b$ can be derived.

12.3 Deformation and settlement of support

In the semianalytical finite strip analysis of bridges, the rigid intermediate supports are included by the flexibility approach introduced in Section 4.7. If the deformation and settlements of the supports are significant, the compatibility equations 4.95 become

$$\begin{bmatrix} f_{11} & f_{12} & \cdots & f_{1n} \\ f_{21} & f_{22} & \cdots & f_{2n} \\ \vdots & & & \\ f_{n1} & f_{n2} & \cdots & f_{nn} \end{bmatrix} \begin{Bmatrix} r_1 \\ r_2 \\ \vdots \\ r_n \end{Bmatrix} + \begin{Bmatrix} \delta_1 \\ \delta_2 \\ \vdots \\ \delta_n \end{Bmatrix}$$

$$= - \begin{bmatrix} f'_{11} & 0 & \ldots & 0 \\ 0 & f'_{22} & \ldots & 0 \\ \vdots & & & \\ 0 & 0 & \ldots & f'_{nn} \end{bmatrix} \begin{Bmatrix} r_1 \\ r_2 \\ \vdots \\ r_n \end{Bmatrix} + \begin{Bmatrix} \Delta_1 \\ \Delta_2 \\ \vdots \\ \Delta_n \end{Bmatrix} \quad (12.10)$$

where f'_{ii} is the flexibility coefficient of column i, which is equal to the deformation of the column under unit redundant force $r_i = 1$, and Δ_i is a prescribed permanent displacement due to the foundation settlement. Equation 12.10 can be rewritten in the form

$$\begin{bmatrix} f_{11} + f'_{11} & f_{12} & \cdots & f_{1n} \\ f_{21} & f_{22} + f'_{22} & \cdots & f_{2n} \\ \vdots & & & \\ f_{n1} & f_{n2} & \cdots & f_{nn} + f'_{nn} \end{bmatrix} \begin{Bmatrix} r_1 \\ r_2 \\ \vdots \\ r_n \end{Bmatrix} = \begin{Bmatrix} \Delta_1 - \delta_1 \\ \Delta_2 - \delta_2 \\ \vdots \\ \Delta_n - \delta_n \end{Bmatrix} \quad (12.11)$$

which can be readily solved for the unknown redundant forces $\{r\}$. Then, the displacement and stress at any point can be obtained from the analysis of released structure under the external loads and the redundant forces.

12.4 Numerical examples

1. Continuous Slab Bridge

The two-span slab bridge shown in Fig. 12.2 was analyzed using the finite strip flexibility technique. Comparisons were made with results obtained from a finite element analysis. In the finite element analysis, advantage was taken of the symmetry by analyzing only one quarter of the bridge (Fig. 12.2(b)). The displacements and moments are in excellent agreement with the finite strip solution as can be seen from Fig. 12.3.

For a comparison of efficiency in computation, the full bridge and half the bridge were analyzed using 16 and eight finite strips, respectively, and eight non-zero harmonics. The CPU times on an IBM 370/158 computer were 14.88 seconds and 6.74 seconds for the full bridge and the half bridges models, respectively, whereas the finite element quarter bridge

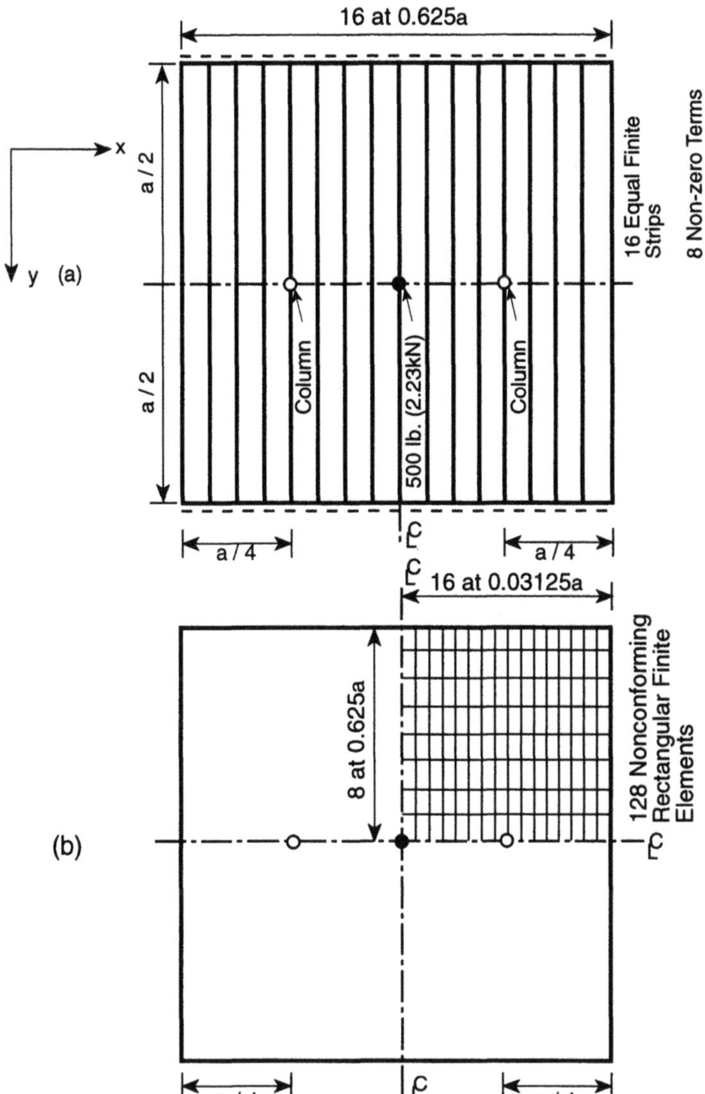

Figure 12.2 *Two-span slab bridge: (a) finite strip idealization; (b) finite element idealization (M. S. Cheung and Chan).*

model required 35.84 seconds. It should be noted that this is the most favorable comparison for the finite element method, because if either the loads or the column reactions were not symmetric, the entire bridge would have to be analyzed.

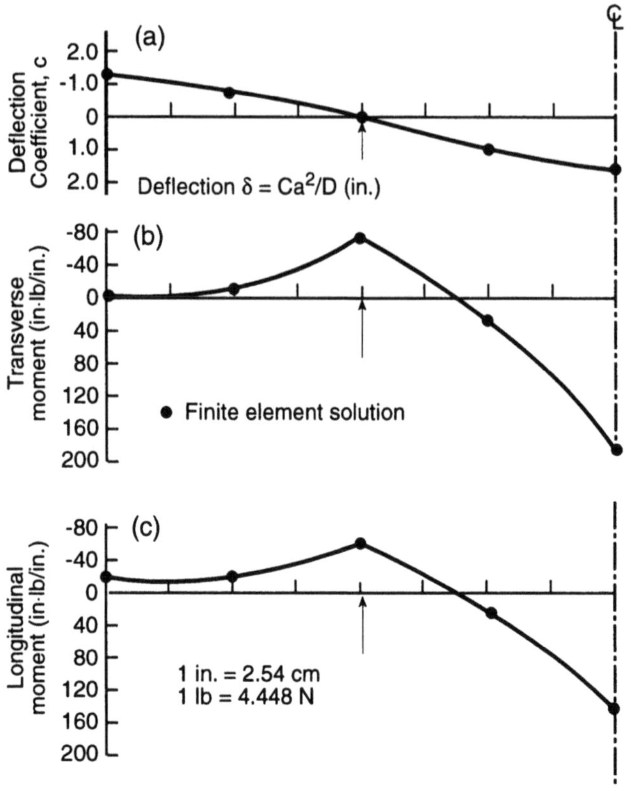

Figure 12.3 *Finite strip element results along transverse center line of the two-span slab bridge: (a) deflection; (b) transverse moment; (c) longitudinal moment (M. S. Cheung and Chan, 1978B).*

2. Continuous slab-on-girder bridge

A three-span continuous slab-on-girder bridge is shown in Fig. 12.4(a). A typical cross-section showing the transverse positions of the wheel loads is depicted in Fig. 12.4(b). The restraints at the supports are as shown in Fig. 12.4(c). Typical values of the calculated moments and stresses at sections under the second axle and over the support are shown in Figs. 12.5–12.7. The deflections and top and bottom longitudinal stresses of the center girder are compared against the beam theory calculation. An effective flange width of 99.26 in (252.12 cm) and a live load distribution factor of 0.37162 (of the total applied load), both determined from the longitudinal stress distribution at the section under the second axle, were used in the beam calculation.

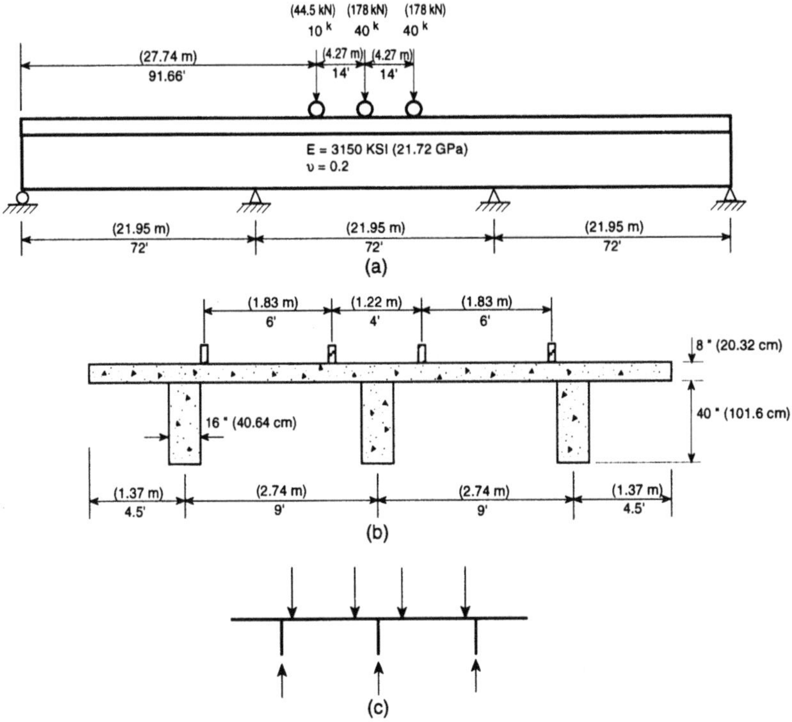

Figure 12.4 *Three-span beam-slab type bridge: (a) longitudinal section; (b) transverse cross-section; (c) support restraints (unless otherwise stated, dimensions are in feet with metric equivalent in parenthesis; M. S. Cheung and Chan, 1978B).*

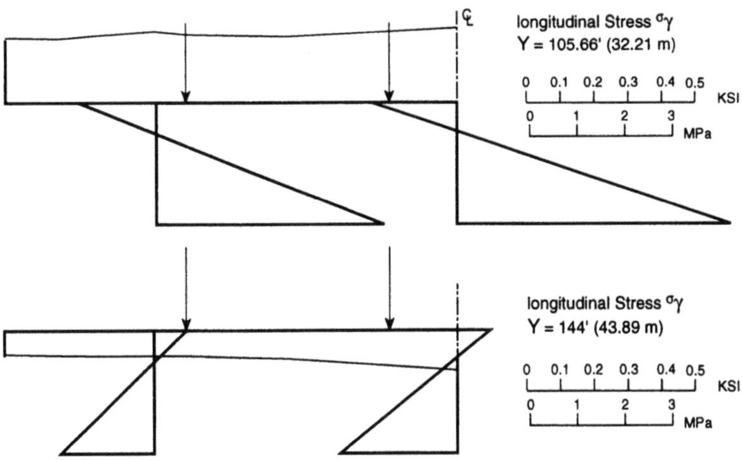

Figure 12.5 *Longitudinal stress in three-span beam-slab bridge at sections under second axis and over second support (M. S. Cheung and Chan).*

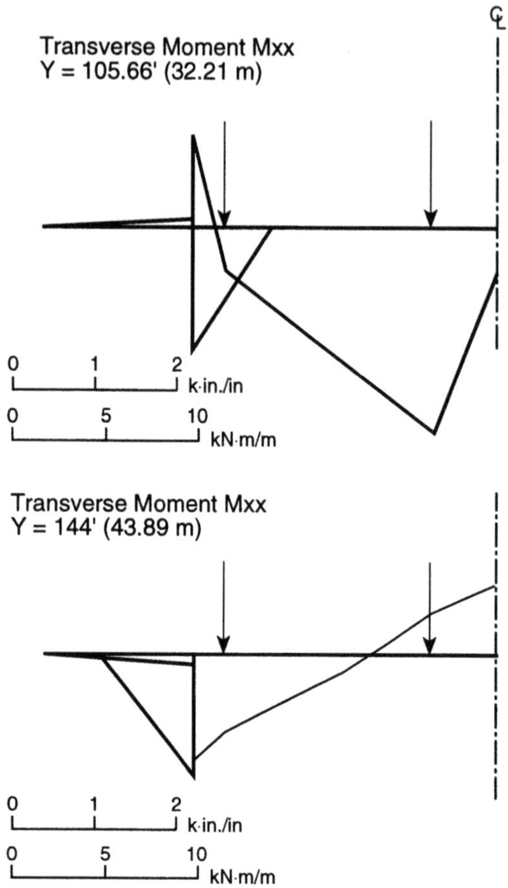

Figure 12.6 *Transverse moment in three-span beam-slab bridge at sections under second axis and over second support (M. S. Cheung and Chan).*

The results of the beam theory calculation are shown in Fig. 12.8 along with the finite strip values. It can be seen that the two sets of values, particularly in the loaded center span, agree reasonably well, and that, in general, the beam theory tends to slightly overestimate both the deflections and the stresses.

Using 18 strips and eight harmonics to model the full bridge, the CPU time required was 1.50 min (the same computer as for the previous example).

3. Slab-on-girder bridge analyzed by compound strip method

Figure 12.9 shows a slab-on-girder bridge under truck loading. The 7 in (1 in = 25.4 mm) deck is supported by four equally spaced plate

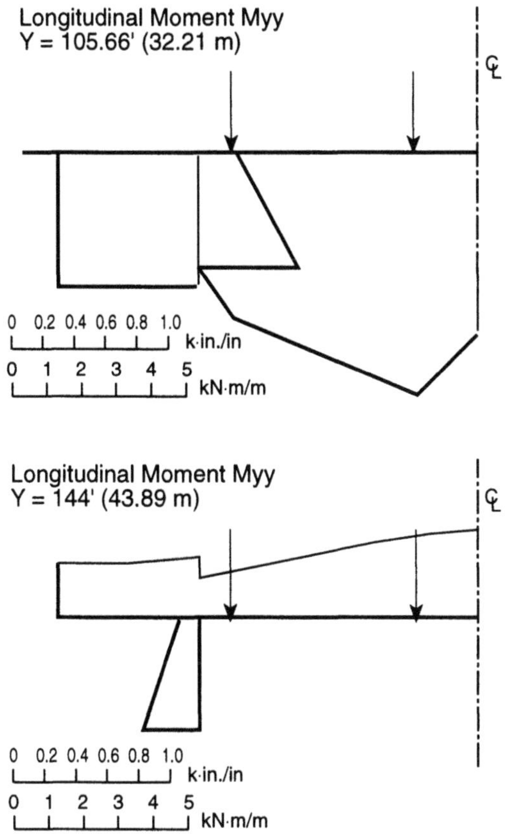

Figure 12.7 *Longitudinal moment in three-span beam-slab bridge at sections under second axis and over second support (M. S. Cheung and Chan).*

girders. The structure was analyzed by Puckett using the compound strip method (CSM) (see Chapter 7) and the finite element method (FEM) in 1986. Because at that time the CMS formulation did not include in-plane deformation, the girders were assumed to be concentric in both CSM and FEM models. Composite behavior was modeled per AASHTO specifications (1977).

The CSM model incorporated axial stiffness elements for the column, and longitudinal and transverse beam elements for the girders and the diaphragms. The compound strip layout is illustrated in Fig. 12.10.

The finite element model is shown in Fig. 12.11. The program ANSYS was employed for FEM analysis (see ANSYS, 1982) using the flat shell and beam elements.

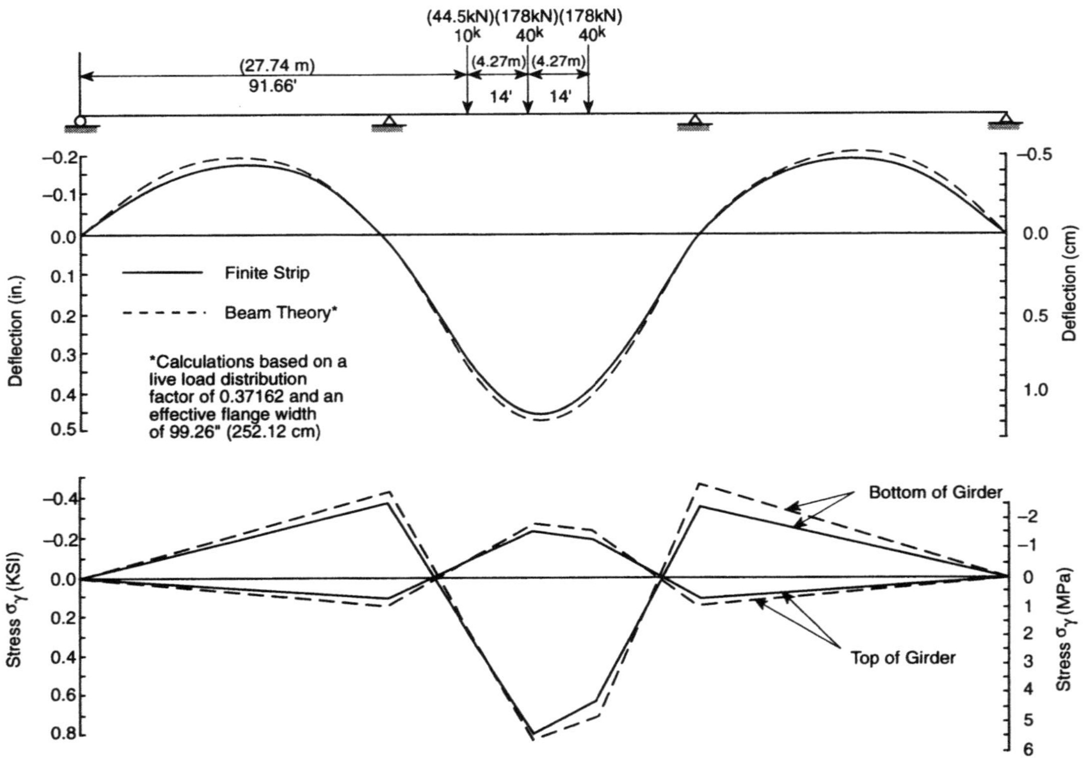

Figure 12.8 *Longitudinal variation of deflection and longitudinal stress of center girder of the three-span beam-slab bridge (dimensions in feet with metric equivalent in parenthesis; M. S. Cheung and Chan, 1978B).*

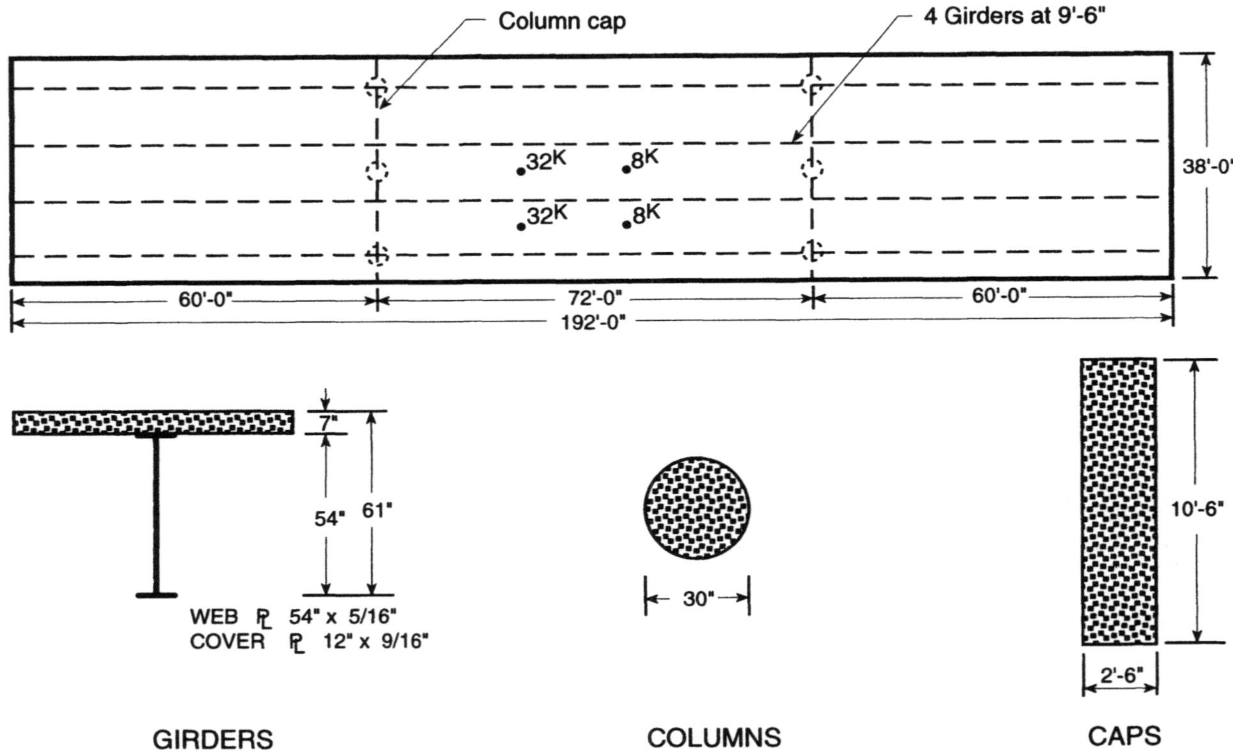

Figure 12.9 *Example slab girder bridge (Puckett, 1986, courtesy of Elsevier Science).*

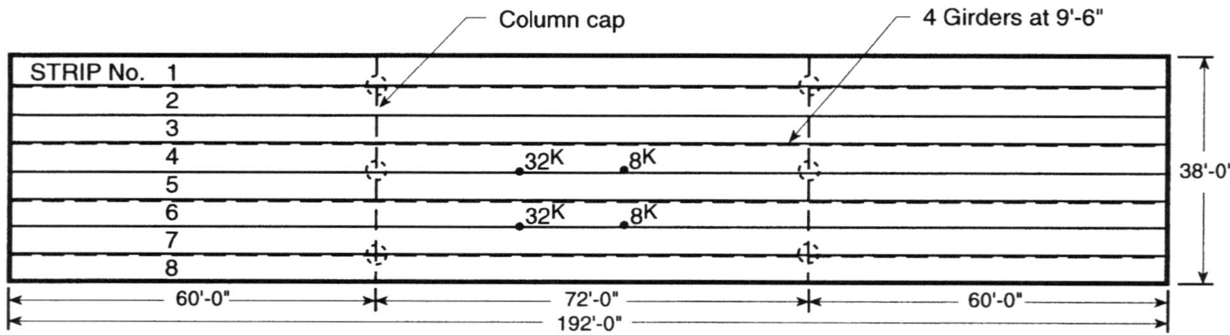

STRIP No.	TYPE
1	STD. (LO2)
2	COMPOUND
3	STD.
4	COMPOUND
5	STD.
6	COMPOUND
7	STD.
8	COMPOUND

Figure 12.10 *Compound strip model (Puckett, 1986, courtesy of Elsevier Science).*

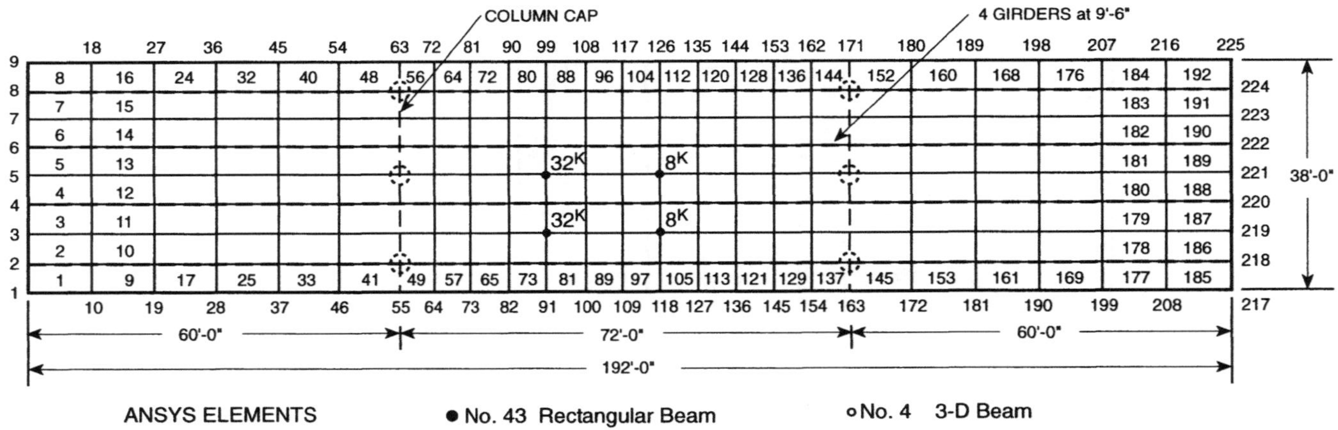

Figure 12.11 *Finite element model (Puckett, 1986, courtesy of Elsevier Science).*

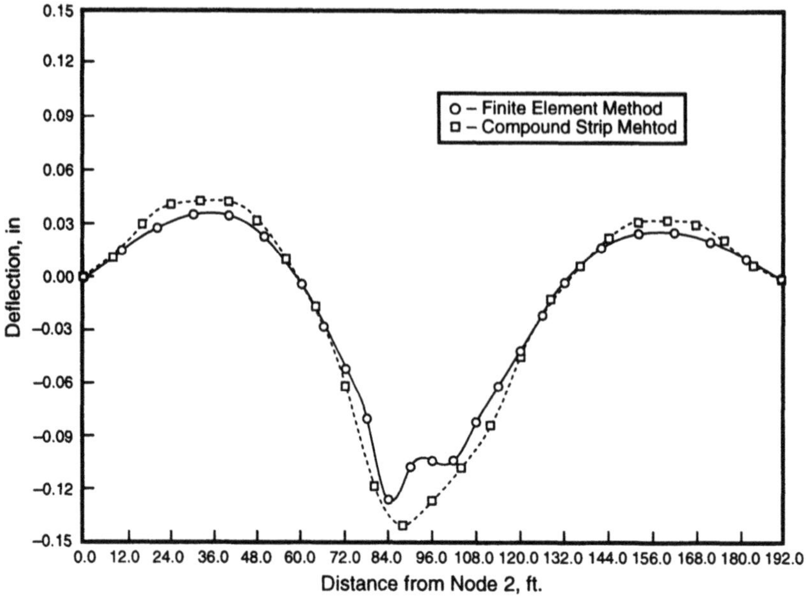

Figure 12.12 *Longitudinal deflection:* (○) = *finite element method,* (□) = *compound strip method (Puckett, 1986, courtesy of Elsevier Science).*

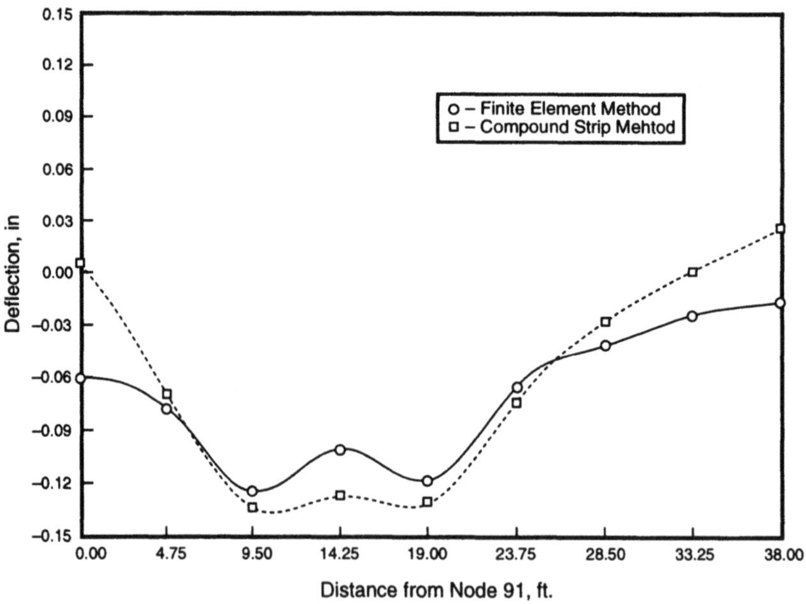

Figure 12.13 *Transverse deflection:* (○) = *finite element method,* (□) = *compound strip method (Puckett, 1986, courtesy of Elsevier Science).*

The two techniques compare very well throughout the structure. For example, Figs. 12.12 and 12.13 show the deflection along the first interior girder and the deflection in the transverse direction at the loaded cross-section, respectively. The maximum deflection in the CSM was 0.128, and 0.126 with the FEM, i.e. an approximate 2% difference. A comparison for the transverse bending moment in the deck is given in Fig. 12.14. The bending moment in the first interior girder is shown in Fig. 12.15. Both the figures reveal the fine agreement between the two methods.

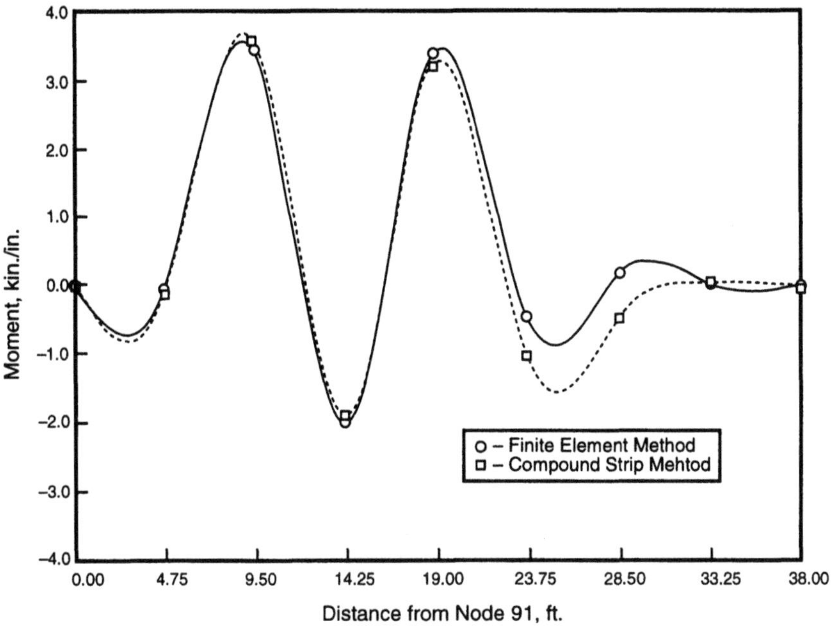

Figure 12.14 *Transverse bending moment in deck:* (○) = *finite element method,* (□) = *compound strip method (Puckett, 1986, courtesy of Elsevier Science).*

The maximum actions are given in Table 12.1. The maximum girder moments differ by 22%, and maximum support reaction by 7%. This difference could be decreased by increasing the number of harmonics considered. Note that FEM required the solution of 675 equilibrium equations, as compared to 270 CSM, with a significant computational saving. In addition, the associated calculations, such as integrations, are significantly less for CSM.

In the above figures and tables, 1 in = 0.0254 m, 1 ft = 0.305 m, K = kip = 4.45 KN, 9'–6" means 9 ft and 6 in.

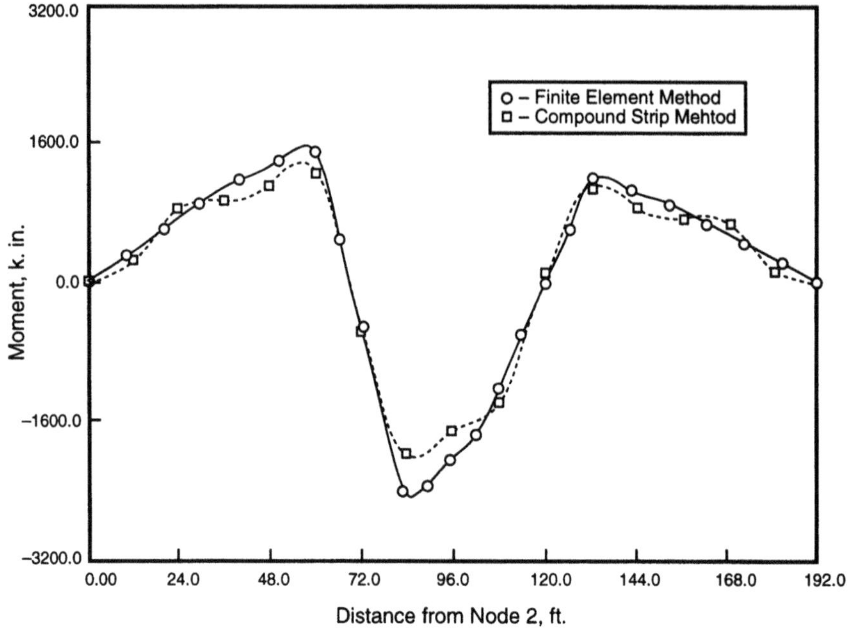

Figure 12.15 *Longitudinal bending moment in the girder: (○) = finite element method, (□) = compound strip method (Puckett, 1986, courtesy of Elsevier Science).*

Table 12.1 *Analysis summary (Puckett, 1986, courtesy of Elsevier Science)*

Parameter	FEM	CSM
Maximum deflection (in)	0.126	0.128
Maximum support reaction (kip)	30.1	32.2
Maximum girder moment (kin)	−2430	−2000
	1500	1270
Maximum transverse moment (kin/in)	3.47	3.59
Degrees of freedom	675	270
Lines of data input	64	21

CHAPTER 13

Curved and skewed bridges

13.1 Introduction

Semi analytical finite strip analysis was extended to circularly curved slab bridges by Y. K. Cheung (1969B), to curved box girder bridges by Meyer and Scordelis (1970) and by Y. K. Cheung and M. S. Cheung (1971), to skewed slab bridges by Brown and Ghali (1972), and to skewed box girder bridges by Brown and Ghali (1975). The compound strip method (Chapter 7) was first applied to the analysis of circularly curved slab girder bridges by Puckett and Lang (1986). The basic procedures in all these applications are presented in this chapter, followed by numerical examples.

The spline finite strip method is applicable not only to circularly curved slab bridges and skewed slab bridges, but also to arbitrarily shaped slab bridges. The related procedure is described in Chapter 6.

13.2 Circularly curved plate strip

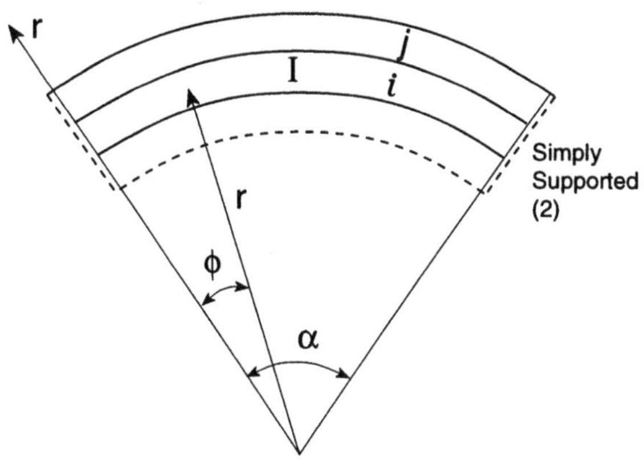

Figure 13.1 *Circularly curved plate.*

In the analysis of a circularly curved slab bridge, the bridge deck is divided into a number of finite strips by concentric circles as shown in Fig. 13.1, and the following displacement function can be adopted:

$$
\begin{aligned}
w(r,\phi) &= \sum_{m=1}^{n}[R_1,R_2,R_3,R_4]\Theta(\phi)\left\{\begin{array}{c} w_{im} \\ \theta_{im} \\ w_{jm} \\ \theta_{jm} \end{array}\right\} \\
&= \sum_{m=1}^{n}[N]_m\{\delta\}_m
\end{aligned}
\tag{13.1}
$$

where w_{im} and w_{jm} are the deflection amplitudes for nodal lines i and j, $\theta_{im} = (\frac{\partial w}{\partial r})_{im}$ and $\theta_{jm} = (\frac{\partial w}{\partial r})_{jm}$ are the corresponding slope amplitudes, α is the spanwise subtended angle in plan, and R_1, R_2, R_3 and R_4 are functions of strip width $2b'$ and the radial distances r, r_i and r_j in the form

$$
\begin{aligned}
R_1 &= 1 - \frac{3R^2}{4} + \frac{R^3}{4} \\
R_2 &= b'(R - R^2 + \frac{R^3}{4}) \\
R_3 &= \frac{3R^2}{4} - \frac{R^3}{4} \\
R_4 &= b'(\frac{R^3}{4} - \frac{R^2}{2})
\end{aligned}
\tag{13.2}
$$

where

$$
R = \frac{r - r_i}{b'} \quad \text{and} \quad b' = \frac{r_j - r_i}{2}
\tag{13.3}
$$

In Equation 13.1 $\Theta_m(\phi)$ is the mth series term of beam eigenfunctions which satisfy the end support conditions of the strip. If both ends are simply supported, $\Theta_m(\phi) = \sin m\pi\phi/\alpha$. If both ends of the strip are otherwise supported, the expression for $\Theta_m(\phi)$ is identical to $Y_m(y)$ and has been explicitly given in Chapter 4.

The curvature displacement relationships are (Timoshenko and Woinowsky–Krieger, 1959)

$$
\{\kappa\} = \left\{\begin{array}{c} -\frac{\partial^2 w}{\partial r^2} \\ \\ -\frac{1}{r}(\frac{1}{r}\frac{\partial^2 w}{\partial \phi^2} + \frac{\partial w}{\partial r}) \\ \\ -\frac{2}{r}(\frac{\partial^2 w}{\partial r \partial \phi} - \frac{1}{r}\frac{\partial w}{\partial \phi}) \end{array}\right\}
\tag{13.4}
$$

Substituting Equation 13.1 into 13.4, the curvatures can be expressed in terms of the displacement parameters as

$$
\{\kappa\} = \sum_{m=1}^{n}[B]_m\{\delta\}_m
\tag{13.5}
$$

where $[B]_m$ is the strain matrix.

The relationships between bending moments and curvatures are

$$\{M\} = \left\{ \begin{array}{c} M_r \\ M_\phi \\ M_{r\phi} \end{array} \right\} = \left[\begin{array}{ccc} D_r & D_1 & 0 \\ D_1 & D_\phi & 0 \\ 0 & 0 & D_{r\phi} \end{array} \right] \{\kappa\} \tag{13.6}$$

i.e.

$$\{M\} = \sum_{m=1}^{n} [D_b][B]_m\{\delta\}_m \tag{13.7}$$

where D_r, D_ϕ, $D_{r\phi}$ and D_1 are the bending rigidities for the r and ϕ directions, the torsional rigidity and the coupling rigidity, respectively.

For an orthotropic plate strip of thickness t

$$\begin{array}{ll} D_r = \dfrac{E_r t^3}{12(1-\nu_r\nu_\phi)}, & D_\phi = \dfrac{E_\phi t^3}{12(1-\nu_r\nu_\phi)} \\ D_{r\phi} = \dfrac{Gt^3}{12}, & D_1 = \nu_r D_r = \nu_\phi D_\phi \end{array} \tag{13.8}$$

For an isotropic plate

$$E_r = E_\phi = E \quad and \quad \nu_r = \nu_\phi = \nu \tag{13.9}$$

Following the procedure described in Chapter 4 gives the stiffness matrix of the curved strip as

$$[k]_m = \int_0^\alpha \int_{r_i}^{r_j} [B]_m^T[D_b][B]_m r dr d\phi \tag{13.10}$$

The expressions to be integrated in Equation 13.10 are fairly complex and, therefore, to work out all the integrals in closed form is time-consuming. Furthermore, due to the presence of $1/r^n$ terms, such closed form integrals are numerically sensitive to very small or very large radii values, and a numerical integration scheme such as Gaussian quadrature should be used instead.

The load vector $\{p\}_m$ of distributed load $q(r,\phi)$ can also be obtained by using the procedure in Chapter 4 as

$$\{p\}_m = \int_0^\alpha \int_{r_i}^{r_j} [N]_m^T q(r,\phi) r dr d\phi \tag{13.11}$$

Next, the assembling and solution procedures are exactly the same as those for right bridges. The final results are summed from individual series terms.

The higher order curved plate strip can also be formed by following the same procedure (Thorpe, 1976).

13.3 Curved compound strip

The aforementioned curved plate strip can be combined with the beams in the circumferential direction, the beams in the radial direction and columns (Fig. 13.2) to form a curved compound strip, which is applicable to the analysis of circularly curved slab girder bridges.

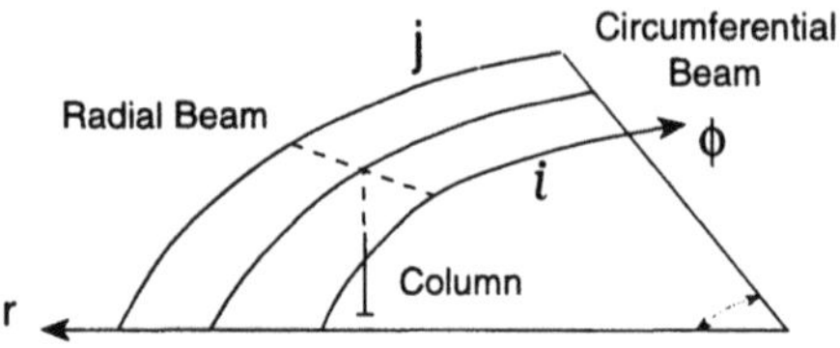

Figure 13.2 *Curved compound strip.*

The stiffness matrices of the attached elements can be derived as follows:

1. Flexural stiffness of a circumferential beam

   The flexural strain energy of a circumferential beam is

$$U^{f\phi} = \frac{EI}{2}\int_0^{\alpha}(\frac{1}{r^2}\frac{\partial^2 w}{\partial\phi^2} + \frac{1}{r}\frac{\partial w}{\partial r})^2 r d\phi \tag{13.12}$$

   Substituting Equation 13.1 into 13.12, the strain energy can be expressed in terms of the displacement parameters as

$$U^{f\phi} = \frac{1}{2}\sum_{m=1}^{r}\sum_{n=1}^{r}\{\delta\}_m^T EI \times \int_0^{\alpha}(\frac{1}{r^2}\frac{\partial^2[N_m]}{\partial\phi^2} + \frac{1}{r}\frac{\partial[N_m]}{\partial r})^T(\frac{1}{r^2}\frac{\partial^2[N_n]}{\partial\phi^2} + \frac{1}{r}\frac{\partial[N_n]}{\partial r})rd\phi\{\delta\}_n \tag{13.13}$$

   Thus, the corresponding stiffness matrix is obtained as

$$[K_{mn}^{f\phi}] = EI\int_0^{\alpha}(\frac{1}{r^2}\frac{\partial^2[N_m]}{\partial\phi^2} + \frac{1}{r}\frac{\partial[N_m]}{\partial r})^T(\frac{1}{r^2}\frac{\partial^2[N_n]}{\partial\phi^2} + \frac{1}{r}\frac{\partial[N_n]}{\partial r})rd\phi \tag{13.14}$$

   The item at the kth row and lth column of the stiffness matrix can then be derived using Equation 13.1 as below

$$(K_{mn}^{f\phi})_{kl} = EI(\frac{1}{r_c^3}R_k R_l I_4 + \frac{1}{r_c^2}R_k R_l' I_2 + \frac{1}{r_c^2}R_k' R_l I_3 + \frac{1}{r_c}R_k' R_l' I_1) \tag{13.15}$$

   in which

$$I_1 = \int_0^{\alpha}\Theta_m\Theta_n d\phi \qquad I_2 = \int_0^{\alpha}\Theta_m''\Theta_n d\phi$$
$$I_3 = \int_0^{\alpha}\Theta_m\Theta_n'' d\phi \qquad I_4 = \int_0^{\alpha}\Theta_m''\Theta_n'' d\phi \tag{13.16}$$
$$R_1' = \frac{3}{4b'}(-2R + R^2) \qquad R_2' = 1 - 2R + \frac{3}{4}R^2$$

$$R'_3 = -R'_1 \qquad R'_4 = \frac{3}{4}R^2 - R \tag{13.17}$$

The terms R_k, R_l, R'_k and R'_l are evaluated at the radial coordinate r_c of the beam.

2. Torsional stiffness of Circumferential beam
The torsional strain energy of a circumferential beam is

$$U^{t\phi} = \frac{GJ}{2}\int_0^{\alpha}(\frac{1}{r}\frac{\partial^2 w}{\partial r\partial\phi} - \frac{1}{r^2}\frac{\partial w}{\partial\phi})^2 r d\phi \tag{13.18}$$

By substituting Equation 13.1 into 13.18, the stiffness matrix and the corresponding item can be obtained as

$$[K^{t\phi}_{mn}] = \frac{GJ}{2}\int_0^{\alpha}(\frac{1}{r}\frac{\partial^2[N_m]}{\partial r\partial\phi} - \frac{1}{r^2}\frac{\partial[N_n]}{\partial\phi})^T(\frac{1}{r}\frac{\partial^2[N_n]}{\partial r\partial\phi} - \frac{1}{r^2}\frac{\partial[N_n]}{\partial\phi})r d\phi \tag{13.19}$$

$$(K^{t\phi}_{mn})_{kl} = GJI_5(\frac{1}{r_c}R'_kR'_l - \frac{1}{r_c^2}R'_kR_l - \frac{1}{r_c^2}R_kR'_l + \frac{1}{r_c^3}R_kR_l) \tag{13.20}$$

in which

$$I_5 = \int_0^{\alpha}\Theta'_m\Theta'_n d\phi \tag{13.21}$$

3. Flexural stiffness of Radial beam
The flexural strain energy of a radial beam is

$$U^{fr} = \frac{EI}{2}\int_{r_i}^{r_j}(\frac{\partial^2 w}{\partial r^2})^2 dr \tag{13.22}$$

Following the same procedure as in the previous paragraphs leads to expressions for the stiffness matrix and the corresponding item as

$$[K^{fr}_{mn}] = EI\int_{r_i}^{r_j}(\frac{\partial^2[N_m]}{\partial r^2})^T(\frac{\partial^2[N_n]}{\partial r^2})^T dr \tag{13.23}$$

$$(K^{fr}_{mn})_{kl} = EI\Theta_m\Theta_n\int_{r_i}^{r_j}R''_kR''_l dr \tag{13.24}$$

in which

$$R''_1 = \frac{3}{2b'^2}(R-1) \qquad R''_2 = \frac{1}{2b'}(3R-4)$$

$$R''_3 = -R''_1 \qquad R''_4 = \frac{1}{2b'}(3R-2) \tag{13.25}$$

The terms Θ_m and Θ_n are evaluated at the ϕ coordinate of the beam.

4. Torsional stiffness of a radial beam
The torsional strain energy of a radial beam is

$$U^{tr} = \frac{GJ}{2}\int_{r_i}^{r_j}(\frac{1}{r}\frac{\partial^2 w}{\partial r\partial\phi} - \frac{1}{r^2}\frac{\partial w}{\partial\phi})^2 dr \tag{13.26}$$

By substituting Equation 13.1 into 13.26, the stiffness matrix and the corresponding item can be obtained as

$$[K^{tr}_{mn}] = GJ\int_{r_i}^{r_j}(\frac{1}{r}\frac{\partial^2[N_m]}{\partial r\partial\phi} - \frac{1}{r^2}\frac{\partial[N_n]}{\partial\phi})^T(\frac{1}{r}\frac{\partial^2[N_n]}{\partial r\partial\phi} - \frac{1}{r^2}\frac{\partial[N_n]}{\partial\phi})dr \tag{13.27}$$

$$(K^{tr}_{mn})_{kl} = GJ\Theta'_m\Theta'_n\int_{r_i}^{r_j}(\frac{1}{r^2}R'_kR'_l - \frac{1}{r^3}R'_kR_l - \frac{1}{r^3}R_kR'_l + \frac{1}{r^4}R_kR_l)dr \tag{13.28}$$

in which Θ_m and Θ_n are evaluated at the ϕ coordinate of the beam.

5. Axial stiffness matrix of a column

The axial strain energy for a column is

$$U^{ac} = \frac{1}{2}K_a w_c^2 \tag{13.29}$$

where K_a is the axial stiffness of the column, and w_c is the axial deflection which can be obtained by evaluating the displacement function N_m at the r_c, ϕ_c coordinates of the column.

Substitution of Equation 13.1 into 13.29 gives the stiffness matrix and the item as

$$[K^{ac}_{mn}] = K_a[N_m]^T[N_n] \tag{13.30}$$

$$(K^{ac}_{mn})_{kl} = K_aR_kR_l\Theta_m\Theta_n \tag{13.31}$$

6. Flexural stiffness of a column

The flexural strain energy in the column has two components, which are respectively related to rotations of the column head in the circumferential and radial directions of the strip.

The strain energy for the radial rotation is

$$U^{rc} = \frac{1}{2}K_r(\frac{\partial w_c}{\partial r})^2 \tag{13.32}$$

where K_r is the radial rotation stiffness of the column head. By means of Equation 13.1 the stiffness matrix and the corresponding item can be derived as

$$[K^{rc}_{mn}] = K_r(\frac{\partial[N_m]}{\partial r})^T\frac{\partial[N_n]}{\partial r} \tag{13.33}$$

$$(K^{rc}_{mn})_{kl} = K_rR'_kR'_l\Theta_m\Theta_n \tag{13.34}$$

The strain energy for the circumferential rotation of the column head is

$$U^{\phi c} = \frac{1}{2}K_\phi(\frac{1}{r_c}\frac{\partial w_c}{\partial\phi})^2 \tag{13.35}$$

where K_ϕ is the circumferential rotation stiffness of the column head.

In the same manner, the related stiffness matrix and the item can be obtained as

$$[K^{\phi c}_{mn}] = K_\phi(\frac{1}{r_c}\frac{\partial[N_m]}{\partial\phi})^T(\frac{1}{r_c}\frac{\partial[N_n]}{\partial\phi}) \tag{13.36}$$

$$(K^{\phi c}_{mn})_{kl} = \frac{K_\phi}{r_c^2} R_k R_l \Theta'_m \Theta'_n \tag{13.37}$$

After forming the stiffness matrices of the plate strip and all support elements attached to the strip, the stiffness matrix of the compound strip is readily obtained as the sum of all the component matrices.

13.4 Curved strip for box bridges

For the analysis of circularly curved box girder bridges, the conical frustrum shell strip is employed (Fig. 13.3), which becomes a cylindrical strip in the vertical position (for $\beta = 0°$), and a flat sector strip in the horizontal position (for $\beta = 90°$). Since the strip is not flat in general, the bending and membrane forces are coupled.

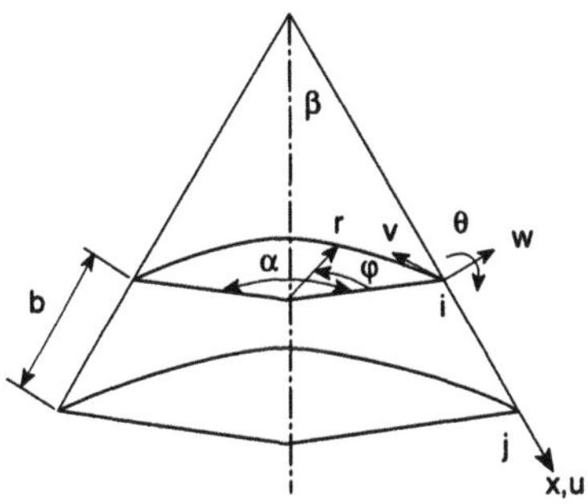

Figure 13.3 *Conical shell strip.*

For simply supported end conditions the displacement functions can be assumed as

$$\begin{aligned} u &= \sum_{m=1}^{n} [(1-X)u_{im} + Xu_{jm}] \sin\frac{m\pi\phi}{\alpha} \\ v &= \sum_{m=1}^{n} [(1-X)v_{im} + Xv_{jm}] \cos\frac{m\pi\phi}{\alpha} \\ w &= \sum_{m=1}^{n} [(1-3X^2+2X^3)w_{im} + x(1-2X+X^2)\theta_{im} \\ &\quad +(3X^2-2X^3)w_{jm} + x(X^2-X)\theta_{jm}] \sin\frac{m\pi\phi}{\alpha} \end{aligned} \tag{13.38}$$

in which $X = x/b$.

The strain–displacement relationship is given by

$$\epsilon_x = \frac{\partial u}{\partial x}$$

$$\begin{aligned}
\epsilon_\phi &= \frac{1}{r}\frac{\partial v}{\partial \phi} + \frac{w\cos\beta + u\sin\beta}{r} \\
\gamma_{x\phi} &= \frac{1}{r}\frac{\partial u}{\partial \phi} + \frac{\partial v}{\partial x} - \frac{v\sin\beta}{r} \\
\kappa_x &= -\frac{\partial^2 w}{\partial x^2} \\
\kappa_\phi &= -\frac{1}{r^2}\frac{\partial^2 w}{\partial \phi^2} + \frac{\cos\beta}{r^2}\frac{\partial v}{\partial \phi} - \frac{\sin\beta}{r}\frac{\partial w}{\partial x} \\
\kappa_{x\phi} &= 2(-\frac{1}{r}\frac{\partial^2 w}{\partial x \partial \phi} + \frac{\sin\beta}{r^2}\frac{\partial w}{\partial \phi} + \frac{\cos\beta}{r}\frac{\partial v}{\partial x} - \frac{\sin\beta\cos\beta}{r^2}v)
\end{aligned} \quad (13.39)$$

Substituting Equation 13.38 into 13.39, the strains at any point in the strip can be expressed in terms of the displacement parameters as

$$\{\epsilon\} = [\epsilon_x, \epsilon_\phi, \gamma_{x\phi}, \kappa_x, \kappa_\phi, \kappa_{x\phi}]^T = \sum_{m=1}^{n} [B]_m \{\delta\}_m \quad (13.40)$$

where $\{\delta\}_m$ is the vector of displacement parameters of the strip defined as

$$\{\delta\}_m = [u_{im}, v_{im}, w_{im}, \theta_{im}, u_{jm}, v_{jm}, w_{jm}, \theta_{jm}]^T \quad (13.41)$$

The strain matrix $[B]_m$ is listed in Table 13.1.

Stress–strain relationships are written as

$$\{\sigma\} = [N_x, N_\phi, N_{x\phi}, M_x, M_\phi, M_{x\phi}]^T = [D]\{\epsilon\} \quad (13.42)$$

where N_x, N_ϕ and $N_{x\phi}$ are the internal tensile and shear membrane forces, respectively, M_x, M_ϕ and $M_{x\phi}$ are the bending and torsional moments, and $[D]$ is the elasticity matrix. For a general orthotropic material, $[D]$ reads as

$$[D] = \begin{bmatrix} K_x & K_1 & 0 & 0 & 0 & 0 \\ K_1 & K_\phi & 0 & 0 & 0 & 0 \\ 0 & 0 & K_{x\phi} & 0 & 0 & 0 \\ 0 & 0 & 0 & D_x & D_1 & 0 \\ 0 & 0 & 0 & D_1 & D_\phi & 0 \\ 0 & 0 & 0 & 0 & 0 & D_{x\phi} \end{bmatrix} \quad (13.43)$$

in which

$$\begin{aligned}
K_x &= \frac{E_x t}{1-\nu_x\nu_\phi}, & K_\phi &= \frac{E_\phi t}{1-\nu_x\nu_\phi} \\
K_{x\phi} &= Gt, & K_1 &= \nu_x K_x = \nu_\phi K_\phi \\
D_x &= \frac{E_x t^3}{12(1-\nu_x\nu_\phi)}, & D_\phi &= \frac{E_\phi t^3}{12(1-\nu_x\nu_\phi)} \\
D_{x\phi} &= \frac{Gt^3}{12}, & D_1 &= \nu_x D_x = \nu_\phi D_\phi
\end{aligned} \quad (13.44)$$

Table 13.1 *Strain matrix* $[B]_m$ *of conical shell strip.*

$-\frac{1}{b}S$	0	0	0	$\frac{1}{b}S$	0	0	0
$\frac{1}{r}(1-X)SS_\beta$	$-\frac{1}{r}(1-X)k_mS$	$\frac{1}{r}N_1SC_\beta$	$\frac{1}{r}N_2SC_\beta$	$\frac{1}{r}XSS_\beta$	$-\frac{1}{r}Xk_mS$	$\frac{1}{r}N_3SC_\beta$	$\frac{1}{r}N_4SC_\beta$
$\frac{1}{r}(1-X)k_mC$	$-\frac{1}{b}C-\frac{1}{r}(1-X)CS_\beta$	0	0	$\frac{1}{r}Xk_mC$	$\frac{1}{b}C-\frac{1}{r}XCS_\beta$	0	0
0	0	$-N_1''S$	$-N_2''S$	0	0	$-N_3''S$	$-N_4''S$
0	$-\frac{1}{r^2}(1-X)k_mSC_\beta$	$\frac{1}{r^2}N_1k_m^2S$ $-\frac{1}{r}N_1'SS_\beta$	$\frac{1}{r^2}N_2k_m^2S$ $-\frac{1}{r}N_2'SS_\beta$	0	$-\frac{1}{r^2}Xk_mSC_\beta$	$\frac{1}{r^2}N_3k_m^2S$ $-\frac{1}{r}N_3'SS_\beta$	$\frac{1}{r^2}N_4k_m^2S$ $-\frac{1}{r}N_4'SS_\beta$
0	$-\frac{2}{r}\frac{1}{b}CC_\beta$ $-\frac{2}{r^2}(1-X)CS_\beta C_\beta$	$-\frac{2}{r}N_1'k_mC$ $+\frac{2}{r^2}N_1k_mCS_\beta$	$-\frac{2}{r}N_2'k_mC$ $+\frac{2}{r^2}N_2k_mCS_\beta$	0	$\frac{2}{r}\frac{1}{b}CC_\beta$ $-\frac{2}{r^2}XCS_\beta C_\beta$	$-\frac{2}{r}N_3'k_mC$ $+\frac{2}{r^2}N_3k_mCS_\beta$	$-\frac{2}{r}N_4'k_mC$ $+\frac{2}{r^2}N_4k_mCS_\beta$

$$k_m = m\pi/\alpha \qquad S = \sin k_m\phi \qquad C = \cos k_m\phi \qquad X = x/b$$

$$N_1 = 1-3X^2+2X^3 \qquad N_2 = x(1-2X+X^2) \qquad N_3 = 3X^2-2X^3 \qquad N_4 = x(X^2-X)$$

$$N_i' = \frac{dN_i}{dx} \qquad N_i'' = \frac{d^2N_i}{dx^2} \qquad S_\beta = \sin\beta \qquad C_\beta = \cos\beta$$

Substituting Equation 13.40 into 13.42, the stresses can be written in terms of the displacement parameters of the strip as

$$\{\sigma\} = \sum_{m=1}^{n} [D][B]_m\{\delta\}_m \tag{13.45}$$

The stiffness matrix of the strip in the local coordinate system can now be obtained as

$$[k]_m = \int_0^b \int_0^\alpha [B]_m^T[D][B]_m r d\phi dx \tag{13.46}$$

Due to the presence of $1/r^n$ terms it is preferable to integrate the above expression numerically both with respect to dr and $d\phi$, e.g. using Gaussian quadrature.

Further, the coordinate transformation, assembling and solution procedures are the same as those for flat shell strips. The final results are summed from individual series terms.

Some higher order conical shell strips for curved box structure such as the HO3 strip have been developed, following the initial work by Thorpe (1976).

13.5 Skewed plate strip

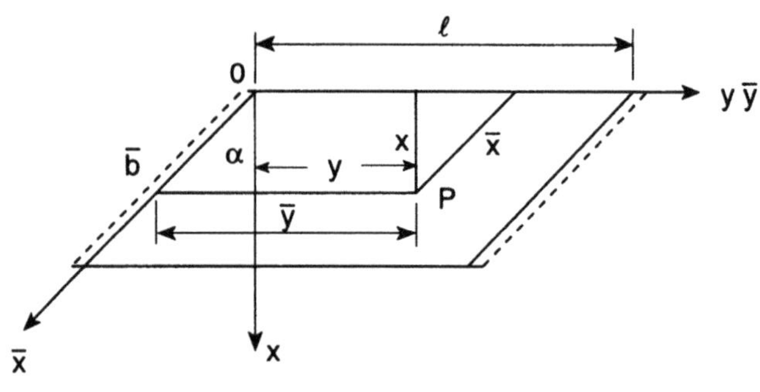

Figure 13.4 *Skewed plate strip.*

A skewed plate strip is shown in Fig. 13.4, in which x and y are the Cartesian coordinates and $\bar{x}$, $\bar{y}$ are the skew coordinates. The two sets of coordinates of the same point are related as

$$\begin{aligned} x &= \bar{x}\cos\alpha \\ y &= \bar{y} - \bar{x}\sin\alpha \end{aligned} \tag{13.47}$$

where α is the skew angle of the bridge. If the both ends of the strip are simply supported, the deflection can be expressed in terms of displacement parameters as follows:

$$\begin{aligned} w &= \sum_{m=1}^{n} [(1-3\bar{X}^2+2\bar{X}^3)\bar{w}_{im} + \bar{x}(1-2\bar{X}+\bar{X}^2)\bar{\theta}_{im} \\ &\quad +(3\bar{X}^2-2\bar{X}^3)\bar{w}_{jm} + \bar{x}(\bar{X}^2-\bar{X})\bar{\theta}_{jm}] \sin\frac{m\pi\bar{y}}{l} \\ &= \sum_{m=1}^{n} [N]\{\delta\}_m \sin\frac{m\pi\bar{y}}{l} \end{aligned} \tag{13.48}$$

where $\bar{X} = \bar{x}/\bar{b}$ and $\bar{\theta} = \partial w/\partial \bar{x}$. The curvatures at any point are

$$\begin{Bmatrix} -\dfrac{\partial^2 w}{\partial x^2} \\ -\dfrac{\partial^2 w}{\partial y^2} \\ 2\dfrac{\partial^2 w}{\partial x \partial y} \end{Bmatrix} = \begin{bmatrix} \dfrac{1}{\cos^2\alpha} & \tan^2\alpha & -\dfrac{\sin\alpha}{\cos^2\alpha} \\ 0 & 1 & 0 \\ 0 & -\dfrac{2\sin\alpha}{\cos\alpha} & \dfrac{1}{\cos\alpha} \end{bmatrix} \begin{Bmatrix} -\dfrac{\partial^2 w}{\partial \bar{x}^2} \\ -\dfrac{\partial^2 w}{\partial \bar{y}^2} \\ 2\dfrac{\partial^2 w}{\partial \bar{x} \partial \bar{y}} \end{Bmatrix} \tag{13.49}$$

or simply

$$\{\kappa\} = [A]\{\bar{\kappa}\} \tag{13.50}$$

Then, the bending moments M_x and M_y, and the twisting moment M_{xy} can be written as

$$\{M\} = [D]\{\kappa\} = [D][A]\{\bar{\kappa}\} \tag{13.51}$$

Thereafter, the stiffness matrix and the load vector of the strip can be obtained by following standard procedure in finite strip analysis.

The boundary conditions at a simply supported edge require that the deflection w and the normal curvature $\partial^2 w/\partial n^2$ should both vanish. However, by using the aforementioned shape function, condition $\partial^2 w/\partial n^2 = 0$ is not satisfied fully because we have only $w = 0$, $\partial^2 w/\partial \bar{x}^2 = 0$ and $\partial^2 w/\partial \bar{y}^2 = 0$ (in general $\bar{x}$ and $\bar{y}$ are not orthogonal). Nevertheless, according to Brown and Ghali (1972), satisfactory results can be obtained using the fifth-order polynomial including the additional curvature compatibility along nodal lines for a uniformly loaded slab with skew angles up to about 45°. For skew angles α greater than 45°, the results begin to deviate more significantly from the results given by other methods. The third-order polynomials give satisfactory results only for skew angle less than 10 – 15° (which, however, has been successfully applied for vibration analysis). It should be noted that although the effect of the assumption in boundary conditions on deflections and stresses at midspan are not always large, values in the important regions near the supports, particularly in the obtuse corner, can be in error. In addition, some possible singularities (Hartmann, 1989) in the solution of bending moments at the corners of the skewed plate, due to deficiencies in the Kirchhoff plate bending theory, may also raise some difficulties in the finite strip analysis.

Mukhopadhyay (1976) has applied the semianalytical finite strip method to the analysis of clamped skew plates under uniform load. In this analysis, the finite strip displacement functions in the skew coordinate system fully satisfy the boundary conditions $w = 0$ and $\frac{\partial w}{\partial \bar{n}}$ at the clamped edges. Abdulla (1977) analyzed the clamped skew plates using the plate strips with internal nodal line and obtained very accurate results in comparison with other solutions. Their results imply that the semianalytical finite strip analysis for clamped skew plates is more accurate than that for simply supported skew plates.

13.6 Analysis of skewed box girder bridges

As mentioned earlier, semianalytical finite strip analysis was extended to skewed box-girder bridges by Brown and Ghali (1975). In the analysis, HO2 higher-order skewed flat shell strips are employed in order to reach higher accuracy. For the ith nodal line and the mth series term, the following displacement parameters are taken:

$$\{\delta\}_{im} = [u, \frac{\partial u}{\partial \bar{x}}, v, \frac{\partial v}{\partial \bar{x}}, w, \frac{\partial w}{\partial \bar{x}}, \frac{\partial^2 w}{\partial \bar{x}^2}]^T_{im} \tag{13.52}$$

Correspondingly, the fifth polynomials of $\bar{x}$ are used for the interpolation of the deflection, and the third ones for the in-plane displacements. After forming the stiffness matrix and load vector of each strip, the higher order parameters, such as $(\frac{\partial u}{\partial \bar{x}})_{im}$, $(\frac{\partial v}{\partial \bar{x}})_{im}$ and $(\frac{\partial^2 w}{\partial \bar{x}^2})_{im}$ can be eliminated by the static condensation (Cook et al., 1989). This procedure not only reduces the size of stiffness matrix, but also eliminates the undesired continuity between the neighboring finite strips in box structures. Then, the final solution can be obtained by following the standard procedure for the right box girder bridges.

For a simply supported bridge the in-plane stress normal to the strip ends is zero. It can be shown that the displacement functions composed of trigonometric functions of $\bar{y}$ and polynomials of $\bar{x}$ satisfy this condition only when the skew angle α (Fig. 13.4) is zero. However, the numerical examples show that the related errors in the finite strip analysis of skewed box bridges are insignificant.

13.7 Numerical examples

1. Clamped semicircular plate under uniform load

A semicircular plate of radius a is clamped along all edges and subjected to a uniformly distributed load. The Poisson's ratio of material is $\nu = 0.3$. The plate is simulated by eight strips of equal width with four series terms (Y. K. Cheung, 1976). The inner radius of the first strip is equal to zero. This did not induce any numerical difficulty since Gaussian quadrature is used in forming the strip stiffness matrix. The displacement parameters of the nodal line forming the outside circular boundary should be eliminated according to the boundary conditions (see Chapter 4).

The resulting maximum center deflection is $w = 0.002021qa^4/D$ at $r = 0.4859a$ and the maximum negative bending moment is $M_r = -0.0697qa^2$ at $r = a$, in a good agreement with the analytical solutions (Woinowsky–Krieger, 1955) $w = 0.002022qa^4/D$ and $M_r = -0.0731qa^2$.

2. Curved three span plate system

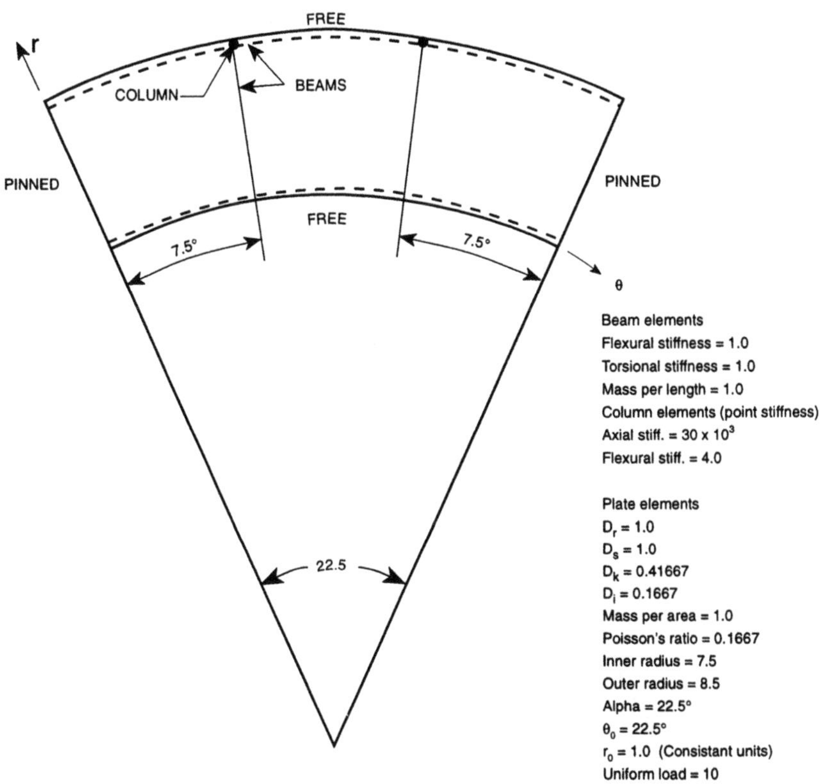

Figure 13.5 *Curved three-span plate (Puckett and Lang, 1986, courtesy of ASCE).*

Figure 13.5 shows a curved three-span plate system, which is simply supported at two ends and elastically supported by beams along two curved edges. The dimensions and elastic properties are also given in this figure. The plate was analyzed by the compound strip method using eight equally spaced strips and nine symmetrical series terms (Puckett and Lang, 1986).

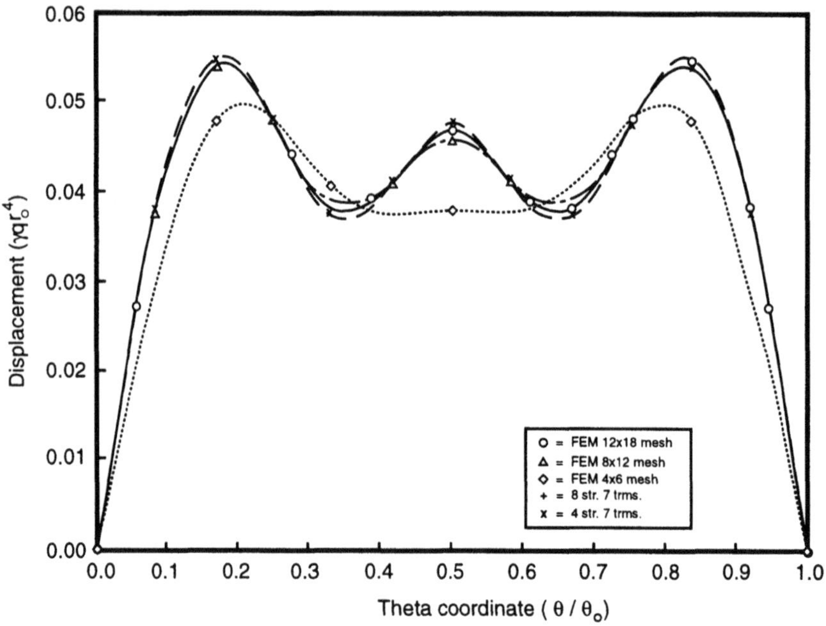

Figure 13.6 *Translation at r=9.0 (Puckett and Lang, 1986, courtesy of ASCE).*

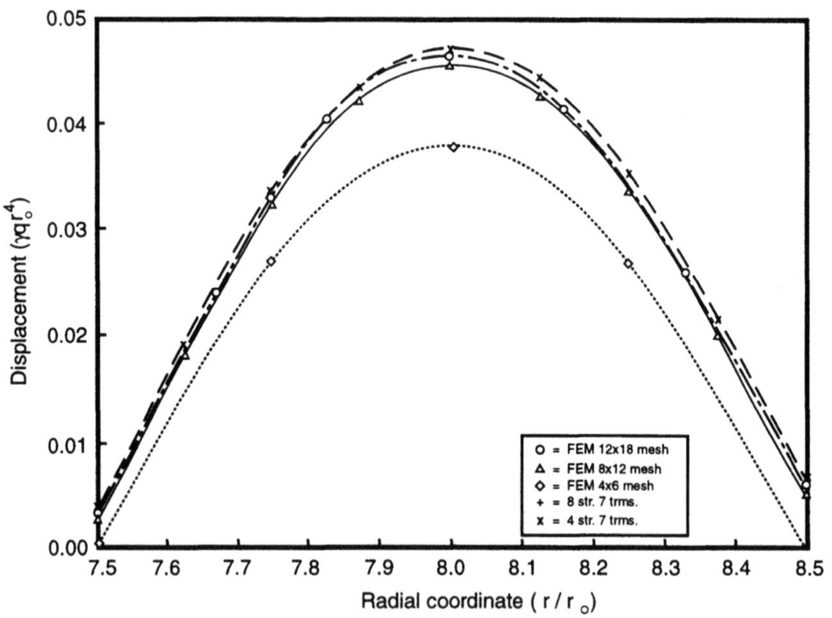

Figure 13.7 *Translation at θ = 11.25° (Puckett and Lang, 1986, courtesy of ASCE).*

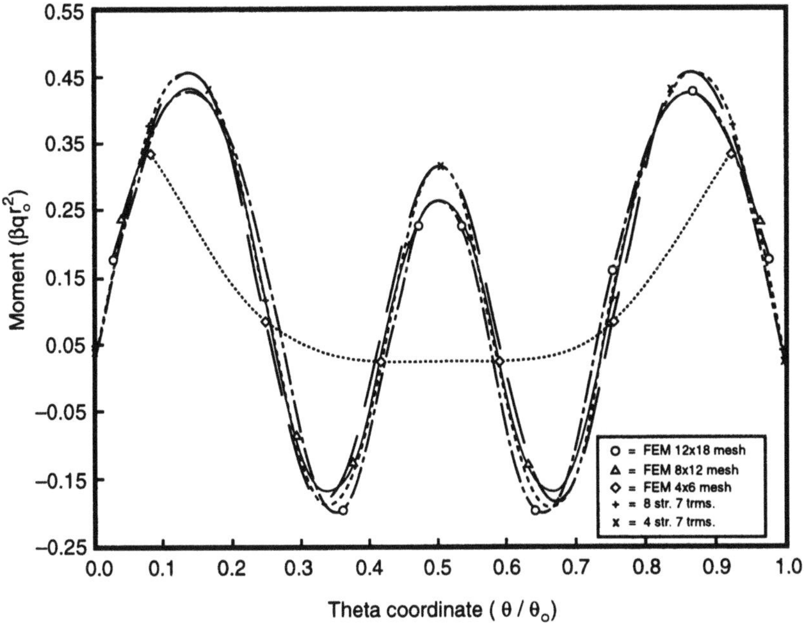

Figure 13.8 *Circumferential moment at r=9.0 (Puckett and Lang, 1986, courtesy of ASCE).*

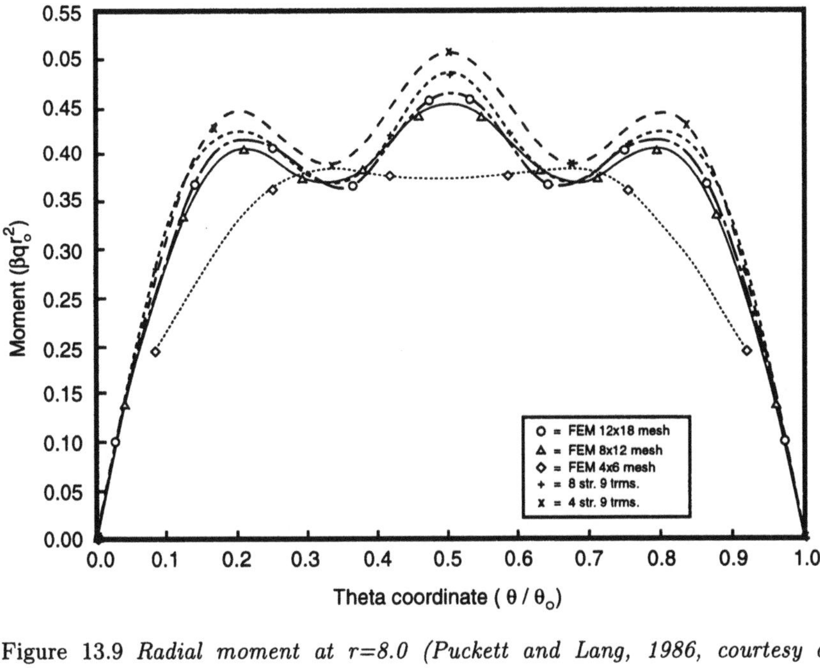

Figure 13.9 *Radial moment at r=8.0 (Puckett and Lang, 1986, courtesy of ASCE).*

The resulting deflections and moments along the center lines of the plate, due to a uniform load of 10, are depicted in Fig. 13.6–13.9. For comparison, this plate system was also analyzed by the finite element method using 12 × 18 quadrilateral plate elements, together with some beam elements and spring elements, using the structural analysis software ANSYS. These results are also shown in the figures. It can be seen that both methods gave the same maximum deflection of 0.0545 m, and the maximum radial and circumferential moments from the two methods compared within 3%.

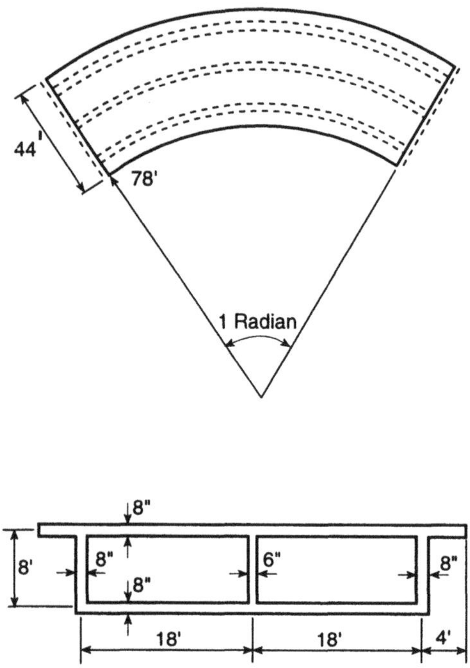

Figure 13.10 *Plan and section of curved box girder bridge* ($E = 1, \nu = 0.16$) *(Y. K. Cheung and M. S. Cheung, 1971). (1'= 0.305 m, 1"= 0.0254m).*

3. Curved box girder bridge under point load

A multicelled, curved box girder bridge (Fig. 13.10) is subjected to a unit central point load at three different radial positions, i.e. outer web, middle web and inner web. The numbering of the nodal lines and strips can be found in Fig. 13.11, in which a very narrow band matrix will result from such a scheme. A total of 20 non-zero terms of the series was used for the analysis (Y. K. Cheung and M. S. Cheung, 1971). The distribution of membrane forces and bending moments for the central section of the bridge are given in Fig 13.12–13.14. For comparison, a straight bridge with the same cross-section and span length equal to the circumferential span

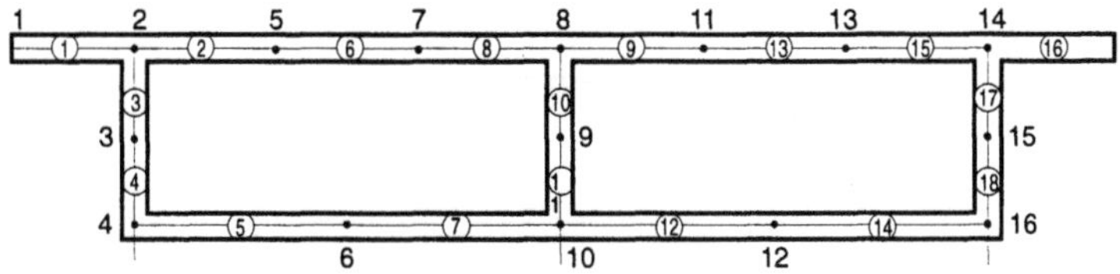

Figure 13.11 *Numbering of nodal lines and strips (for Example 2 – Y. K. Cheung and M. S. Cheung, 1971).*

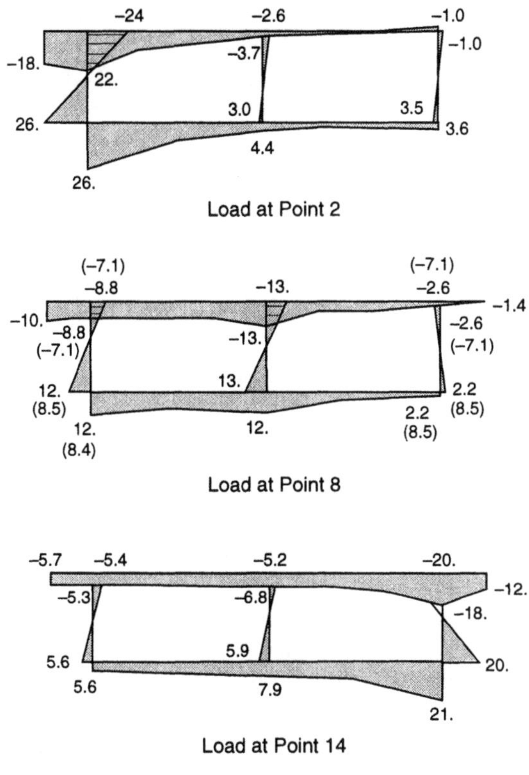

Figure 13.12 $N_\theta(10^{-2})$ *distribution in the midsection of the bridge due to load at three different positions (Y. K. Cheung and M. S. Cheung, 1971).*

of the curved bridge at the middle web is also analyzed for a central point load, and the results are given in parentheses inside the same figure.

4. Clamped skew plate

A skew plate is clamped along its four edges of length 2b each and is subjected to uniform load of intensity q. The Poisson's ratio of material is $\nu = 0.3$. This plate was analyzed by Abdulla (1977) using the higher order

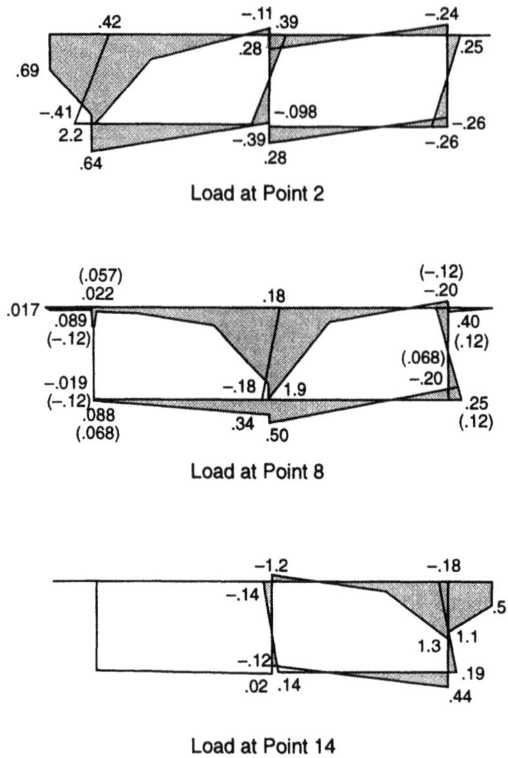

Figure 13.13 $M_\theta(10^{-2})$ *distribution in the midsection of the bridge due to load at three different positions (Y. K. Cheung and M. S. Cheung, 1971).*

skewed plate strips with one internal nodal line. Eleven series terms are used in the analysis. The results are compared in Table 13.2 with values given by Morley (1964), Hamada (1966), Rushton (1964) and Iyengar and Srinivasan (1967).

5. Skewed box-girder bridge with sloping webs

Figure 13.15 shows a skew box-girder bridge with sloping webs. In this case not only the flanges but also the webs are parallelogrammic. The bridge was analyzed by Brown and Ghali (1975) using 12 skewed flat shell strips with simply supported ends and seven series terms. The results obtained for the three load cases are depicted in Figs 13.16 and 13.17 in comparison with the finite element analysis using the finite elements QLC3 described by Sisodiya et al. (1972). The finite element mesh divisions in the longitudinal direction are shown in Fig. 13.15(a). The divisions in the transverse direction are identical to the strip divisions and the Poisson's ratio is 0.2. It is apparent that the agreement between the two methods is excellent.

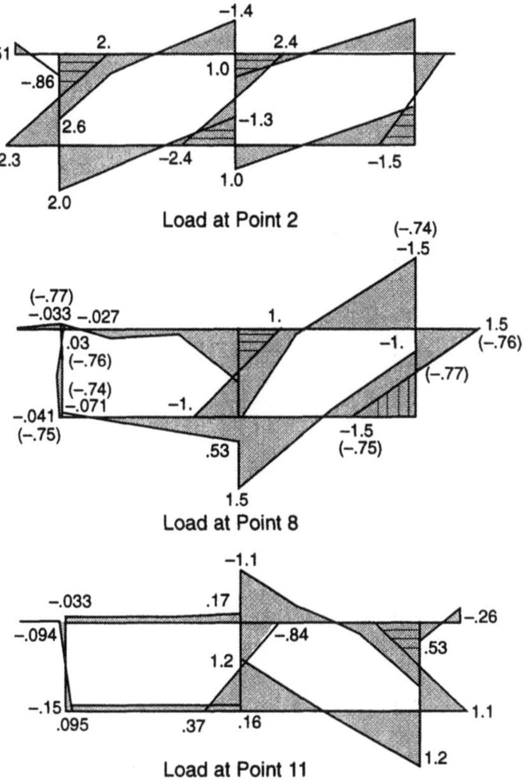

Figure 13.14 $M_r(10^{-2})$ *distribution in the midsection of the bridge due to load at three different positions (Y. K. Cheung and M. S. Cheung, 1971).*

Table 13.2 *Results for clamped skew plates (Loo and Cusens, 1978)*

Skew angle	Method	Maximum deflection	Maximum central bending moment	Maximum edge bending moment
30°	ANL finite strip	1.244	7.998	– 16.30
	Morley	1.230	8.000	– 16.24
	Hamada	1.230	–	– 16.20
	Iyengar	1.230	–	–
45°	ANL finite strip	0.608	5.552	– 11.85
	Morley	0.603	5.760	– 11.90
	Hamada	0.603	–	– 11.40
	Rushton	0.622	5.280	– 11.48
60°	ANL finite strip	0.173	–	– 7.23
	Hamada	0.174	–	– 6.20
Multipliers		$\frac{qb^4}{D}10^{-2}$	$(qb^2)10^{-2}$	$(qb^2)10^{-1}$

Basis of methods: Morley, finite differences; Hamada, Fourier series; Iyengar and Srinivasan Fourier series; Rushton, electric analogue; ANL finite strip: H03 strip.

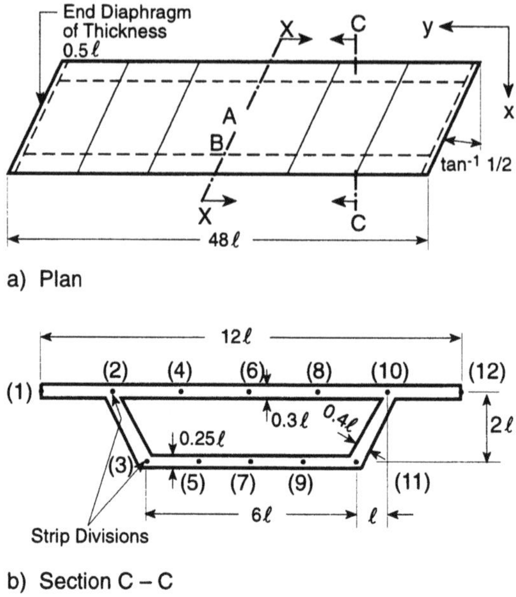

Figure 13.15 *Skew box girder bridge: (a) plan, (b) section C–C (Brown and Ghali, 1975).*

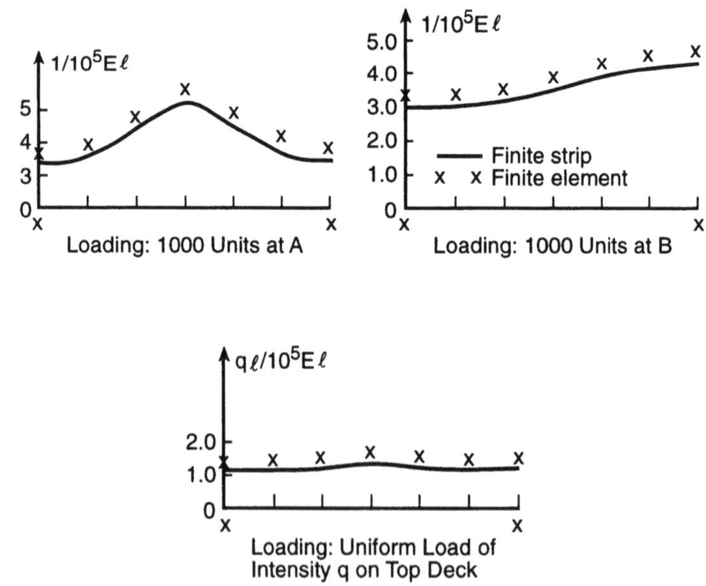

Figure 13.16 *Downward diplacements along section X–X (Brown and Ghali, 1975).*

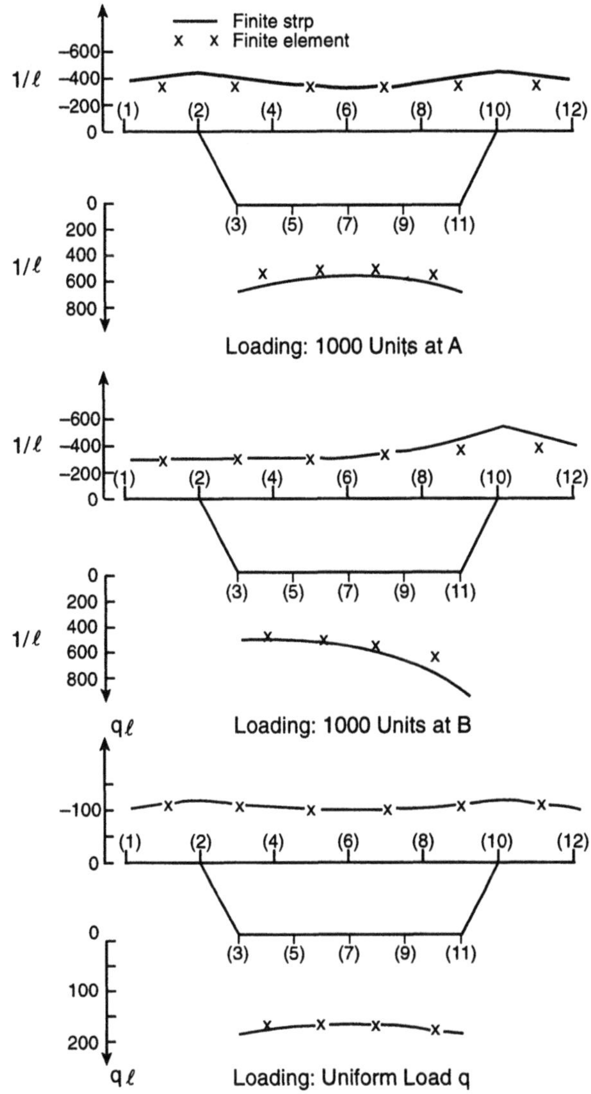

Figure 13.17 *In-plane forces in the y direction along X–X (Brown and Ghali, 1975).*

CHAPTER 14

Box girder bridges

14.1 Introduction

Box girder bridges are very popular for highways because of their high torsional rigidity and good appearance. However, the structural analysis of box girder bridges is a very difficult undertaking because of their complex deformation pattern and stress distribution, especially if the structure is continuous over intermediate supports and includes unsupported stiffening diaphragms. Although both the finite strip method and finite element method can be employed to analyze such bridges, the analysis cost is greater for the latter. In this chapter, the implementation of the finite strip procedure coupled with the flexibility approach is presented as an economical solution for the analysis of box girder bridges.

In the analysis of the bridges, the flanges and webs are usually considered as orthotropic plates with equivalent elastic properties. Therefore, the flat shell strips (Chapter 4) or the conical frustrum shell strips (Chapter 13) can be employed to model the right and curved box girders. Each end support of the bridge is assumed to be a diaphragm which is infinitely stiff in its own plane but infinitely flexible out of plane. Consequently, an idealized simply supported boundary condition is achieved for both in-plane and out-of-plane behaviors. The intermediate supports and stiffening diaphragms can be included by means of the flexibility approach.

Many concrete bridges are prestressed. The prestressing effects, which are affected by the prestressing cable on the concrete, can be simulated by applying the force system, as an external loading. The analysis of simple and continuous prestressed bridges is discussed in this chapter. The same analogy can be easily applied to pre-tension and post-tension prestressed girders.

At the immediate vicinity of wheel loads, the stresses and moments obtained by the summation of a Fourier series tend to converge very slowly because of the discontinuity in the stress field at the location of a point load. Therefore, if the local bending effect of wheel loads is of concern, a dense mesh and many series terms are required for satisfactory accuracy. This chapter offers recommendations for this problem, based on a comparison study between numerical and experimental results. This is followed by some numerical examples illustrating details in the finite strip simulation of concrete and steel box girder bridges.

14.2 Elastic properties of constituent plates

In steel box girder bridges, the webs and flanges are usually constructed from steel plates. Each constituent plate is an assembly of steel plates and stiffeners (Fig. 14.1). In the overall analysis of a steel bridge, the constituent plate is replaced by an orthotropic plate, i.e. a uniform plate with orthotropic material properties. Before the finite strip simulation is carried out, the elastic properties of the plate must be determined.

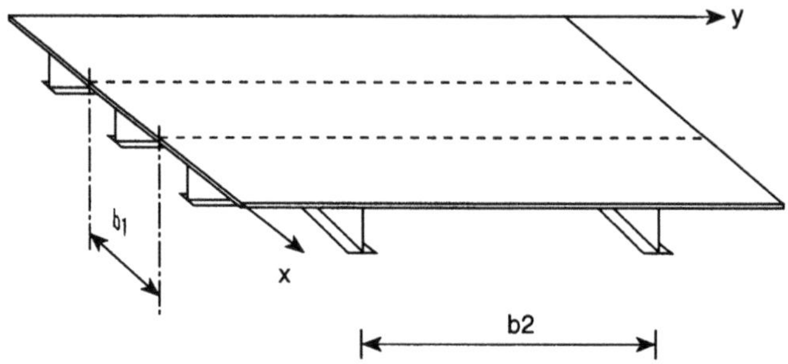

Figure 14.1 *Steel plate with stiffeners.*

14.2.1 Bending rigidities

The bending rigidity matrix has the form

$$[D] = \begin{bmatrix} D_x & D_1 & 0 \\ D_1 & D_y & 0 \\ 0 & 0 & D_{xy} \end{bmatrix} \tag{14.1}$$

For the plate strip in Fig. 14.1, D_y is the longitudinal bending rigidity per unit width:

$$D_y = I_y E / b_1 \tag{14.2}$$

Similarly,

$$D_x = I_x E / b_2 \tag{14.3}$$

The torsional rigidity of the top plate is $Gt_t^3/12$, in which t_t is the thickness of the top plate only. The equivalent torsional rigidity of a longitudinal stiffener D'_{xy} can be evaluated from the energy principle. Under pure torsion, the strain energy of the stiffener should be identical to that of the orthotropic plate with the equivalent torsional rigidity as follows:

$$U = \frac{1}{2} G J_1 (\frac{\partial^2 w}{\partial x \partial y})^2 b_2 = \frac{1}{2} \int_0^{b_1} \int_0^{b_2} D'_{xy} (2 \frac{\partial^2 w}{\partial x \partial y})^2 dx dy$$

where J_1 is the torsional constant of the longitudinal stiffener. Furthermore, under pure torsion, $\frac{\partial^2 w}{\partial x \partial y}$ is a constant. Therefore, the above integral becomes a simple multiplication of the integrand by the area $b_1 b_2$, and D'_{xy} can then be obtained as

$$D'_{xy} = \frac{GJ_1}{4b_1}$$

Similarly, the equivalent torsional rigidity of a transverse stiffener can be written as

$$D''_{xy} = \frac{GJ_2}{4b_2}$$

where J_2 is the torsional constant of the transverse stiffener.

Thus, for the whole constituent plate, D_{xy} can be calculated as

$$D_{xy} = \frac{Gt_t^3}{12} + \frac{GJ_1}{4b_1} + \frac{GJ_2}{4b_2} \tag{14.4}$$

For a conservative assessment, the coupling rigidity may be evaluated by considering only the top plate as effective, thus

$$D_1 = \frac{\nu E t_t^3}{12(1-\nu^2)} \tag{14.5}$$

14.2.2 In-plane elastic properties

The in-plane elastic matrix has the form

$$[D_p] = \frac{1}{1-\nu_x\nu_y}\begin{bmatrix} E_x & E_1 & 0 \\ E_1 & E_y & 0 \\ 0 & 0 & (1-\nu_x\nu_y)E_{xy} \end{bmatrix} \tag{14.6}$$

where

$$E_1 = \nu_x E_x = \nu_y E_y \tag{14.7}$$

Since the longitudinal in-plane stress σ_y is a major stress in a box bridge, the equivalent thickness t of a stiffened plate strip (Fig. 14.1) can be assumed as:

$$t = A_y/b_1 \tag{14.8}$$

where A_y is the net cross-sectional area of the top plate and stiffener, and b_1 is the stiffener spacing. Since the area of cross-section is used in this direction, the elastic modulus E_y is equal to the Young's modulus of steel E. Then, the elastic modulus in the x direction can be calculated as

$$E_x = A_x E/tb_2 \tag{14.9}$$

where A_x is the net cross-sectional area of the top plate and transverse stiffener, and b_2 is the spacing between transverse stiffeners.

The shear modulus can be approximated from

$$E_{xy} = t_t G/t \tag{14.10}$$

as a conservative estimate. Similarly, only the top plate is considered as the effective structural member in evaluating E_1, and

$$E_1 = \nu t_t E/t \tag{14.11}$$

Then, ν_x and ν_y can be calculated from Equation 14.7.

14.3 Intermediate supports and diaphragm

Box girder bridges are often supported by intermediate columns, and diaphragms are normally provided at these supported sections. If the deflection of the columns and the in-plane deformation of the diaphragms are considered negligible, the flexibility approach presented in Section 4.7 is suitable for the analysis of this type of structure. However, if the deformation of the supports and the settlement of foundations are significant, the flexibility matrix of the bridge should be modified according to the procedure given in Section 12.2.

Some box girders are also stiffened by internal rigid diaphragms or deformable plane frames, which are movable in their own plane with the girders. For these features, the flexibility approach can also be employed, and the details are described here.

14.3.1 Rigid movable diaphragm

The adaptation of the flexibility approach for the analysis of box bridges with movable diaphragms requires that each diaphragm be considered as the redundant member, and assumed to be connected to the girder only at a number of joints (Fig. 14.2). In most cases, only the joints at the cell corners are to provide an adequate simulation of the diaphragm connection, except for the diaphragm under truck loading, where one or two joints should be inserted on the top edge of a cell. The joint forces at stiffened section i forms the vector

$$\{X\}_i = [H_1, V_1, M_1, H_2, V_2, M_2, \ldots, H_N, V_N, M_N]_i^T \tag{14.12}$$

where N is the number of joints, and H, V and M are the horizontal, vertical and rotational joint forces, respectively.

Since the diaphragm is in equilibrium, the following equilibrium equations hold:

$$\sum_{j=1}^{N} H_j = 0$$

$$\sum_{j=1}^{N} V_j = 0$$

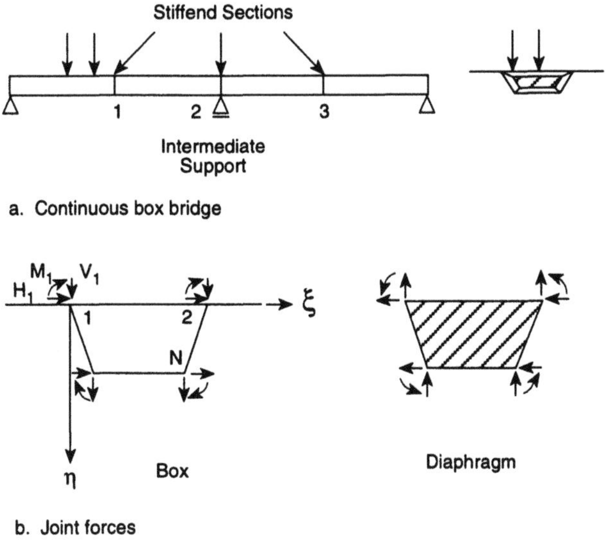

Figure 14.2 *Box bridge with movable diaphragms: (a) continuous box bridge; (b) joint forces.*

$$\sum_{j=1}^{N} Moment_j = 0 \tag{14.13}$$

Therefore, of all the 3N redundant forces, only $3N-3$ are independent, and the redundant force vector $\{X\}_i$ can be expressed in terms of independent redundant force vector $\{\bar{X}\}_i$ with $3N-3$ components as

$$\{X\}_i = [T]_i\{\bar{X}\}_i \tag{14.14}$$

For instance, we may chose the three forces at joint 1 as the dependent forces (Fig. 14.2(b)), then

$$\{\bar{X}\}_i = [H_2, V_2, M_2, \ldots, H_N, V_N, M_N]^T \tag{14.15}$$

and

$$[T]_i = \begin{bmatrix} -1 & 0 & 0 & -1 & 0 & 0 & \ldots & -1 & 0 & 0 \\ 0 & -1 & 0 & 0 & -1 & 0 & \ldots & 0 & -1 & 0 \\ \eta_2 & -\xi_2 & -1 & \eta_3 & -\xi_3 & -1 & \ldots & \eta_N & -\xi_N & -1 \\ 1 & & & & & & & & 0 & \\ & 1 & & & & & & & & \\ & & \cdot & & & & & & & \\ & & & & & & & & \cdot & \\ & 0 & & & & & & & & 1 \end{bmatrix} \tag{14.16}$$

in which ξ–η is the coordinate system with the origin at joint 1.

The box girder, released from all the redundant members, deforms under both external loads and redundant forces. The displacements of all the joints on section i constitute the vector

$$\{a\}_i = [u_1, w_1, \theta_1, u_2, w_2, \theta_2, \ldots, u_N, w_N, \theta_N]_i^T \tag{14.17}$$

which can be obtained as

$$\{a\} = [f]\{X\} + \{\delta\} \tag{14.18}$$

where

$$\{X\} = [[X]_1, [X]_2, ...[X]_L]^T \tag{14.19}$$

with L being the total number of stiffened and supported sections,

$$[f] = \begin{bmatrix} [f_{11}] & [f_{12}] & \cdots & [f_{1L}] \\ [f_{21}] & [f_{22}] & \cdots & [f_{2L}] \\ & & \vdots & \\ [f_{L1}] & [f_{L2}] & \cdots & [f_{LL}] \end{bmatrix} \tag{14.20}$$

is the flexibility matrix obtained by finite strip analysis of the released structure under unit redundant forces, and

$$\{\delta\} = [[\delta]_1, [\delta]_2, ...[\delta]_L]^T \tag{14.21}$$

is the joint displacements of the released structure due to external loading only.

Now, it is assumed that the diaphragm i is fixed on the box girder at joint 1 (the forces at this joint were assumed to be dependent). During the deformation, each joint on the box girder moves relatively to the diaphragm. The energy principle indicates that the matrix $[T]$ is also applicable for the transformation between the relative displacements $\{\bar{a}\}$ and absolute displacements $\{a\}$[1]:

$$\{\bar{a}\}_i = [T]_i^T \{a\} \tag{14.22}$$

Taking into account all the stiffened and supported sections (Fig. 14.2(a)), the following relationships are obtained:

$$\{X\} = [T]\{\bar{X}\} \tag{14.23}$$

[1] The work done in the two systems should be identical, i.e.

$$\{\bar{a}\}_i^T \{\bar{X}\}_i = \{a\}_i^T \{X\}_i$$

or

$$\{\bar{a}\}_i^T \{\bar{X}\}_i = \{a\}_i^T [T]_i \{\bar{X}\}_i$$

Transposing,

$$\{\bar{X}\}_i^T \{\bar{a}\}_i = \{\bar{X}\}_i^T [T]_i^T \{a\}_i$$

Since the relationship is valid for any set of redundant forces, the following relationship must hold:

$$\{\bar{a}\}_i = [T]_i^T \{a\}$$

$$\{\Delta\} = [T]^T\{\delta\} \tag{14.24}$$

in which $\{\bar{X}\}$ includes the independent redundant forces at all the stiffened and supported sections, $\{\Delta\}$ is the vector of relative displacements due to external loads, and

$$[T] = \begin{bmatrix} [T]_1 & & & \\ & [T]_2 & & \\ & & \ddots & \\ & & & [T]_L \end{bmatrix} \tag{14.25}$$

Because the diaphragms are assumed to be rigid, the relative displacement of every joint at any stiffened section should be zero. Thus, the compatibility conditions become

$$[T]^T\{a\} = \{0\} \tag{14.26}$$

Substituting Equation 14.18 into 14.26 and taking into consideration Equations 14.23 and 14.24 yields

$$[F]\{\bar{X}\} + \{\Delta\} = \{0\} \tag{14.27}$$

where

$$[F] = [T]^T[f][T] \tag{14.28}$$

is the modified flexibility matrix.

For a supported section, the displacements of all the joints appear in the compatibility equation directly and, therefore, $[T]_i$ is a unit matrix.

For a hinge-type connection, there are only two force components, i.e. H_j and V_j, at each joint. It is also possible to chose three redundant forces at any diaphragm section, e.g. H_1, V_1 and V_N, as the linear dependents and a similar matrix $[T]_i$ can be derived using the static equilibrium equations.

After solving Equation 14.27 for $\{\bar{X}\}$ and using Equation 14.23 to obtain $\{X\}$, the final solution of the indeterminate structure is the combination of the effects of external loading and the redundant forces on the released structure.

14.3.2 Deformable stiffening frames

Box girders are often stiffened by plane frames at a regular interval. These frames may have considerable deformation under wheel loads. Due to the deformation of the frame, the relative displacements of the box at any joint are not zero, in general, and must be equal to those of the frames. Thus, the compatibility conditions 14.27 become

$$[F]\{\bar{X}\} + \{\Delta\} = [f'](-\{\bar{X}\}) \tag{14.29}$$

or

$$([F] + [f'])\{\bar{X}\} + \{\Delta\} = \{0\} \tag{14.30}$$

where

$$[f'] = \begin{bmatrix} [f'_{11}] & & & 0 \\ & [f'_{22}] & & \\ & & \cdot & \\ 0 & & & [f'_{LL}] \end{bmatrix} \tag{14.31}$$

and $[f'_{ii}]$ is the flexibility matrix of the stiffening frame at section i and is null matrix if the frame is rigid.

Comparing Equations 14.30 and 14.27 shows that the deformation of frames can be included if only the flexibility matrix $[F]$ is substituted by $[F] + [f']$.

The matrix $[f'_{ii}]$ can be formed through a finite element analysis of the frame under unit redundant forces, with the frame being supported in a statically determinate manner. If the three redundant forces at joint 1 are chosen to be linearly dependent in forming the matrix $[T]_i$ (Equation 14.16), it is necessary to support the frame at built-in joint 1.

14.4 Prestressing forces

In the finite strip simulation, the forces applied by the prestressing cable to the concrete are treated as external loads. For single-span bridges, the friction loss is negligible, and the prestressing forces consists of the anchorage forces, i.e. point forces at two ends, and the induced forces, i.e. a distributed pressure along the span (Fig. 14.3).

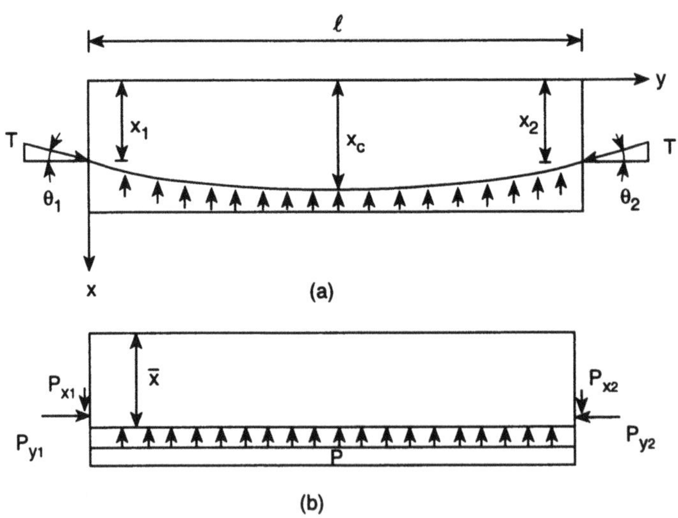

Figure 14.3 *Prestressing forces. (a) Forces induced by prestressing, (b) Simplified forces.*

The anchorage force can be resolved into the transverse and longitudinal components

$$\begin{aligned} P_x &= T\sin\theta \\ P_y &= T\cos\theta \end{aligned} \tag{14.32}$$

where T is the prestressing force along the cable. Thus, the corresponding load vector can be readily calculated.

In practice, the induced forces can be adequately modeled by a uniform in-plane pressure p along a longitudinal line as shown in Fig. 14.3(b), and

$$p = (P_{x1} + P_{x2})/l \tag{14.33}$$

with the line being located at

$$\bar{x} = (x_1 + x_2 + x_c)/3 \tag{14.34}$$

in which x_1, x_2 and x_c are the transverse coordinates of cable ends and center, respectively. Then, the load vector can be formed accordingly.

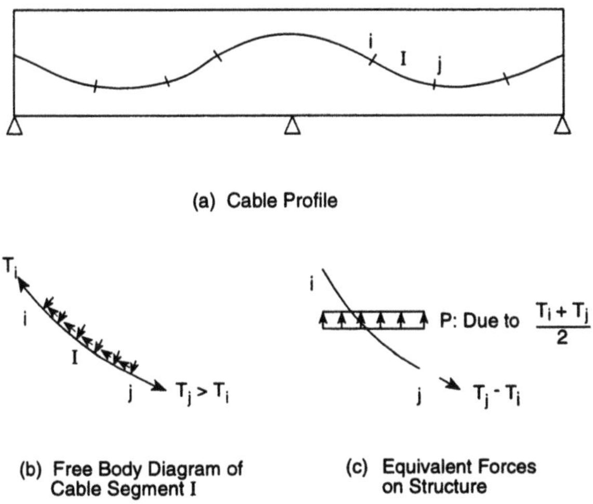

Figure 14.4 *Simulation of curved cable in multispan structure: (a) cable profile; (b)free-body diagram of cable segment I; (c) equivalent forces on the structure.*

For continuous bridges, the profile of the prestressing cable may be complicated and the effect of frictional losses should not be neglected. In this case, the cable can be modeled by a number of small, simple segments (Fig. 14.4). In each segment, the curvature is assumed to be constant and the frictional loss can be estimated. The induced pressure is evaluated using the average prestressing force in the segment. The frictional loss over each segment is applied at one end in an opposite direction, as shown in Fig. 14.4(c). Thus, the force system induced by a complex curved cable is simulated by a number of uniform pressures, two end-anchorage forces and a series of corrective forces representing frictional losses.

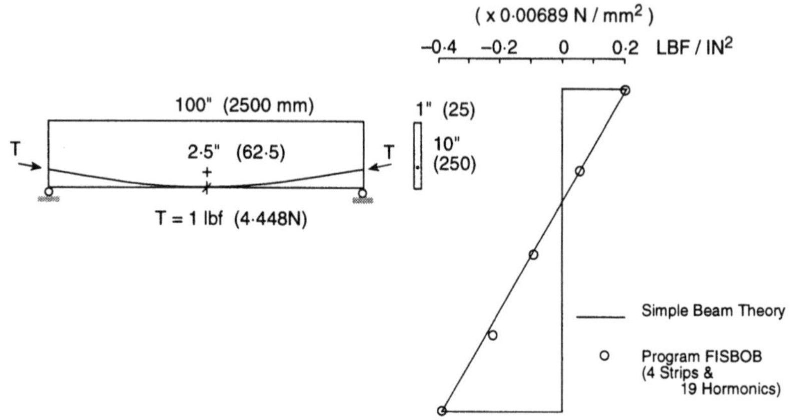

Figure 14.5 *Midspan longitudinal stress profiles for a beam with a curved post tensioned cable (Loo and Cusens, 1978).*

Figure 14.5 illustrates a simple prestressed deep beam that is simulated by four equal in-plane strips with 19 symmetrical series terms. The resulting mid-span longitudinal stresses are shown in the same figure. The correlation with the beam theory solution is excellent.

14.5 Local bending moment

According to thin plate theory, the bending moments at the application point of a concentrated force are infinity. Therefore, in bridge decks, the local bending moments induced by wheel loading are often significant, and the finite strip simulation of this effect normally requires a dense mesh and many series terms. For a better understanding of convergence behavior in this simulation, an asbestos-cement slab-on-girder bridge model was built and tested under point loads (Loo, 1973). The experimental results were compared with those of finite strip simulations employing many different meshes with different numbers of series terms, and the following recommendations were made for simulating local bending moments of wheel loading in continuous box-girder bridges (Loo and Cusens, 1978):

1. The end conditions have little influence on the distribution of local bending moments around concentrated loads. Therefore, if the wheels are located in the region between two supports, it is only necessary to analyze a simply supported bridge with a span equal to the spacing of the two related supports.

2. For local bending moment distribution in the vicinity of an intermediate support, it is only necessary to analyze a two-span bridge including only the adjacent supports, which are assumed to be simple.

3. Around the load, the top slab should be divided into strips in the following manner: each longitudinal line of wheels coincides with a nodal line; at least one additional nodal line should be added at each side of the vehicle and also between the lines of wheels.

4. Transverse moment output at 25 harmonics and longitudinal moments at 75 harmonics respectively are recommended for the finite strip HO3 with one internal nodal line. The numbers of harmonics recommended for low-order finite strip simulations are 45 and 95, respectively.

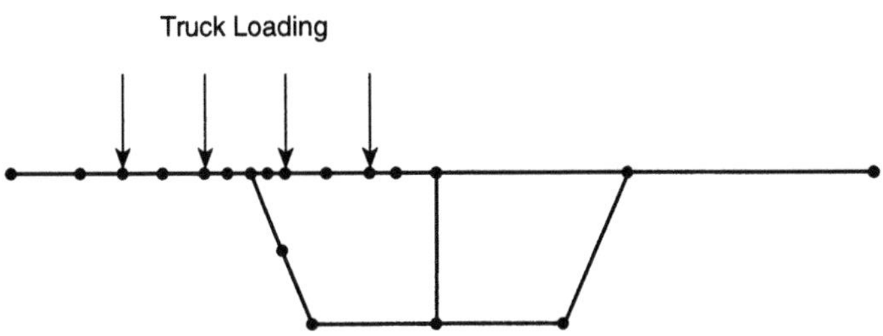

Figure 14.6 *Finite strip dvision for a box girder bridge under truck loading.*

Figure 14.6 shows a finite strip division for a box girder bridge under HB loading. If the top slab has variable cross-sections, more strips may be required for the simulation.

14.6 Numerical examples

14.6.1 Three-span concrete box girder bridge

Figure 14.7 illustrates a three-span concrete box girder bridge under truck loading. A typical cross-section of the bridge along with the wheel loads is shown in Fig. 14.7(b). For symmetry, only half the bridge was analyzed and only the symmetrical series terms were required. For simulating overall behavior of the bridge, eight strips and eight harmonics were employed (M. S. Cheung and Chan, 1979). The intermediate supports and diaphragms were all assumed to be rigid. Therefore, the flexibility approach in Section 4.7 was applied. Three types of support restraints shown in Fig. 14.7(c) were examined. Results of the three analyses are given in Figs 14.8 and 14.9, and in Tables 14.1 and 14.2.

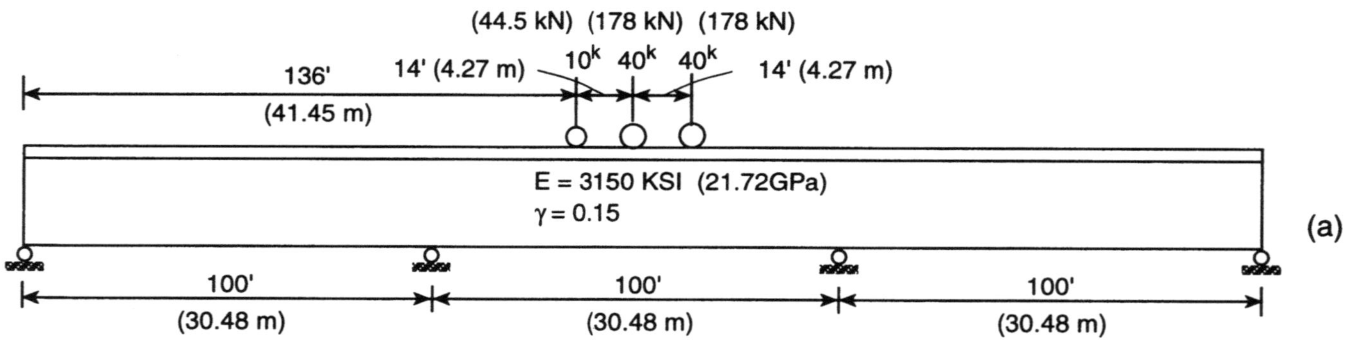

Figure 14.7 *Three-span box girder bridge: (a) longitudinal section; (b) transverse cross-section; (c) support restraints (dimensions in feet with metric equivalent in parenthesis; M.S. Cheung and Chan, 1979).*

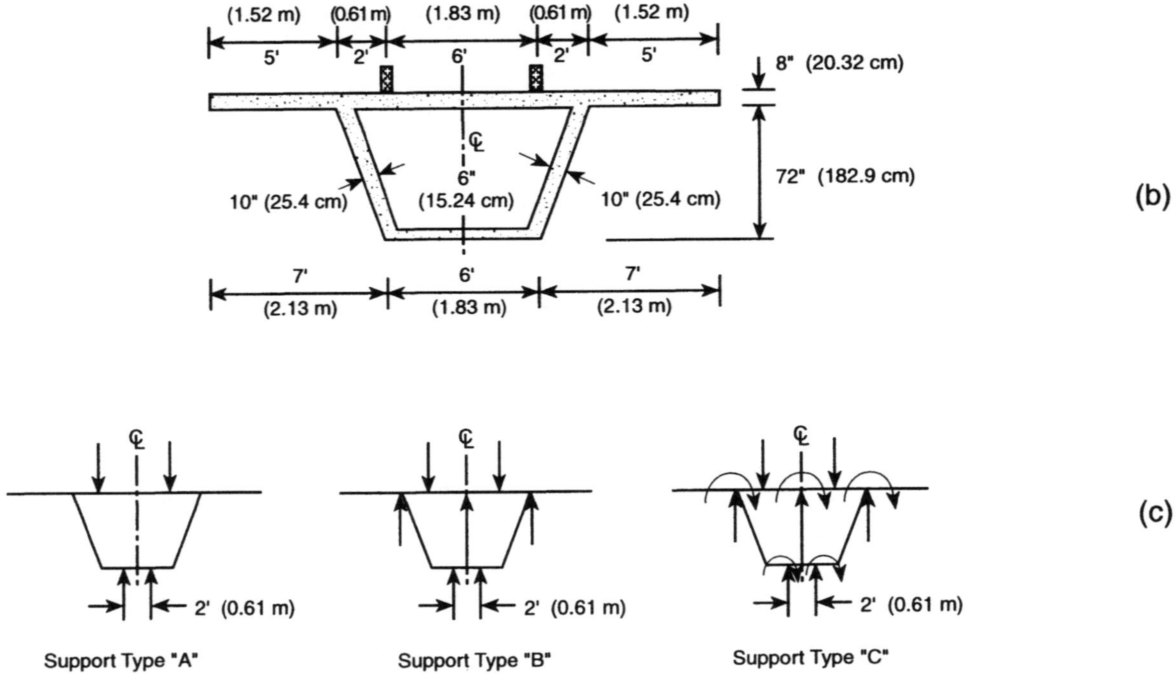

Figure 14.7 *(Continued)*

Table 14.1 *Deflection δ at top and bottom of center line for support types A, B and C (M. S. Cheung and Chan, 1979)*

Y (ft)	δ at top (in)			δ at bottom (in)		
	A	B	C	A	B	C
50	−0.0656	−0.0672	−0.0671	−0.0672	−0.0679	−0.0678
100	0.0026	0.0	0.0	0.0	0.0	0.0
150	0.1990	0.1960	0.1961	0.1930	0.1887	0.1877
200	0.0045	0.0	0.0	0.0	0.0	0.0
250	−0.0765	−0.0785	−0.0783	−0.0761	−0.0789	−0.0787

Table 14.2 *Longitudinal stresses σ_y at top and bottom of center line for support types A, B and C (1 ksi=6.8947 MPa, 1 ft=0.3048 m)(M. S. Cheung and Chan, 1979)*

Y (ft)	σ_y at top (ksi)			σ_y at bottom (ksi)		
	A	B	C	A	B	C
50	0.0215	0.0215	0.0216	−0.0476	−0.0475	−0.0475
100	0.0440	0.0437	0.0437	−0.1023	−0.1026	−0.1025
150	−0.1036	−0.1035	−0.1036	0.2394	0.2393	0.2390
200	0.0408	0.0405	0.0406	−0.0948	−0.0954	−0.0953
250	0.0269	0.0269	0.0269	−0.0598	−0.0599	−0.0598

From Table 14.1, it can be seen that the deflections along the center line of the bridge are not greatly affected by the support types. Table 14.2 and Fig. 14.8 indicate that the longitudinal stresses remain essentially the same regardless of the type of support restraints. Figure 14.9 reveals that away from the support diaphragm, the longitudinal bending moment M_y remain more or less unchanged while the transverse moments are slightly altered due to the change in support types. Figure 4.10 shows that at the location of the support diaphragm, support types B and C produced approximately the same results, while significant different results can be observed for support type A. Consequently, if the overall rather than the local behavior of bridge is the overriding consideration, a support simulation such as that for type A or type B should be accurate enough for most practical purposes.

14.6.2 Five-span steel box girder bridge

The elevation and cross-section of a five-span steel box girder bridge are shown in Figs 14.11(a) and 14.12, respectively. The dimensions of the transverse and longitudinal stiffeners are given in Figure 14.13.

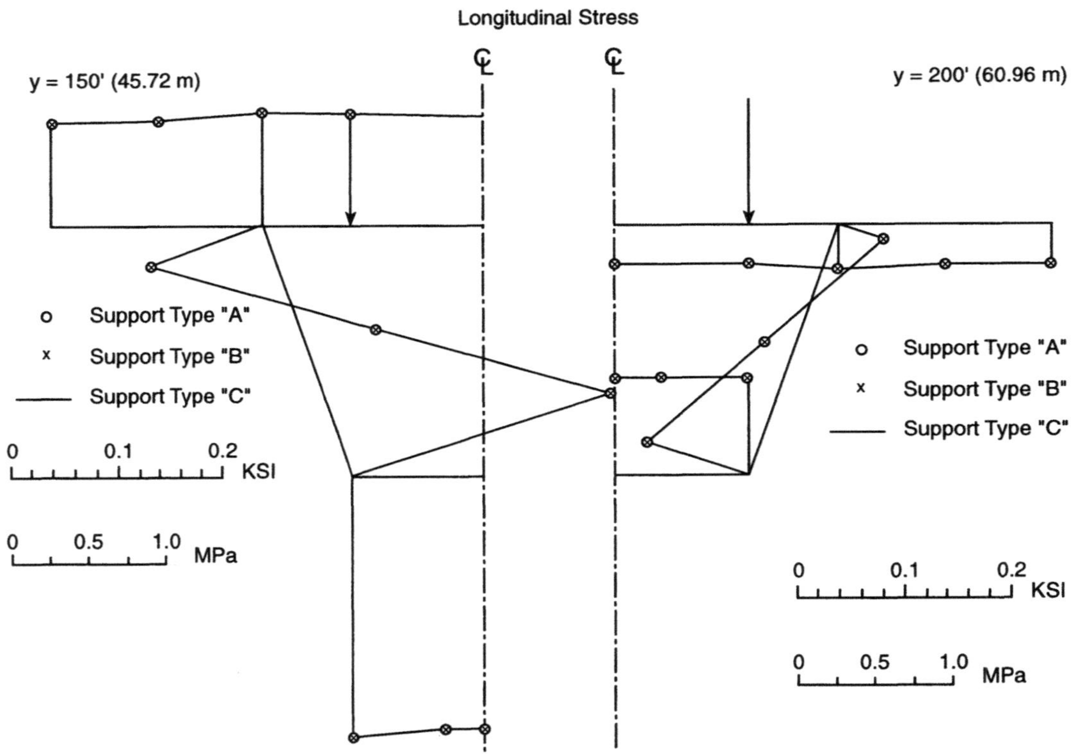

Figure 14.8 *Longitudinal stress in three-span box girder bridge at sections under a second axle and over second support (dimensions in feet with metric equivalent in parenthesis; M. S. Cheung and Chan, 1979).*

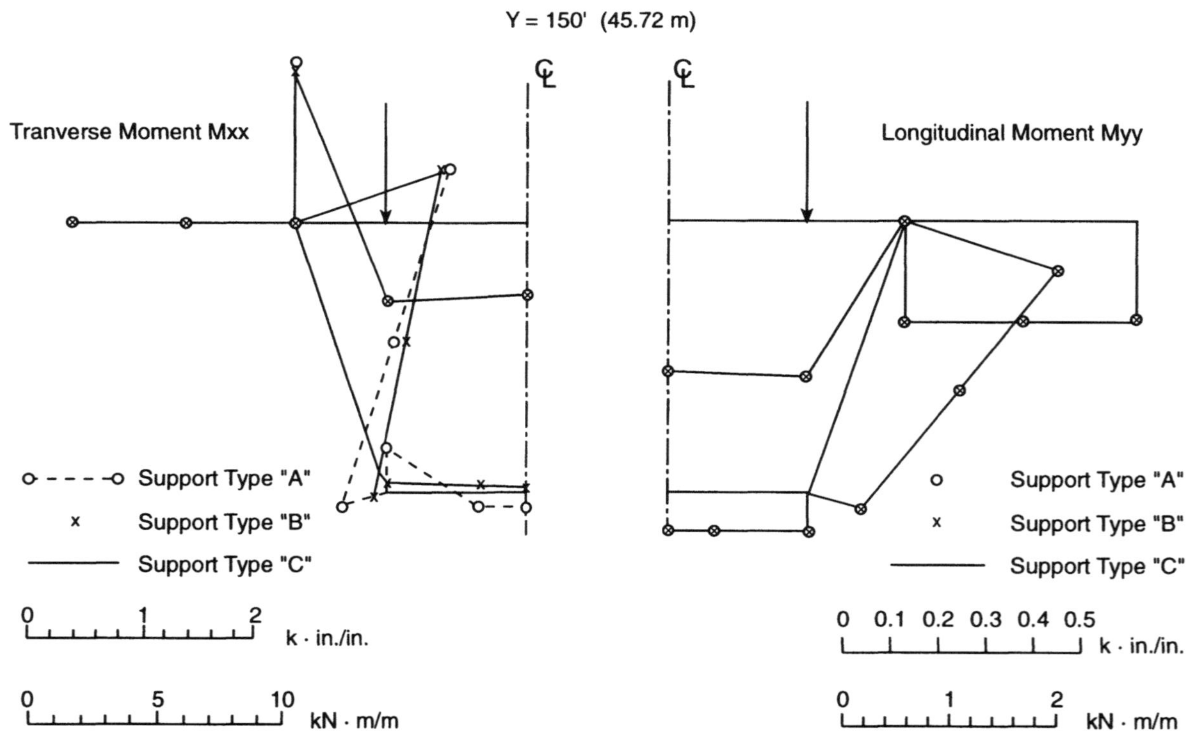

Figure 14.9 *Transverse moment and longitudinal moment in three-span box girder bridge at section under second axle (M. S. Cheung and Chan, 1979).*

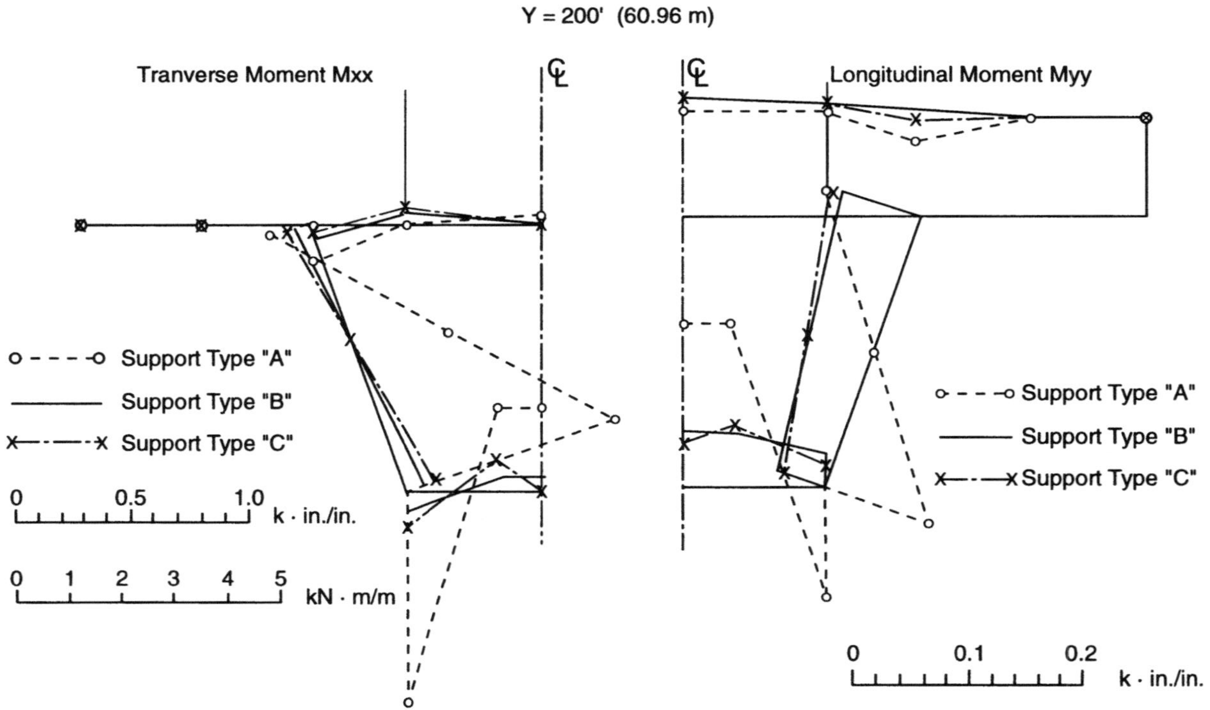

Figure 14.10 *Transverse moment and longitudinal moment in three-span box girder bridge at section over second support (M. S. Cheung and Chan, 1979).*

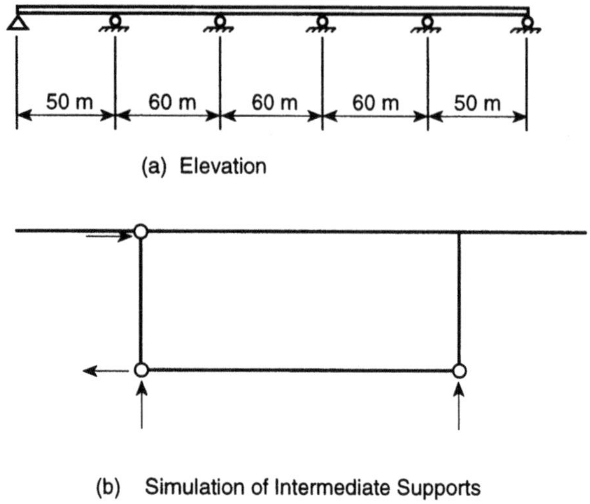

Figure 14.11 *Elevation and positions of restrained points for five-span continuous steel box bridge: (a) elevation; (b) simulation of intermediate supports (Loo and Cusens, 1978).*

This bridge was analyzed using low order finite strips (Loo and Cusens, 1978). For each intermediate support, two vertical and two horizontal restraints were introduced (Fig. 14.11b).

Overall analysis

For simulating the overall behavior of the bridge, the flanges and webs of the box structure were modeled by orthotropic plates, whose elastic properties were calculated following the procedures given in Section 14.2. Because the cross-beam for the top flange has variable height, E_x, D_x and D_{xy} were evaluated at several positions along the beam. Then interpolation curves were plotted to give E_x, D_x and D_{xy} at the position of concern. For each constituent plate, the transverse section and the longitudinal one have different neutral axes. In that analysis, the neutral axis of transverse section was assumed to coincide with the mid-plane of finite strips.

The loading and the finite strip model are illustrated in Fig. 14.14. The variations of the resulting longitudinal in-plane stresses and longitudinal local moments at the sections under the third and fourth axles of the HB vehicle are given in Figs 14.15 and 14.16.

Detailed analysis

If the accurate distributions of in-plane stresses and local bending moments in the top plates and the longitudinal stiffeners are required, a detailed analysis becomes necessary. For this purpose, the top plates and the walls of the longitudinal stiffeners should be modeled by finite strips, and the cross-members by deformable stiffening frames (Section 14.3.2). Consequently,

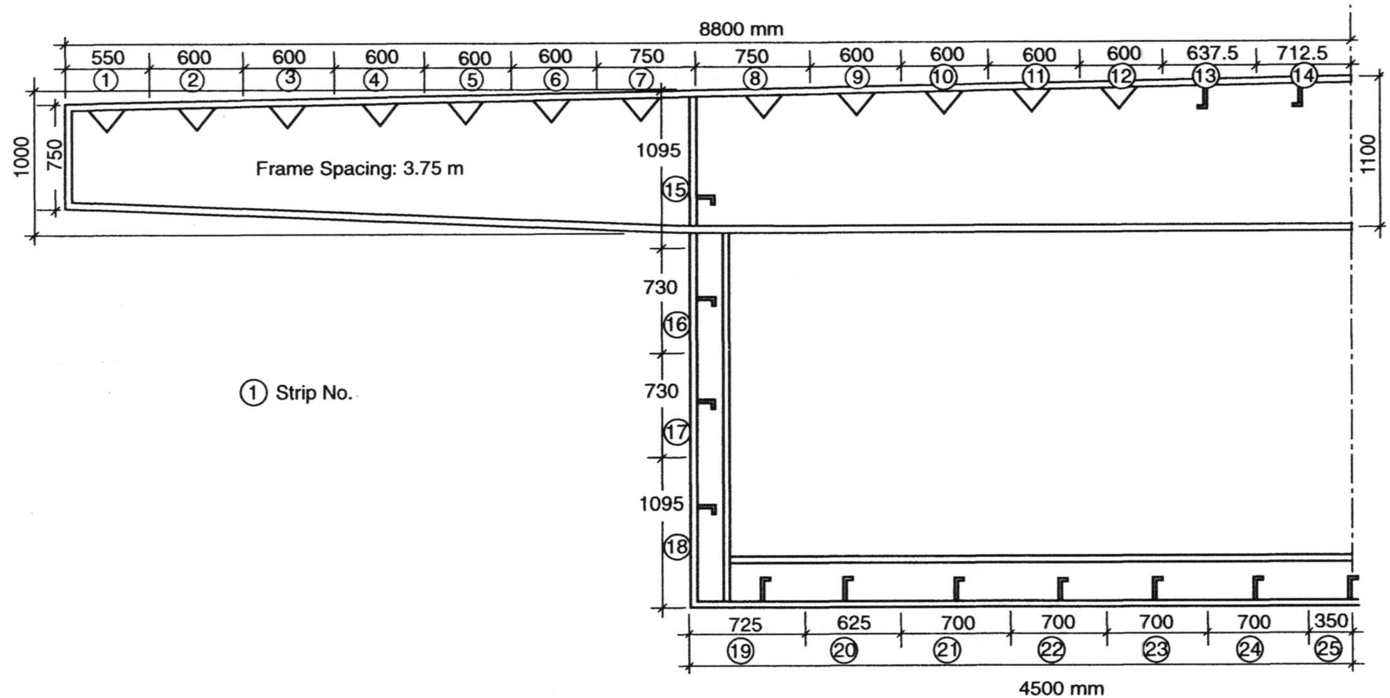

Figure 14.12 *Cross-section details of five-span continuous steel box bridge (Loo and Cusens, 1978).*

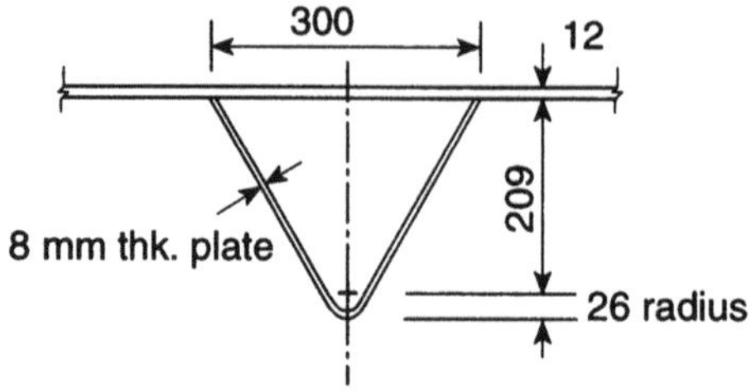

Longitudinal stiffeners	
Location (type)	Stiffeners
Top flange (trough)	As above drawing
Top flange (bulb flat)	203 × 10.2 × 20.76 kg
Web (bulb flat)	152 × 8.4 × 12.56 kg
Bottom flange (bulb flat)	203 × 10.2 × 20.76 kg

Transverse stiffness (T-section)			
Location	Web (thickness)	Flange (width)	Flange (thickness)
Top flange	12 mm	400 mm	25 mm
Webs	10 mm	178 mm	128 mm
Bottom flange	10 mm	200 mm	20 mm

Figure 14.13 *Dimensions of stiffeners for five-span continuous steel box bridge (Loo and Cusens, 1978).*

the finite strip model in Fig. 14.14 becomes that in Fig. 14.17; it is necessary to use a large number of finite strips and to simulate many frames, at a considerable cost. However, if only a particular part of the bridges is of concern, such as the portion under the HB vehicle, the detailed simulation may be necessary only for that part. The remainder of the constituent plates may still be simplified as orthotropic plates. Moreover, only four or five cross-frames around the HB truck would give adequate simulation for the stiffening effect of the nearby cross-members. Normally, this procedure yields a conservative solution since other frame stiffening effects are neglected.

Thus, the detailed finite strip simulation can be reduced to that shown in Fig. 14.18, in which only the left cantilever of the box is modeled in details because an HB vehicle operates over this part (Fig. 14.14). This finite strip model produces accurate distributions of in-plane stresses and local bending moments in the cantilever portion of the bridge around the third axle of the HB vehicle.

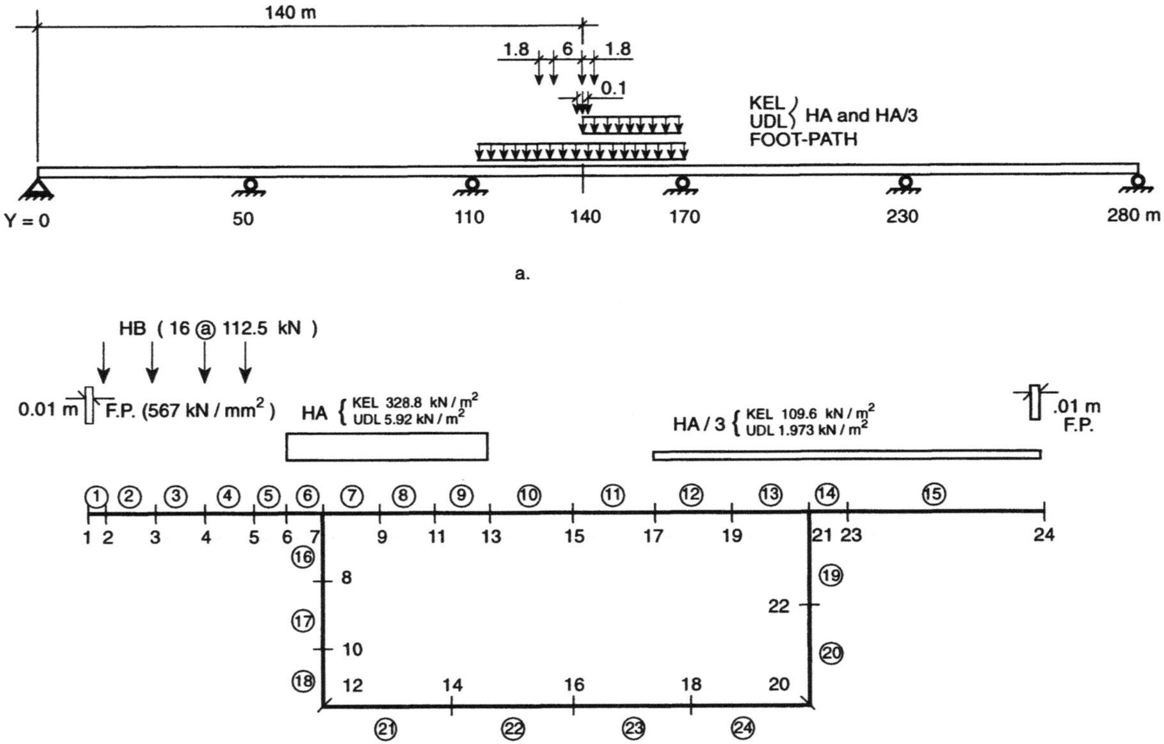

Figure 14.14 *Finite strip layout and loading details for five-span continuous steel box bridge (Loo and Cusens, 1978).*

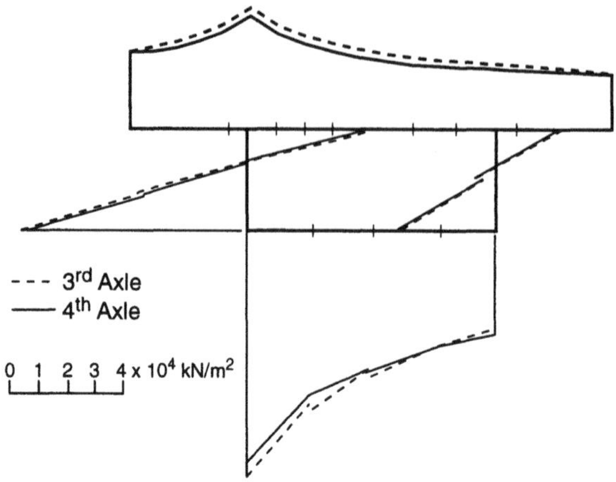

Figure 14.15 *Longitudinal in-plane stress for five-span continuous steel box bridge (overall analysis) (Loo and Cusens, 1978).*

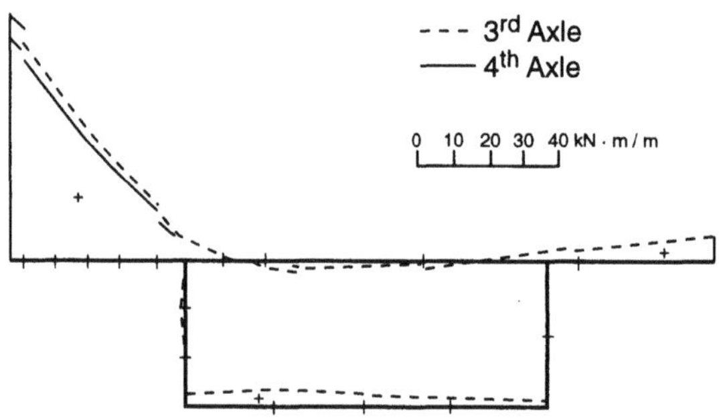

Figure 14.16 *Longitudinal local bending moment profiles for five-span steel box bridge (overall analysis) (Loo and Cusens, 1978).*

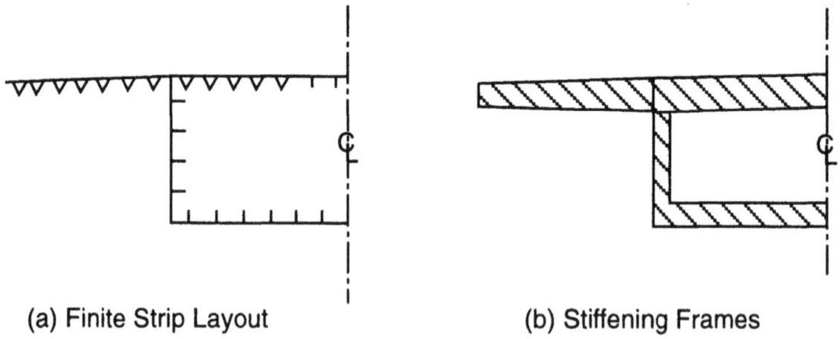

Figure 14.17 *Detailed finite strip simulation of steel box-girder bridge: (a) Simulations of supports and stiffening; (b) finite strip layout (Loo and Cusens, 1978).*

The corresponding analysis (Fig. 14.18) was implemented (Loo and Cusens, 1978), and the distribution of the resulting longitudinal stress in the midplane of top plate along the left cantilever is depicted in Figs 14.19 and 14.20 for the sections under the third and fourth axles of the HB truck. The third axle is right above a cross-beam, while the fourth is between two cross-members. In comparison with the results from the overall analysis, it can be observed that the overall analysis gives an acceptable maximum compressive stress. However, since the overall analysis could not simulate the stiffening effects of cross-beam, this analysis yields a compressive stress distribution, even though almost the entire section is under tension because of a heavy stiffening beam (Fig. 14.19).

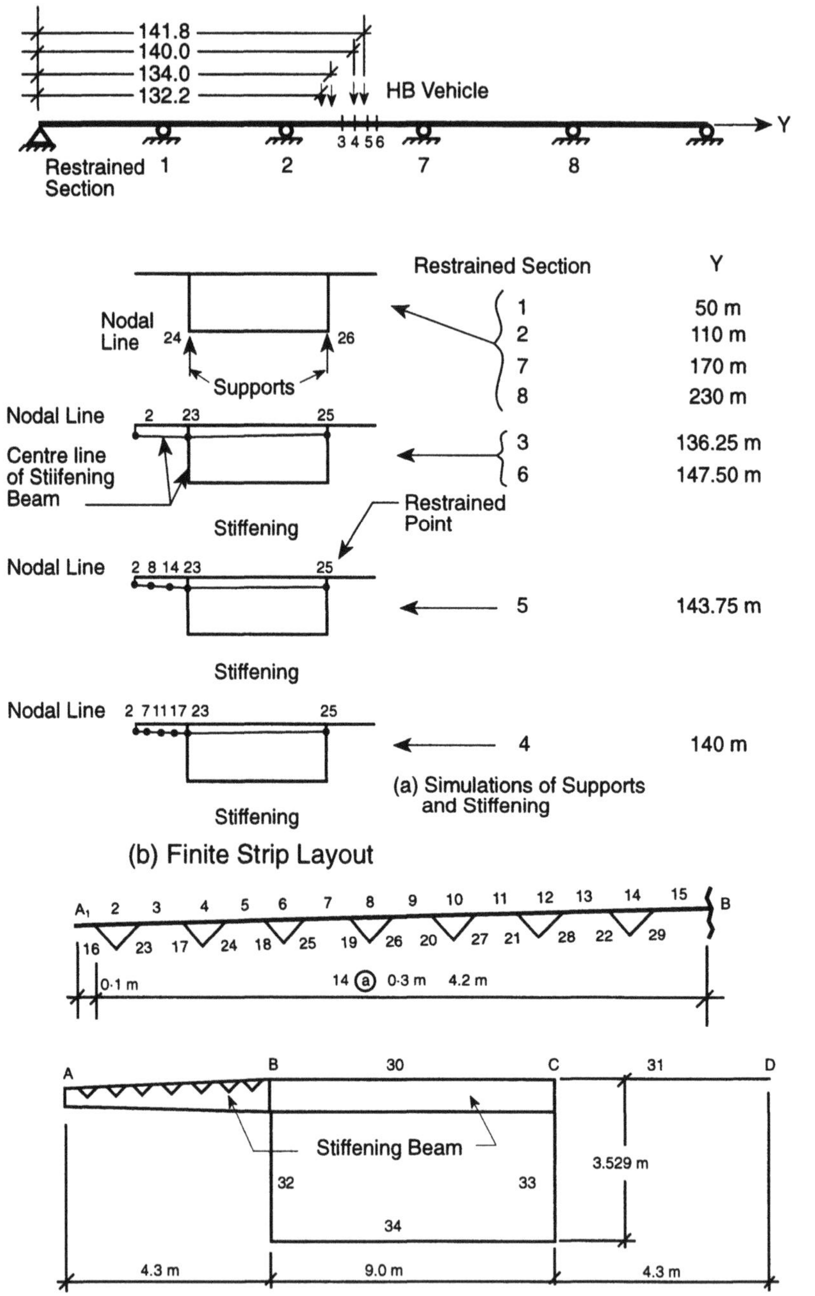

Figure 14.18 *Detailed analysis of five-span continuous steel bridge (approximate scheme) (Loo and Cusens, 1978).*

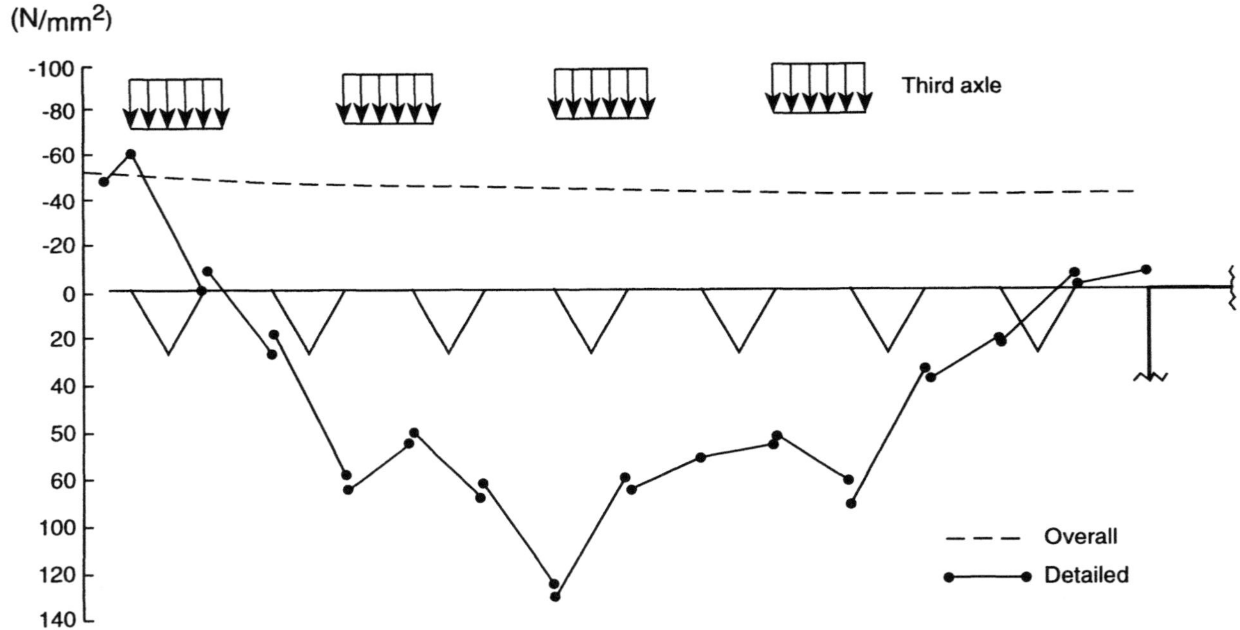

Figure 14.19 *Longitudinal in-plane stress profile at neutral plane of top plate (at section* $y = 140.0$ *m) (Loo and Cusens, 1978).*

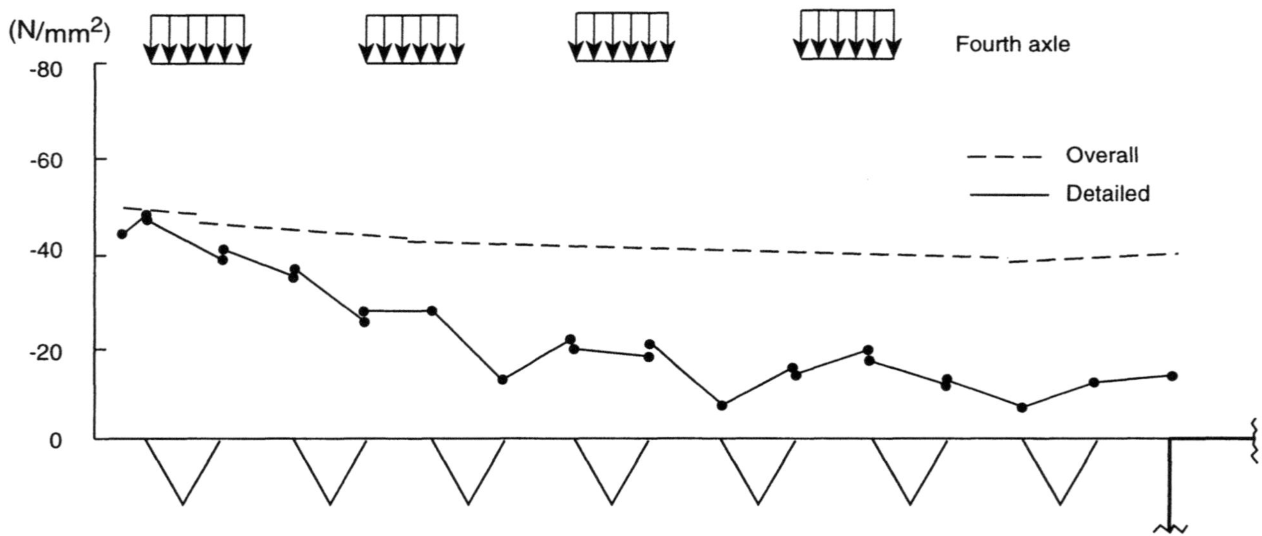

Figure 14.20 *Longitudinal in-plane stress profiles at neutral plane of top plate (at section $y = 141.8m$) (Loo and Cusens, 1978).*

CHAPTER 15

Continuous haunched bridges

15.1 Introduction

Recently, the finite strip method and the spline finite strip method have been extended to the analysis of continuous haunched slab-on-girder and box-girder bridges (M. S. Cheung and Li, 1988B,1989A, 1990B; M. S. Cheung, Li and Jaeger, 1992).

For bridges under uniform loads, the finite strip method produces satisfactory results for the deflections and longitudinal stresses using very few series terms. However, as mentioned in Chapter 6, for simulating the behavior of a bridge subjected to point loads, the spline finite strip method becomes more efficient. Both the methods are introduced in the following sections.

15.2 Finite strip analysis

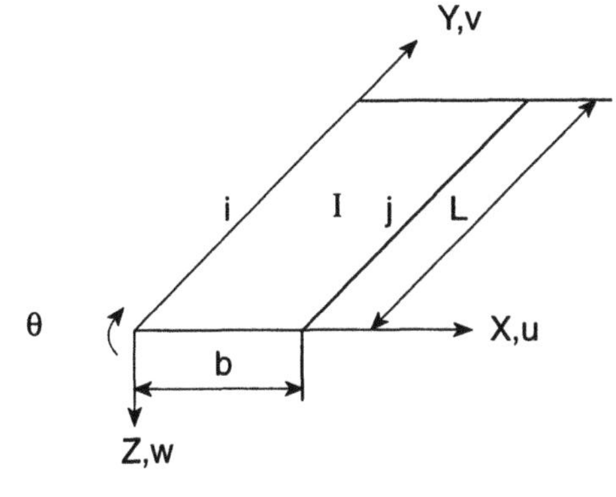

Figure 15.1 *Top flange plate strip.*

For analysis of a haunched bridge, three types of finite strip are employed, the top flange plate strip (Fig. 15.1), the vertical web strip (Fig. 15.2) and the bottom flange shell strip (Fig. 15.3).

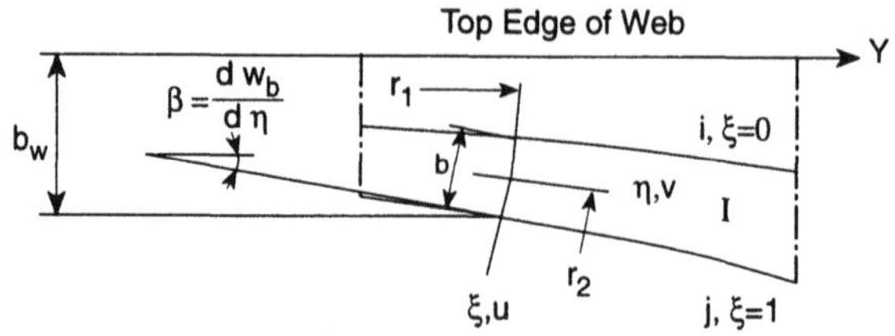

Figure 15.2 *Web strip.*

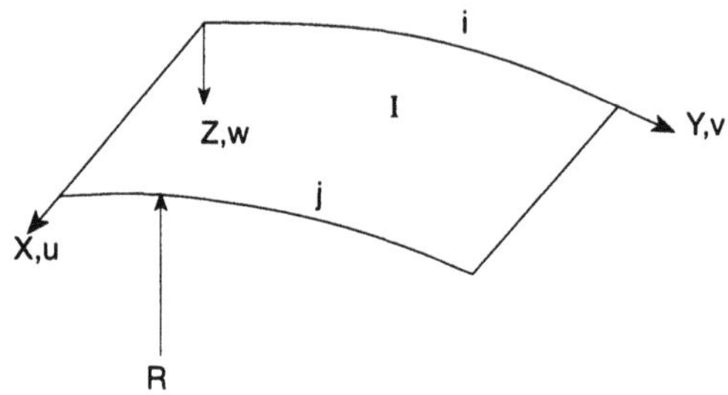

Figure 15.3 *Shell strip.*

15.2.1 Strain–displacement relationships

For the top flange plate strip, the Cartesian coordinate system is used and the following strain–displacement relationships are applicable:

$$
\begin{aligned}
\epsilon_x &= \frac{\partial u}{\partial x} \\
\epsilon_y &= \frac{\partial v}{\partial y} \\
\gamma_{xy} &= \frac{\partial u}{\partial y} + \frac{\partial v}{\partial x} \\
\chi_x &= -\frac{\partial^2 w}{\partial x^2} \\
\chi_y &= -\frac{\partial^2 w}{\partial y^2} \\
\chi_{xy} &= 2\frac{\partial^2 w}{\partial x \partial y}
\end{aligned}
\tag{15.1}
$$

For the web strip of variable depth, the curvilinear orthogonal coordinate system (ξ–η) is more convenient, in which η has units of length. The web can be divided into a number of equal or unequal width strips. The width of each strip is taken as $b = c \times b_w$, where b_w is the depth of web, with $0 < c \leq 1$. In the web, the most important deformation is in-plane bending, and the dominant strain components are the longitudinal normal strain ϵ_η and the in-plane shear strain $\gamma_{\xi\eta}$. According to the actual proportion of ordinary haunched box girder bridges, the relationship between displacements and strains can be written as

$$\begin{aligned}
\epsilon_\xi &= \frac{1}{b}\frac{\partial u}{\partial \xi} \\
\epsilon_\eta &= -\frac{u}{r_2} + \frac{\partial v}{\partial \eta} \\
\gamma_{\xi\eta} &= \frac{\partial u}{\partial \eta} + \frac{1}{b}\frac{\partial v}{\partial \xi} - \frac{u}{r_1} \\
\chi_\xi &= -\frac{1}{b^2}\frac{\partial^2 w}{\partial \xi^2} \\
\chi_\eta &= -\frac{\partial^2 w}{\partial \eta^2} \\
\chi_{\xi\eta} &= \frac{2}{b}\frac{\partial^2 w}{\partial \xi \partial \eta}
\end{aligned} \tag{15.2}$$

where r_1 and r_2 are the radii of curvature of coordinate lines (Fig. 15.2), and the following approximate geometrical relationship can be taken with reasonable accuracy:

$$\frac{1}{r_2} = \frac{x'}{b_w}\frac{d^2 b_w}{d\eta^2} \tag{15.3}$$

$$r_1 = \frac{b_w}{\frac{db_w}{d\eta}}. \tag{15.4}$$

in which x' is the distance from the top edge of web to the point under consideration.

For the bottom flange shell strip, cylindrical surface coordinates are used, and the relationships between displacements and strains (Timoshenko & Woinowsky–Krieger, 1959) are:

$$\begin{aligned}
\epsilon_x &= \frac{\partial u}{\partial x} \\
\epsilon_y &= \frac{\partial v}{\partial y} - \frac{w}{R} \\
\gamma_{xy} &= \frac{\partial u}{\partial y} + \frac{\partial v}{\partial x} \\
\chi_x &= -\frac{\partial^2 w}{\partial x^2} \\
\chi_y &= -\frac{\partial^2 w}{\partial y^2} - \frac{w}{R^2} \\
\chi_{xy} &= 2\frac{\partial^2 w}{\partial x \partial y} + \frac{2}{R}\frac{\partial v}{\partial x}
\end{aligned} \tag{15.5}$$

where R is the radius of curvature of the middle surface in the longitudinal direction, this is usually a function of y. In ordinary haunched box girder bridges it can be assumed that

$$\frac{1}{R} = \frac{d^2 b_w}{d\eta^2} \tag{15.6}$$

15.2.2 Displacement functions

The choice of suitable displacement functions is the most crucial and difficult part of the analysis. The right choice should lead to satisfactory results for maximum deflection and maximum longitudinal stress using only very few series terms if the bridge is subjected to uniform load (according to the experience of the authors, the required number of series terms should be less than five times the number of spans).

After comprehensive research and experimentation, the following formulations have been derived.

The displacement parameters of a strip for the mth term are taken as

$$\{\delta\}_{im} = [u_{im}\ v_{im}\ \bar{v}_{im}\ w_{im}\ \theta_{im}\ u_{jm}\ v_{jm}\ \bar{v}_{jm}\ w_{jm}\ \theta_{jm}]^T$$

where v_{im} and v_{jm} mainly represent the longitudinal motion due to in-plane bending and $\bar{v}_{im}$ and $\bar{v}_{jm}$ mainly represent displacements in the y direction caused by the longitudinal elongation or compression of the strip. The displacements within a top flange plate strip are

$$\begin{aligned} u &= \sum_{m=1}^{r}((1-X)u_{im} + Xu_{jm})Y_m(y) \\ v &= \sum_{m=1}^{r}((1-X)v_{im} + Xv_{jm})Y_m^*(y) + \sum_{m=1}^{r}((1-X)\bar{v}_{im} + X\bar{v}_{jm})\bar{Y}_m(y) \\ w &= \sum_{m=1}^{r}\Big((1-3X^2+2X^3)w_{im} + x(1-2X+X^2)\theta_{im} \\ &+ (3X^2-2X^3)w_{jm} + x(X^2-X)\theta_{jm}\Big)\,Y_m(y) \end{aligned} \tag{15.7}$$

where Y_m is the eigenfunction of a continuous beam which has the span lengths of the actual bridge (Section 4.8), and

$$Y_m^*(y) = (b_w \frac{dY_m}{dy} - \frac{db_w}{dy} Y_m) \tag{15.8}$$

$$\bar{Y}_m(y) = \int_y (Y_m(y)/R)\,dy \tag{15.9}$$

$$X = x/b.$$

In Equation 15.9 the interval of integration is from the origin of the first span to the current coordinate y. If the bottom line of the web in each span is a parabolic curve, R can be assumed to be a constant (see Equation 15.6) and the integration can be implemented analytically.

For the web strip, similar equations can be obtained by replacing X by ξ, x by bξ and y by η in the above equations.

These equations can also represent the displacements in the bottom flange shell strip without any change. However, in this case, x, y and z are cylindrical surface coordinates, and u, v and w are the corresponding displacement components in this coordinate system (Fig. 15.3).

The expressions for u and w meet the boundary conditions at the end and intermediate supports which are assumed to be diaphragms that are infinitely stiff in their own planes, so that $u = 0$ and $w = 0$ there.

By substituting the displacement functions 15.7 for the web strip into the expression for $\gamma_{\xi\eta}$ in Equation 15.2 and taking into consideration Equations 15.3 and 15.4, $\gamma_{\xi\eta}$ can be expressed in terms of displacement parameters as follows:

$$\begin{aligned}
\gamma_{\xi\eta} &= \frac{\partial u}{\partial \eta} + \frac{1}{b}\frac{\partial v}{\partial \xi} - \frac{u}{r_1} \\
&= \sum_{m=1}^{r}((1-\xi)u_{im} + \xi u_{jm})\frac{dY_m}{d\eta} \\
&\quad + \sum_{m=1}^{r}\{(-v_{im} + v_{jm})Y_m^*/b + (-\bar{v}_{im} + \bar{v}_{jm})\bar{Y}_m/b\} \\
&\quad - \sum_{m=1}^{r}((1-\xi)u_{im} + \xi u_{jm})Y_m/\frac{b_w}{\frac{db_w}{d\eta}} \\
&= \sum_{m=1}^{r}((1-\xi)u_{im} + \xi u_{jm})(\frac{dY_m}{d\eta} - \frac{db_w}{d\eta}Y_m/b_w) \\
&\quad + \sum_{m=1}^{r}\{(-v_{im} + v_{jm})Y_m^*/cb_w + (-\bar{v}_{im} + \bar{v}_{jm})\bar{Y}_m/b\} \\
&= \sum_{m=1}^{r}((1-\xi)u_{im} + \xi u_{jm})Y_m^*/b_w \\
&\quad + \sum_{m=1}^{r}\{(-v_{im} + v_{jm})Y_m^*/cb_w + (-\bar{v}_{im} + \bar{v}_{jm})\bar{Y}_m/b\}
\end{aligned}$$

From the above expression, it can be seen that $\gamma_{\xi\eta} = 0$ if $u_{im} = u_{jm}$, $u_{im} = (v_{im} - v_{jm})/c$ and $\bar{v}_{im} = \bar{v}_{jm}$. This means that the displacement functions 15.7 can simulate the pure in-plane bending of webs at the term level.

By substituting the displacement functions 15.7 for the shell strip into the expression for ϵ_y in Equation 15.5, ϵ_y can be expressed as

$$\begin{aligned}\epsilon_y =& \frac{\partial v}{\partial y} - \frac{w}{R}\\ =& \sum_{m=1}^{r}((1-X)v_{im} + Xv_{jm})\frac{\partial Y_m^*}{\partial y} + \sum_{m=1}^{r}((1-X)\bar{v}_{im} + X\bar{v}_{jm})\frac{\partial \bar{Y}_m(y)}{\partial y}\\ &- \sum_{m=1}^{r}((1-3X^2+2X^3)w_{im} + x(1-2X+X^2)\theta_{im}\\ &+(3X^2-2X^3)w_{jm} + x(X^2-X)\theta_{jm})Y_m(y)/R\end{aligned} \tag{15.10}$$

Using Equation 15.9, we have

$$\frac{d\bar{Y}_m(y)}{dy} = \frac{d}{dy}\int_y (Y_m(y)/R)\,dy = Y_m(y)/R.$$

Substituting this expression into Equation 15.10 yields

$$\begin{aligned}\epsilon_y =& \sum_{m=1}^{r}((1-X)v_{im} + Xv_{jm})\frac{\partial Y_m^*}{\partial y} + \sum_{m=1}^{r}((1-X)\bar{v}_{im} + X\bar{v}_{im})Y_m(y)/R\\ &- \sum_{m=1}^{r}((1-3X^2+2X^3)w_{im} + x(1-2X+X^2)\theta_{im}\\ &+(3X^2-2X^3)w_{jm} + x(X^2-X)\theta_{jm})Y_m(y)/R\end{aligned}$$

From this expression, it is not difficult to find that all the terms of $Y_m(y)/R$ can be cancelled out if $\theta_{im} = 0$, $\theta_{jm} = 0$, and $w_{im} = w_{jm} = \bar{v}_{im} = \bar{v}_{jm}$. Similar derivations can also prove that the longitudinal normal strains on any given curve $\xi = constant$ in the web strip and the entire top flange strip due to the arching effect w/R may be released under proper combinations of displacement parameters. This means that the displacement shape functions 15.7 are capable of reflecting that the arching effect can affect the distribution of longitudinal stress on the cross-section, but may not cause any longitudinal stress at the neutral axis of the cross-section, because in any continuous bridge only the supports at one cross-section are fixed in the longitudinal direction.

If the displacement functions cannot simulate the pure in-plane bending of the web ($\gamma_{\xi\eta} = 0$) or the stress-free longitudinal expansion of the bridge, the structure will seem much stiffer than it really is, and the correct answer may never be obtained.

15.2.3 Solution procedures

By substituting the displacement functions 15.7 into Equations 15.1, 15.2 and 15.5 in turn, the strains in each type of strip can be expressed in terms

Table 15.1 *Number of segments and Gauss points*

	b_w constant		b_w variable	
N_μ	N_{seg}	N_G	N_{seg}	N_G
1,2	1	8	1	8
3,4	1	10	1	10
5,6	2	8	2	8
7,8	2	10	2	10
9	2	10	3	10
10,11	3	8	3	10
12-14	3	10	3	10
15-17	4	10	4	10
18	4	10	5	10
19-24	5	10	5	10
25-30	6	10	6	10

$N_\mu = \mu_{max} l_i / \pi$, N_{seg}= number of segments in Span i, N_G= number of Gauss points in each segment

of displacement parameters in the following matrix form:

$$\{\epsilon\} = \sum_{m=1}^{r} [B]_m \{\delta\}_m$$

The stiffness matrix of the strip corresponding to the mth and nth series terms can then be expressed as:

$$[k]_{mn} = \int_L \int_b [B]_m^T [D][B]_n \, dx \, dy$$

In the x direction , the integration can be implemented numerically or analytically. However, in the y direction a numerical integration method such as Gaussian quadrature has to be used, because of the complexity of the integration in the stiffness matrix. Since $\int_L Y_m' Y_n' \, dy$ and $\int_L Y_m'' Y_n dy$ are not equal to zero when m and n are different, the different terms of the series are coupled. In evaluating the stiffness matrix by Gaussian quadrature, a proper number of Gaussian integration points should be chosen in order to achieve the desired accuracy and efficiency. When necessary, every span can be divided into several segments. In each segment, 8 to 10 integration points may be used. The selection of segment size and Gaussian integration points depends mainly on the number of highest order harmonics in a given span, $n_\mu = \mu_{max} l_i / \pi$. Based on the authors' experience, the selection which will achieve a reasonable balance between accuracy and efficiency of the solution is given in Table 15.1.

Following commonly used procedures, it is possible to obtain the final solutions required, including the distribution of displacements and stresses within the whole structure.

15.3 Spline finite strip analysis

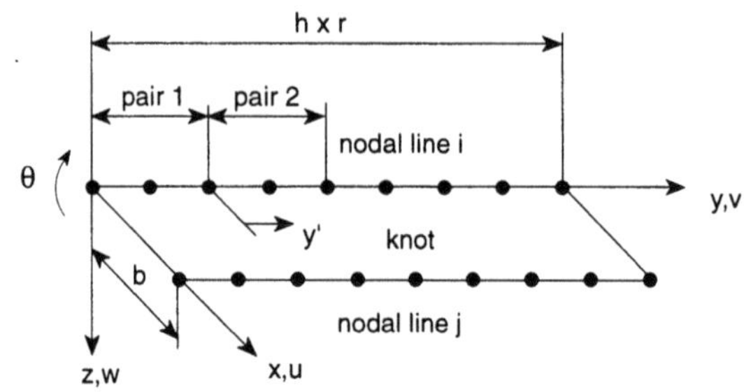

Figure 15.4 *Top flange spline strip.*

The spline finite strip analysis of haunched bridges also requires three types of strip, i.e. the top flange plate strip (Fig. 15.4), the vertical web strip (Fig. 15.5) and the bottom flange shell strip (Fig. 15.6).

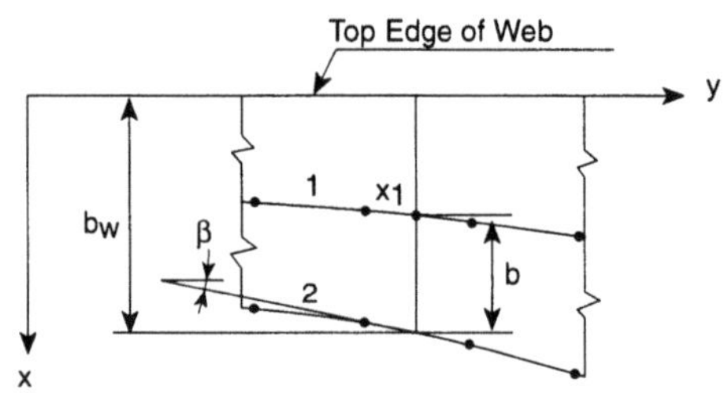

Figure 15.5 *Web spline strip.*

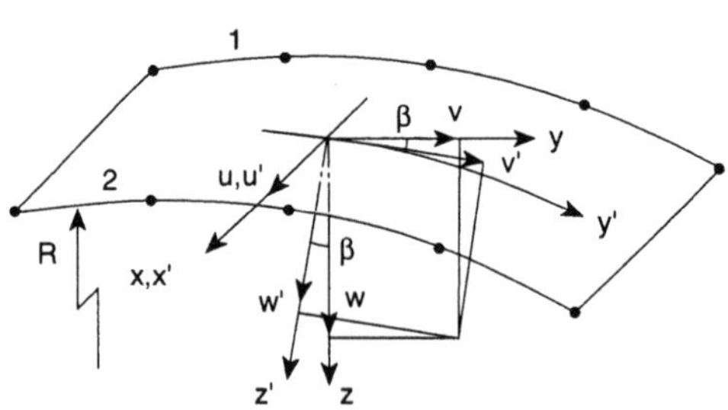

Figure 15.6 *Shell spline strip.*

15.3.1 Strain–displacement relationships

For the top flange plate strip and the vertical web strip, the strain–displacement relationships in Equation 15.1 are applicable. However, in the strip, x, y and z refer to the axes of an individual local Cartesian coordinate system, and u, v and w are the displacement components in the directions x, y and z respectively.

In the previous section, for the bottom flange shell strip, the strain–displacement relationships were written in the cylindrical surface coordinate system. However, in the present analysis a global Cartesian coordinate system is taken. In order to facilitate the assembly of structure matrices, the following coordinate transformations are required (Fig. 15.6):

$$\begin{aligned} u' &= u \\ v' &= v\cos\beta + w\sin\beta \\ w' &= -v\sin\beta + w\cos\beta \\ \frac{\partial}{\partial x'}f(x,y) &= \frac{\partial}{\partial x}f(x,y) \\ \frac{\partial}{\partial y'}f(x,y) &= \cos\beta\frac{\partial}{\partial y}f(x,y) \end{aligned} \tag{15.11}$$

where x, y and z are the global Cartesian coordinates, u, v and w are the corresponding displacements, while the primed ones refer to the cylindrical surface coordinate system, $f(x, y)$ represents an arbitrary function of x and y, and

$$\beta = \tan^{-1}(\frac{db_w}{dy}) \tag{15.12}$$

Starting from Equation 15.5, performing the above coordinate transformations and carrying out some simplifications according to the theory of shallow shells, the relationships between the strains in the bottom flange shell strip and its displacement components expressed in the global Cartesian coordinate system can be obtained in the following form:

$$\begin{aligned} \epsilon'_x &= \frac{\partial u}{\partial x} \\ \epsilon'_y &= \frac{\partial v'}{\partial y'} - \frac{w'}{R} \\ &= \cos\beta\frac{\partial}{\partial y}(v\cos\beta + w\sin\beta) - \frac{w'}{R} \\ &= \cos\beta\{\frac{\partial v}{\partial y}\cos\beta + \frac{\partial w}{\partial y}\sin\beta) + (-v\sin\beta + w\cos\beta)\frac{\partial\beta}{\partial y}\} - \frac{w'}{R} \\ &= \cos\beta(\frac{\partial v}{\partial y}\cos\beta + \frac{\partial w}{\partial y}\sin\beta) + w'\frac{\partial\beta}{\partial y'} - \frac{w'}{R} \\ &= \frac{\partial v}{\partial y}\cos^2\beta + \frac{\partial w}{\partial y}\cos\beta\sin\beta \end{aligned}$$

$$\begin{aligned}
\gamma'_{xy} &= \frac{\partial u'}{\partial y'} + \frac{\partial v'}{\partial x'} \\
&= \frac{\partial u}{\partial y}\cos\beta + \frac{\partial v}{\partial x}\cos\beta + \frac{\partial w}{\partial x}\sin\beta \\
\chi'_x &= -\frac{\partial^2 w}{\partial x^2} \\
\chi'_y &= -\frac{\partial^2 w}{\partial y^2} \\
\chi'_{xy} &= 2\frac{\partial^2 w}{\partial x \partial y}
\end{aligned} \tag{15.13}$$

15.3.2 Displacement functions

In analysis, each nodal line is divided into an even number of sections by knots, and these sections are grouped pair by pair (Fig. 15.4). The B_3-spline expressions introduced in Chapter 6 are employed as the longitudinal displacement functions for displacement components u and w, and quadratic interpolation in each pair of sections is used as the longitudinal displacement function for v.

The displacement parameters of the strip corresponding to knot m are

$$\{\delta\}_m = [u_{im}, v_{im}, w_{im}, \theta_{im}, u_{jm}, v_{jm}, w_{jm}, \theta_{jm}]^T \tag{15.14}$$

The displacement field within a top flange plate strip or a bottom flange shell strip is:

$$\begin{aligned}
u &= \sum_{m=-1}^{r+1} ((1-X)u_{im} + Xu_{jm})\,\Phi_m(y) \\
v &= \sum_{k=1}^{3} ((1-X)v_{ik} + Xv_{jk})\,L_k(y) \\
w &= \sum_{m=-1}^{r+1} ((1-3X^2+2X^3)w_{im} + x(1-2X+X^2)\theta_{im} \\
&\quad +(3X^2-2X^3)w_{jm} + x(X^2-X)\theta_{jm})\,\Phi_m(y)
\end{aligned} \tag{15.15}$$

where $X = x/b$, r is the number of sections, Φ_m is the B_3-spline expression with $y = y_m$ as the center (see Equations 6.1 and 6.4), k is the local number of knot m in the corresponding pair of sections, and L_k are quadratic interpolations in the following form:

$$\begin{aligned}
L_1(y) &= (1-Y)(1-0.5Y) \\
L_2(y) &= Y(2-Y) \\
L_3(y) &= 0.5Y(Y-1)
\end{aligned} \tag{15.16}$$

in which, $Y = y'/h$ (see Fig. 15.4).

Equation 15.15 is also applicable for the vertical web strip. However, for this strip, x, y and z are local Cartesian coordinates, and u, v and w are the corresponding displacement components. In addition, X is not only the function of x, but also varies with y in the form

$$X = \frac{x - x_i}{b} = \frac{x}{cb_w(y)} - \frac{c_i}{c} \tag{15.17}$$

where $b_w(y)$ is the variable depth of the web, $b = cb_w$ is the width of the strip and $x_i = c_i b_w$ is the coordinate of nodal line i, with c and c_i being constant (Fig. 15.5).

Substituting the displacement functions of the web strip into the expression for ϵ_y in Equation 15.1 yields the relationship

$$\begin{aligned} \epsilon_y &= \frac{\partial v}{\partial y} \\ &= \sum_{k=1}^{3}\{\frac{\partial}{\partial y}((1-X)v_{ik} + Xv_{jk})L_k(y) + ((1-X)v_{ik} + Xv_{jk})\frac{dL_k}{dy}(y)\} \\ &= \sum_{k=1}^{3}\{(v_{ik} - v_{jk})\frac{x}{cb_w^2}\frac{db_w}{dy}L_k(y) + ((1-X)v_{ik} + Xv_{jk})\frac{dL_k}{dy}(y)\} \end{aligned} \tag{15.18}$$

It can be seen that ϵ_y depends not only on the first derivative of the longitudinal displacement function $\frac{dL_j}{dy}$, but also on the slope of bottom line of the web, $\frac{db_w}{dy}$. In actual structures, the slope of the bottom line of web may have abrupt changes, especially at intermediate supports. This may cause discontinuities of the resulting strain ϵ_y and significant error in the results of the most important stress component σ_y if the B_3-spline functions (with undesired continuity of their first derivatives) are used as the longitudinal displacement function for v. The use of quadratic interpolations as the longitudinal displacement functions of v can solve this problem as long as the abrupt changes of the slope are located at the ends of the pairs, because the first derivatives of $L_j(y)$ can be discontinuous across these locations. An alternative method of treating this problem is to use the spline function with only C_0 continuity as the longitudinal displacement functions of v; however, this will produce constant longitudinal strain and, consequently, poor convergence.

Once the displacement functions for the strips have been chosen, the stiffness matrix, load vector etc., can be obtained by following standard finite strip formulation procedures (see Chapter 4).

15.3.3 Penalty function method

A number of methods for imposing boundary conditions have been proposed (see Y. K. Cheung and Fan, 1983). In the present analysis, the penalty function method is chosen to treat end and intermediate support conditions of the strips.

As an example, it is assumed that the displacement component w at point (x, y) is prescribed to be zero, that is

$$w = \sum_m N_m(x,y)w_m = 0 \tag{15.19}$$

where $N_m(x, y)$ and w_m are the associated displacement function and displacement parameter, respectively.

In order to impose this constraint in analysis, a fictitious spring with large stiffness α might be introduced in the corresponding direction at the point (x, y). Its strain energy or penalty function is

$$U = \frac{1}{2}\alpha w^2 = \frac{1}{2}\alpha \sum_m \sum_n N_m(x,y)N_n(x,y)w_m w_n \tag{15.20}$$

From this expression it can be seen that the stiffness matrix of this imaginary spring is

$$[k_{mn}] = [\alpha N_m(x,y)N_n(x,y)] \tag{15.21}$$

If the first derivative of deflection, say $\frac{\partial w}{\partial y}$, at point (x, y) is prescribed to be zero, then

$$\frac{\partial w}{\partial y} = \sum_m \frac{\partial N_m(x,y)}{\partial y} w_m = 0 \tag{15.22}$$

Following the same procedure, the required stiffness matrix for this constraint can be obtained as

$$[k_{mn}] = [\alpha \frac{\partial N_m(x,y)}{\partial y} \frac{\partial N_n(x,y)}{\partial y}] \tag{15.23}$$

Assembling the stiffness matrices of all point constraints with the structural stiffness matrix will accomplish the treatment of boundary conditions.

Any elastic support can also be included by the same method, as long as the α is the actual stiffness of this support.

15.4 Numerical examples

15.4.1 Continuous beam with variable depth

The beam is continuous over two equal spans $l_1 = l_2 = 10$ m (Fig. 15.7). The depth of beam is 1.0 m at both ends, 2.0 m over the intermediate support, and varies linearly in-between. The thickness of the beam is 1.0 m. The properties of material are $E = 10000$ MPa and $\nu = 0$. A vertical

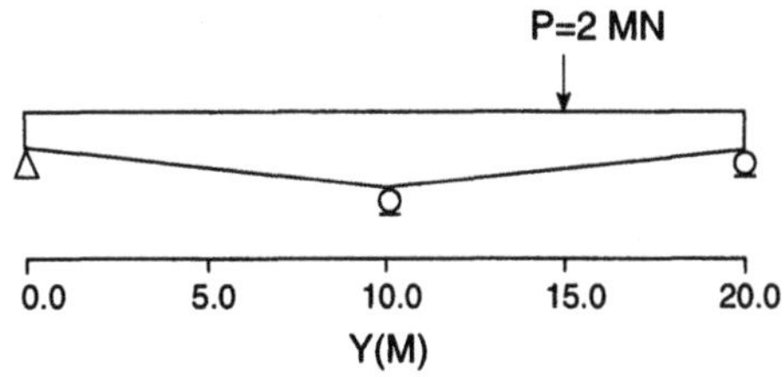

Figure 15.7 *Continuous beam.*

Table 15.2 *Deflection and longitudinal stresses in continuous beam*

Number of sections	w (m)	σ_y (MPa)		
	$y = 15$ m	$y = 5$ m	$y = 10$ m	$y = 15$ m
8	0.011279	−3.520	−4.111	9.998
16	0.011480	−3.370	−3.826	9.995
24	0.011524	−3.347	−3.777	10.002
Beam theory	0.011265	−3.339	−3.756	9.994

point force $P = 2.0$ MN is acting at the center of the second span. The beam is analyzed by the spline finite strip method using only one web strip. The deflection at the loaded cross-section and the longitudinal stresses at several points along the bottom nodal line are listed in Table 15.2, in which the values from beam theory (shear deformation included) are also given for comparison. It can be seen that this method is accurate and that it converges very rapidly.

15.4.2 Two-span haunched concrete slab-on-girder bridge

The bridge with span $l_1 = 12$ m and $l_2 = 16$ m is shown in Fig. 15.8. The end of shorter span is simply supported, and the other end is fixed. The thickness of the slab is 0.2 m and the thickness of all webs is 0.4 m. The spacing of webs is 2.8 m. The bottom line of the web is a parabola in each span. The depth of the web (from its bottom to the middle plane of slab) is 0.5 m at the simply supported end, 1.1 m over the intermediate support, 0.5 m at the center of longer span, and 0.8 m at the fixed end. The material properties are $E = 27000$ MPa, $\nu = 0.15$. In loading case 1, a unit vertical point load $P = 1.0$ MN acts on the top of the exterior girder at the center of the longer span; in case 2, the same load moves to the top of the central girder at the same cross-section. In the spline finite strip analysis, the structure is divided into 12 slab strips, seven web strips

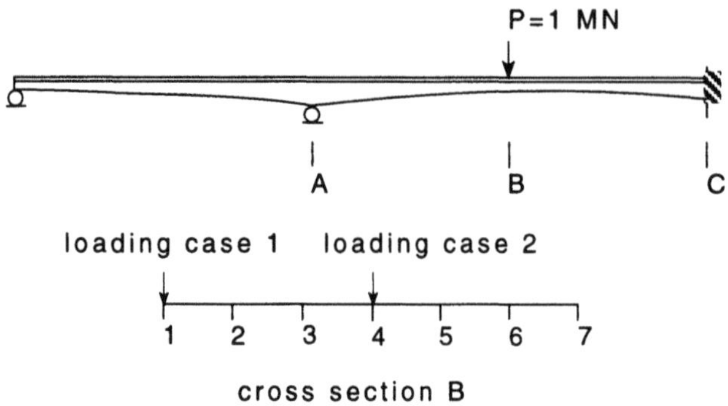

Figure 15.8 *Two-span concrete bridge.*

Table 15.3 *Longitudinal stresses in a two-span slab-on-girder bridge*

Loading Case	Cross-section	σ_y (MPa) on the top and bottom			
		Girder 1	Girder 2	Girder 3	Girder 4
1	A	5.24	1.22	0.01	–
		−10.75	−2.76	0.06	–
	B	−15.64	−2.41	−0.37	–
		45.51	10.75	1.67	–
	C	11.66	3.51	0.55	–
		−29.36	−10.06	−1.15	–
2	A	–	0.14	1.03	1.86
		–	−0.21	−2.87	−6.20
	B	–	−0.42	−1.59	−6.42
		–	1.84	8.62	30.19
	C	–	0.62	2.49	3.95
		–	−1.71	−8.83	−16.10

and 28 longitudinal sections. The resulting longitudinal stresses on the top and bottom of webs at a number of cross-sections are listed in Table 15.3.

If the load is put on the second or third girder, the maximum longitudinal stress occurs at the bottom of the loaded girder at the loaded cross-section and its value is little different from that in loading case 2.

15.4.3 Two-span haunched steel box girder bridge

A multi cell box-girder bridge is continuous over two equal spans as shown in Fig. 15.9. It is haunched over the interior supports, with $c = 2.40$ m. The bridge span l is chosen as 40 m, the spacing between webs $b_2 = 2.80$ m;

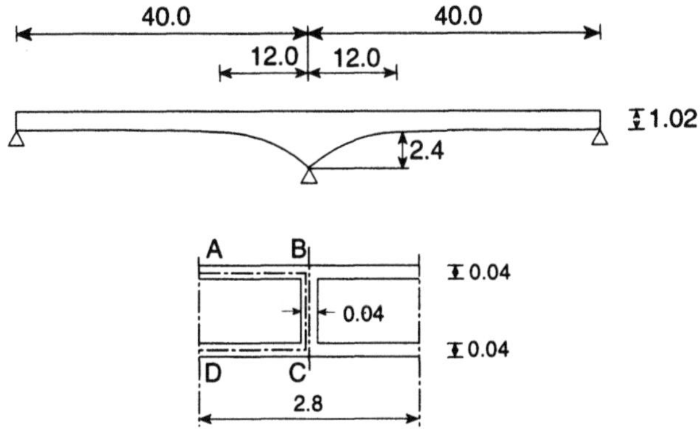

Figure 15.9 *Two-span haunched box girder bridge (dimensions in meters).*

the thickness of the flanges and the webs is 4.0 cm; the haunched portion covers a distance of 12 m. In this portion, the generating line of the bottom flange is a parabola. The depth of the box girder (excluding the haunch) is 1.04 m. The load is self-weight, corresponding to a specific weight of material of 76.4 KN m^{-3}. The material properties are $E = 200000$ MPa, $\nu = 0.3$.

Due to the symmetry, only half a cell and one span needed to be analyzed.

For comparison, this structure was first analyzed by the finite element package ADINA. Each flange was divided into one row of 16-node shell elements; the web was divided into two rows, with each row including eight elements (Fig. 15.10). Using an AMDAHL-5860 mainframe, a CPU time of 33.6 s was required for this analysis.

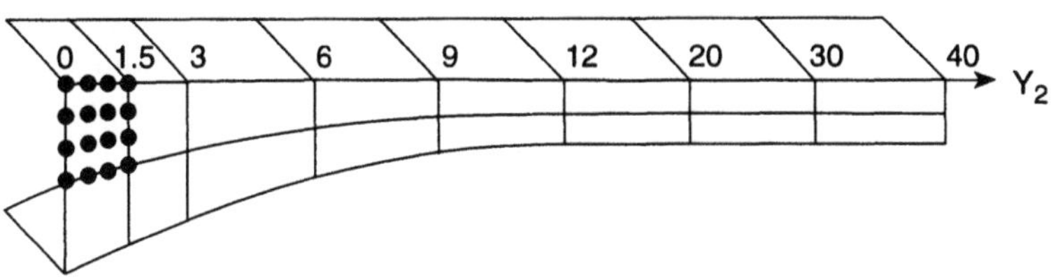

Figure 15.10 *Mesh of shell elements.*

Next, the finite strip method was employed to analyze the same structure. The top flange, web and bottom flange were modeled by one strip each; 10 symmetrical terms were taken in the longitudinal displacement function series. The CPU time needed was 19.3 s.

Finally, the spline finite strip method was used. The flanges and web were modeled by one strip each with 20 longitudinal sections each. The CPU time needed was reduced to 11.7 s.

Table 15.4 *Longitudinal stresses in a two-span box girder bridge*

y_2 (m)	Method	σ_y (MPa) along nodal line B	C
0	ADINA	13.37	−11.75
	FSM	11.72	−11.16
	SFSM	12.46	−11.95
3	ADINA	13.70	−15.11
	FSM	13.58	−16.69
	SFSM	13.88	−14.44
6	ADINA	13.92	−16.85
	FSM	13.95	−18.04
	SFSM	14.03	−16.01
12	ADINA	3.53	−3.69
	FSM	4.36	−4.46
	SFSM	5.02	−4.00
20	ADINA	−10.71	11.19
	FSM	−10.23	10.38
	SFSM	−10.65	10.65
30	ADINA	−14.26	14.26
	FSM	−13.80	13.72
	SFSM	−13.57	13.57

The resulting longitudinal stresses of the middle surface at a number of cross sections are listed in Table 15.4, in which FSM denotes the finite strip method and SFSM represents the spline finite strip method.

In the analysis of this structure, the time spent on constructing stiffness matrices was the largest component of total CPU time consumed. The spline function is relatively simpler than the beam eigenfunction, and it is this feature that results in its higher efficiency.

15.4.4 Five-span composite box girder bridge

The bridge is a two-lane steel–concrete composite box girder structure and is continuous over five spans. There are two steel boxes with a cast-in-place reinforced concrete deck. Figure 15.11 shows the elevation of the bridge, the depth of the webs and its typical cross sections. The loads are two, five-axle trucks. Their wheel weights and positions are shown in Figure 15.12. For steel, the material properties are $E = 210$ GPa, $\nu = 0.3$ and, for reinforced concrete, $E = 25$ GPa, $\nu = 0.2$.

The structure is simulated by 22 finite strips with 21 nodal lines (Fig. 15.13). Ten terms of the eigenfunction series are used in the analysis.

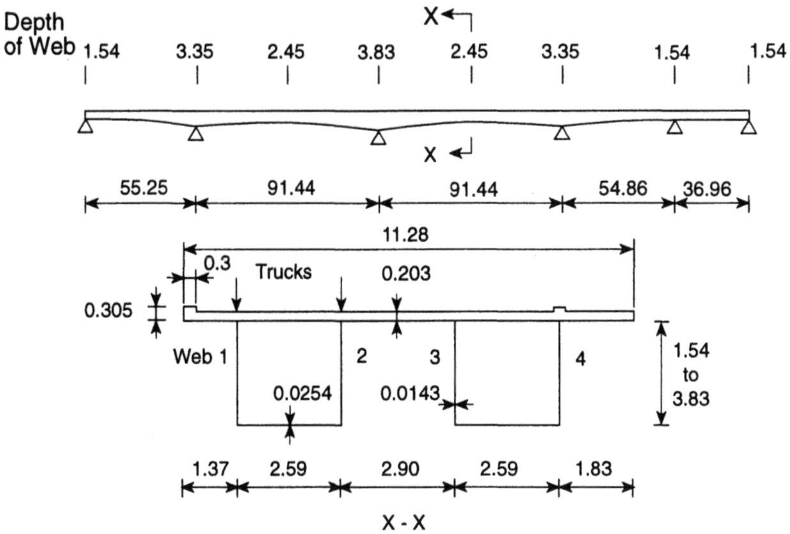

Figure 15.11 *Five-span box girder bridge (dimensions in meters).*

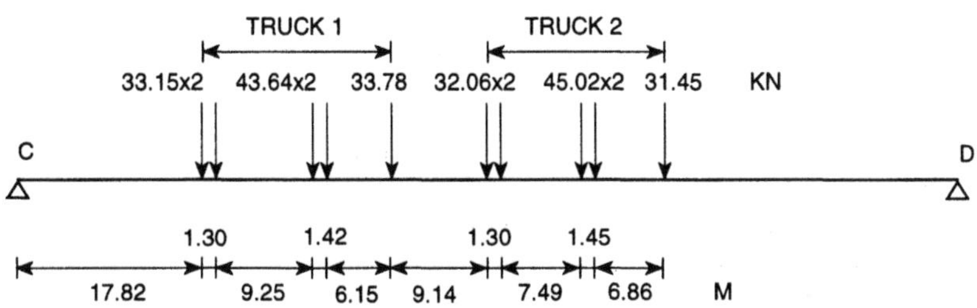

Figure 15.12 *Wheel weight of two trucks.*

The distribution of the longitudinal stresses in web 1, web 2 and the bottom flange in-between at cross-section x–x is depicted in Fig. 15.14. The experimental values (M. S. Cheung and Chan, 1978A) for the same stress are also shown in this figure for comparison purpose. The maximum longitudinal stress of the web is at the bottom corner of web; its numerical

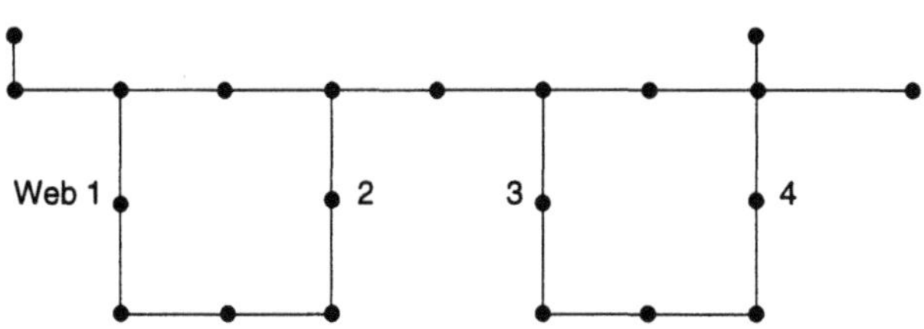

Figure 15.13 *Division of strips.*

value is 26.25 MPa, while the experimental value is 24.47 MPa. The maximum longitudinal stress of the flange appears in the bottom flange adjacent to web; its numerical value is 25.67 MPa while the experimental value is 29.64 MPa.

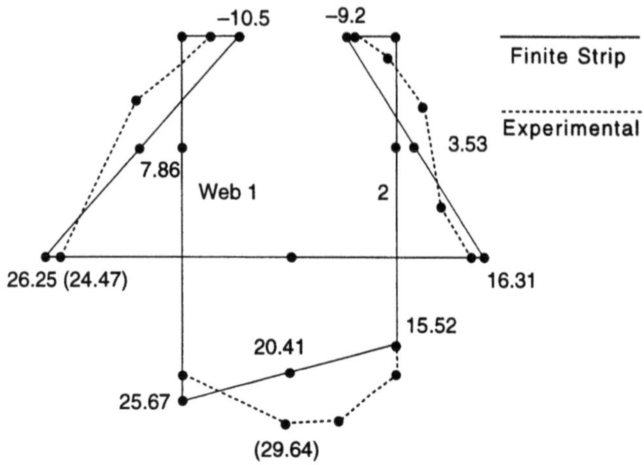

Figure 15.14 *Longitudinal stress at section X–X (MPa).*

It may be concluded that the numerical results are in good agreement with the experimental results for all high-stress regions. Stress comparisons for other regions seem to be less favorable. However, this may be attributed to the accuracy of the experimental measurements themselves. Several irregularities of the experimental results, as shown in Fig. 15.14, clearly indicate this possibility.

CHAPTER 16

Cable-stayed bridges

16.1 Introduction

The erection of cable-stayed bridges for medium to long spans has been increasing worldwide since 1955. This trend is due to several outstanding advantages that these structures possess, such as economy, stiffness, superior aesthetic qualities, ease of erection without falsework, and freedom in selecting the structural arrangement. The principal components of a cable-stayed bridge at superstructure level are girders, pylons and cables, (Fig. 16.1). The girder is subjected to bending and axial forces. In the pylons, the axial forces are significant but bending also contributes to the deformations. The cables can carry only axial loads.

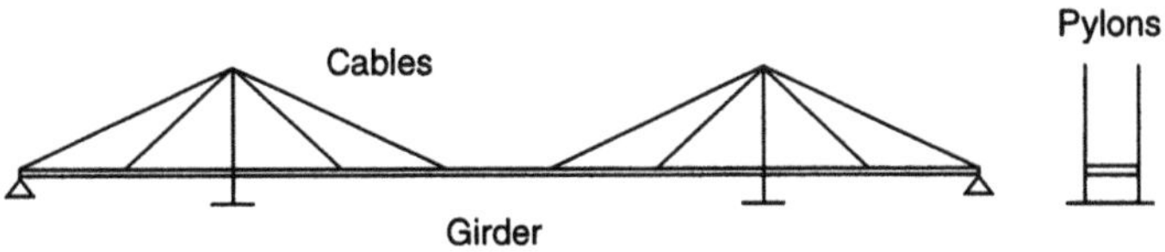

Figure 16.1 *Cable-stayed bridge.*

A major departure in the behavior of cable-stayed bridges from conventional structures is the effect of deformation on the three principal components. Especially in cables, even though only axial forces are present, the magnitude of the forces is highly dependent on the end displacements and their weight effect, namely, sag. In the analysis of girders and pylons, due to the presence of both axial and bending loads, the beam-column effect may become significant for some slender structures. Therefore, nonlinear analysis of this type of bridge often becomes necessary.

A number of techniques have been developed for the analysis of cable-stayed bridges (e.g. Tang, 1971; Podolny *et al.*, 1976; Seif and Dilger, 1990). However, in most publications the girder of a cable-stayed bridge is treated as a beam. When greater accuracy is required, the girder should be considered as a plate or a three-dimensional structure, and analyzed by the finite element method (Kajita and Cheung, 1973). The nonlinear behavior of the cables necessitates the use of an iterative method for the analysis, and if the entire girder has to be analyzed in every iteration, the computer time requirement is likely to become unacceptably large.

In most cable-stayed bridges, the girders are continuous over two or three spans. For the three-dimensional analysis of this girder, the finite strip method using the eigenfunctions of a continuous beam which has span lengths equal to those of the actual bridge (Section 4.8), is preferable to finite element analysis, because of its higher efficiency.

Because the second-order derivatives of every beam function have a smooth variation within each span, it is difficult to simulate the peak value of longitudinal bending moment at any cable attachment point. In fact, the deflection of a girder at any cable attachment point is much smaller than the deflection of the same girder without cables under the same load. Hence, if an additional support is added under every cable attachment point, the bending moment of the girder will not change very much, which means that the bending moment of a cable-stayed bridge is very similar to that of the continuous beam with additional supports. The bending moment of such a continuous beam is readily represented by its eigenfunctions. Therefore, during the course of analysis, the eigenfunctions of such a continuous beam are added into the series of longitudinal displacement functions. By following this procedure, the accuracy of the bending moments of a girder at the cable attachment points is improved significantly.

Another measure with which to improve the efficiency of the iterative process is the substructure technique. This transforms the whole girder by means of the flexibility approach into a substructure with only very few nodes at the points of cable attachments. Hence, only the substructure with very few displacement parameters is involved in every iteration and the solution time is reduced significantly.

The theory of the catenary is utilized in calculating the cable tension from the positions of the cable ends. In this way, nonlinearity due to sag and the angle change of cables is taken into consideration.

The initial-stiffness method is taken into the nonlinear iterations. Hence, the stiffness matrix of the bridge is formed and inverted only once. Because of the high tensions of the cables, the nonlinearity is not very marked, and the convergence can be achieved within very few iterations.

In accordance with the ASCE Task Committee recommendation (Task Committee, 1977), nonlinearities of the girder and pylons generally are not taken into account. The effect of axial displacements of girder and pylons on cable tensions is also neglected. It is further assumed that all cables are fixed to the pylons, and that the girder is simply supported on piers as it passes through the pylon legs, which are fixed to the piers.

16.2 Girder substructure

In the girder substructure, each point of cable attachment is taken as a node, and the deflection of this point is the only nodal displacement parameter. The girder, released from all stays, is first analyzed by the above-mentioned finite strip method under external loads. The deflections at all the nodes due to the external loads can be easily found and expressed in terms of a column vector $\{\delta\}$. Then a unit vertical force is applied at node j alone, with j being 1–n, where n is the number of cable attachment points, and the deflection f_{ij} of every node $i (i = 1\text{–}n)$ is obtained. This procedure gives the flexibility matrix $[F_g]$ of the substructure:

$$[F_g] = [F_{ij}] \tag{16.1}$$

The inverse matrix of $[F_g]$ is the stiffness matrix $[K_g]$ of the substructure,

$$[K_g] = [F_g]^{-1} \tag{16.2}$$

In order to avoid possible singularities in $[F_g]$ and consequent failure to invert it, the number of series terms used in the analysis of the girder must not be less than the number of nodes on any nodal line.

Multiplying the stiffness matrix $[K_g]$ by the displacement vector $\{\delta\}$ due to external loads gives the equivalent nodal forces $\{P\}$ as follows:

$$\{P\} = [K_g]\{\delta\} \tag{16.3}$$

16.3 Formulas for cables

In this section, the initial-stiffness matrix of the cable is discussed together with basic catenary theory. This is introduced since the shape of a cable under self-weight and end tensions is similar to a catenary. Then, the solution procedure for the cable tension is briefly presented.

16.3.1 Initial-stiffness matrix

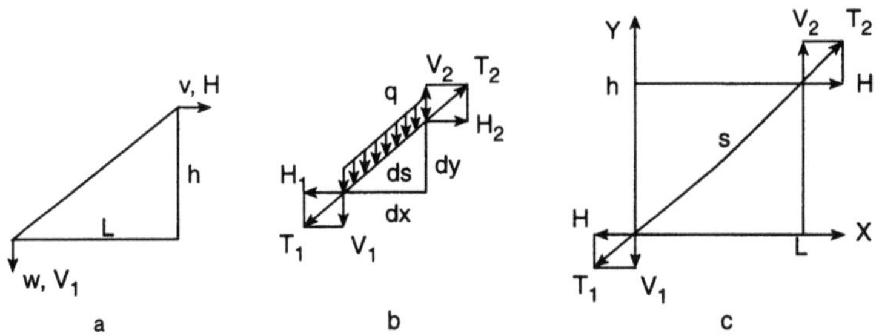

Figure 16.2 *Forces and tensions in a cable.*

The displacement vector of a cable is $[w, v]^T$ (Fig. 16.2(a)), and the linear elastic stiffness matrix is represented by

$$[K_c] = \frac{EA}{S_0}\begin{bmatrix} \sin^2\alpha & \cos\alpha\sin\alpha \\ \cos\alpha\sin\alpha & \cos^2\alpha \end{bmatrix} \tag{16.4}$$

in which E is the modulus of elasticity, A is the area of cross-section, S_0 is the free length of the cable, and α is the angle of inclination of the cable.

16.3.2 Basic catenary theory

If q is the self-weight per unit lengths, and T is the cable tension with horizontal component H and vertical component V, the equilibrium conditions of an infinitesimal cable element (Fig. 16.2(b)) are

$$\sum X = 0, \qquad H_2 - H_1 = 0 \tag{16.5}$$

$$\sum Y = 0, \qquad V_2 - V_1 - qds = 0 \tag{16.6}$$

From Equation 16.5 it can be concluded that $H_2 = H_1 = H$, which means that H is constant along the entire cable.

Taking into consideration that

$$V_2 - V_1 = H \tan\alpha_2 - H \tan\alpha_1 = H(\frac{dy}{dx})_2 - H(\frac{dy}{dx})_1 = H\frac{d^2y}{dx^2}dx$$

and

$$ds = \sqrt{dx^2 + dy^2} = \sqrt{1 + (\frac{dy}{dx})^2}dx$$

Equation 16.6 can be rewritten as

$$H\frac{d^2y}{dx^2} = q\sqrt{1 + (\frac{dy}{dx})^2} \tag{16.7}$$

It should be noted that Equation 16.7 is the differential equation of a catenary and the general solution is given by

$$y = a\cosh\frac{x - x_0}{a} + c \tag{16.8}$$

in which x_0 and c are constants that are determined from the end conditions, and

$$a = H/q \tag{16.9}$$

For a cable with horizontal projection l and vertical projection h (Fig. 16.2(c)), the end conditions can be assumed to be

$$y(0) = 0 \tag{16.10}$$

$$y(l) = h \tag{16.11}$$

Substituting Equation 16.8 into 16.10 gives

$$a\cosh\frac{-x_0}{a} + c = 0$$

from which the following expressions can be obtained:

$$c = -a\cosh\frac{x_0}{a}$$

and

$$y = a(\cosh \frac{x - x_0}{a} - \cosh \frac{x_0}{a}) \tag{16.12}$$

Substitution of Equation 16.12 into 16.11 yields

$$a(\cosh \frac{l - x_0}{a} - \cosh \frac{x_0}{a}) = h$$

which can then be rewritten in the form

$$2a \sinh \frac{l}{2a} \sinh \frac{l - 2x_0}{2a} = h$$

whence

$$\sinh \frac{l - 2x_0}{2a} = \frac{h}{2a \sinh \frac{l}{2a}} \tag{16.13}$$

This equation can be used to find x_0, giving

$$x_0 = \frac{l}{2} - a \sinh^{-1} \frac{h}{2a \sinh \frac{l}{2a}}$$

Because $\sinh^{-1}(t) = \ln(t + \sqrt{1 + t^2})$, the above equation can also be expressed as

$$x_0 = \frac{l}{2} - a \ln(t + \sqrt{t^2 + 1}) \tag{16.14}$$

in which

$$t = \frac{h}{2a \sinh \frac{l}{2a}} \tag{16.15}$$

Once x_0 is defined, Equation 16.12 can be utilized to find any values needed in the analysis, as summarized below:

1. The cable tension T and its vertical component V are computed from

$$V = H\frac{dy}{dx} = H \sinh \frac{x - x_0}{a} \tag{16.16}$$

$$T = \sqrt{H^2 + V^2} = H \cosh \frac{x - x_0}{a} \tag{16.17}$$

2. The curved length of the cable is given by

$$\begin{aligned} S &= \int_0^l \sqrt{1 + (\frac{dy}{dx})^2} dx \\ &= \int_0^l \sqrt{1 + \sinh^2 \frac{x - x_0}{a}}\, dx \\ &= \int_0^l \cosh \frac{x - x_0}{a}\, dx \\ &= a \sinh \frac{l - x_0}{a} - a \sin \frac{-x_0}{a} \end{aligned}$$

$$= 2a\sinh\frac{l}{2a}\cosh\frac{l-2x_0}{2a}$$

$$= 2a\sinh\frac{l}{2a}\sqrt{\sinh^2\frac{l-2x_0}{2a}+1}$$

Substituting Equation 16.13 into the above eventually gives the curved length S of the cable as

$$S = \sqrt{h^2 + 4a^2\sinh^2\frac{l}{2a}} \tag{16.18}$$

3. Elongation of the cable due to tension is calculated as follows:

$$\begin{aligned}
S - S_0 &= \int_0^l \frac{Tds}{EA} \\
&= \frac{1}{EA}\int_0^l T\sqrt{1+(\frac{dy}{dx})^2}\,dx \\
&= \frac{1}{EA}\int_0^l H\cosh\frac{x-x_0}{a}\sqrt{1+\sinh^2\frac{x-x_0}{a}}dx \\
&= \frac{H}{EA}\int_0^l \cosh^2\frac{x-x_0}{a}\,dx \\
&= \frac{H}{2EA}(l+\frac{a}{2}(\sinh\frac{2(l-x_0)}{a}+\sinh\frac{2x_0}{a})) \\
&= \frac{H}{2EA}(l+a\sinh\frac{l}{a}\cosh\frac{l-2x_0}{a}) \\
&= \frac{H}{2EA}(1+a\sinh\frac{l}{a}(2\sinh^2\frac{l-2x_0}{2a}+1))
\end{aligned}$$

Using Equation 16.13 again and taking into account

$$\sinh\frac{l}{a} = 2\sinh\frac{l}{2a}\cosh\frac{l}{2a}$$

and

$$\coth\frac{l}{2a} = \cosh\frac{l}{2a}/\sinh\frac{l}{2a},$$

the above equation can be rewritten in the form

$$S - S_0 = \frac{H}{2EA}(l+\frac{h^2}{a}\coth\frac{l}{2a}+a\sinh\frac{l}{a}) \tag{16.19}$$

16.3.3 Solution procedure for cable tension

Substituting Equation 16.18 into Equation 16.19 yields

$$\sqrt{h^2+4(\frac{H}{q})^2\sinh^2\frac{lq}{2H}} - S_0 = \frac{H}{2EA}(l+\frac{qh^2}{H}\coth\frac{ql}{2H}+\frac{H}{q}\sinh\frac{ql}{H}) \tag{16.20}$$

From Equations 16.14–16.16, the vertical force V_1 applied by the cable on the girder and the horizontal force H applied by the cable on the pylon can be related to each other as follows:

$$t = \frac{qh}{2H} / \sinh \frac{ql}{2H}$$
$$x_0 = \frac{l}{2} - \frac{H}{q} \ln(t + \sqrt{t^2 + 1})$$
$$V_1 = H \sinh \frac{qx_0}{H} \tag{16.21}$$

Initially, q, l and h are known; thence, from the given initial value of V_1, the horizontal force H can be computed by solving Equation 16.21 numerically, using the first-order Reguli–Falsi iteration. In this iteration, for any trial value of H, t can be obtained by means of the first expression of Equation 16.21. Then, substituting t into the second expression gives x_0. Thereafter, V_1 can be computed using the third expression. After H is obtained, substituting its value into Equation 16.20 then gives the free length S_0 of the cable.

At any deformed position, after w and v are found, it is easy to obtain h and l. Using a numerical approach again, H can be obtained from Equation 16.20. Then V_1 is readily evaluated by substituting H into Equation 16.21. Both V_1 and H reflect the nonlinear resistance of the cable to deformation.

16.4 Stiffness matrix for pylons

In the present analysis, the pylon is treated as a cantilever beam, and only its deflection, not its rotation, is taken into account. For this reason, it is convenient to form the flexibility matrix first, and then let the computer invert it to obtain the stiffness matrix.

Figure 16.3 *Forces in a pylon.*

The flexibility coefficient f_{ij} of the pylon (Fig. 16.3) is defined by

$$f_{ij} = \frac{1}{2EI_p} h_j^2 (h_i - \frac{1}{3} h_j) \tag{16.22}$$

in which EI_p is the bending stiffness of the pylon and $h_i \geq h_j$. Hence, by assembling the f_{ij}, the flexibility matrix is obtained as

$$[F_p] = [f_{ij}] \tag{16.23}$$

The inverse matrix of $[F_p]$ is the stiffness matrix $[K_p]$ of pylon,

$$[K_p] = [F_p]^{-1} \tag{16.24}$$

16.5 Initial-stiffness iteration

Assembling the linear stiffness matrices of the girder substructure, cables and pylons together produces the initial linear stiffness matrix $[K_0]$ of the whole structure. $[K_0]$ is formed and inverted only once and then is used in every iteration for the nonlinear solution without any change. Because of the relatively high cable tension in ordinary cable-stayed bridges, the nonlinearity is usually not very significant, and hence convergence can be achieved within very few iterations.

During the course of the analysis, the structure is divided into a linear part, the girder and pylons, and a nonlinear part, the cables. After kth iteration, the resistance of the linear part to displacement $\{\delta\}^k$ is given by

$$\{R_l\}^k = [K_l]\{\delta\}^k \tag{16.25}$$

where $[K_l]$ is the linear stiffness matrix of this part. The resistance $\{R_{nl}\}$ of the nonlinear part to the displacement $\{\delta\}^k$ is the cable tension and can be found from $\{\delta\}^k$ by following the procedure described in Section 16.3. The resistance of the entire structure to this deformation is the combination of the resistances of both parts:

$$\{R\}^k = \{R_l\}^k + \{R_{nl}\}^k = [K_l]\{\delta\}^k + \{R_{nl}\}^k \tag{16.26}$$

If the external load $\{P\}$ is not equal to the resistance $\{R\}^k$, the structure will deform further under the action of the so-called 'unbalanced force' $\{F\}^k$ to the next position $\{\delta\}^{k+1}$, with the initial linear stiffness matrix $[K_0]$ assumed to be unchanged (Fig. 16.4). Hence,

$$[K_0](\{\delta\}^{k+1} - \{\delta\}^k) = \{F\}^k = \{P\} - \{R\}^k \tag{16.27}$$

If the second term in the bracket on the left side of above equation is shifted to the right side, this equation can be rewritten as

$$\begin{aligned} [K_0]\{\delta\}^{k+1} &= [K_0]\{\delta\}^k + \{P\} - \{R\}^k \\ &= ([K_l] + [K_{nl}])\{\delta\}^k + \{P\} - ([K_l]\{\delta\}^k + \{R_{nl}\}^k) \\ &= \{P\} + [K_{nl}]\{\delta\}^k - \{R_{nl}\}^k \end{aligned} \tag{16.28}$$

where $[K_{nl}]$ is the initial linear stiffness matrix of the nonlinear part (cables), and $[K_0] = [K_l] + [K_{nl}]$. Thus, Equation 16.28 can be used to find an improved approximation $\{\delta\}^{k+1}$ from $\{\delta\}^k$. This procedure is repeated until convergence is obtained. Convergence may be assumed to occur if

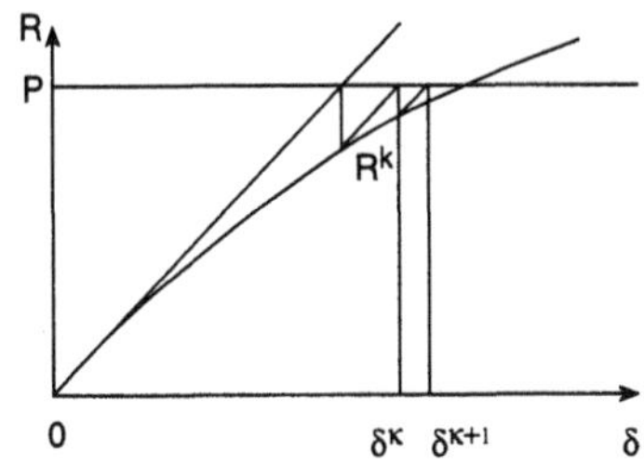

Figure 16.4 *Initial-stiffness method.*

the difference in the value of every displacement parameter of every cable, determined from two successive iterations, does not exceed a chosen small percentage (say 0.1%).

16.6 Numerical examples

16.6.1 Single-plane cable-stayed bridge

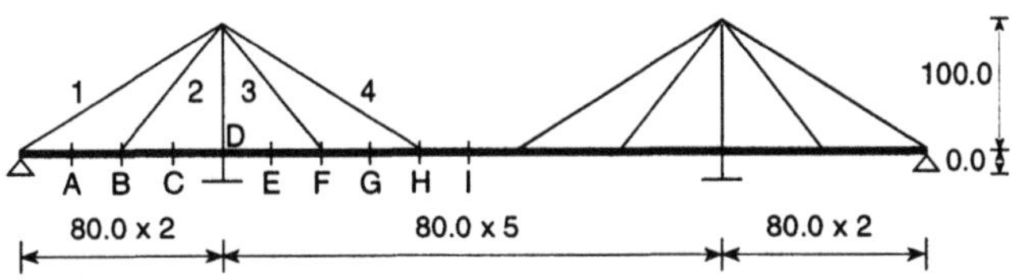

Figure 16.5 *Single-plane cable-stayed bridge (dimensions in meters).*

A single-plane cable-stayed bridge over three spans is shown in Fig. 16.5. In order to compare the results of the finite strip analysis with beam theory more conveniently, it is assumed that the girder has a narrow width in comparison with its height. For the girder, the Young's modulus $E = 2.0 \times 10^5$ MPa, the Poisson's ratio $\nu = 0.3$, and bending moment of inertia $I_g = 1.0$ m^4. For the pylons, $EI_p = 1.0 \times 10^6$m^2 MN . For the cables, $E_c = 1.52 \times 10^5$ MPa, the specific weight $\gamma_c = 0.0764$ MN m^{-3}. The areas of the cross-sections are $A_1 = A_4 = 0.10$m^2, $A_2 = A_3 = 0.05$ m^2.

In load case 1, the girder is subjected to dead load $q_d = 0.1$ MN m^{-1} over all spans, and the deflections at all the cable attachment points are set to zero by adjusting the cable tensions. In this situation, the vertical forces applied to the girder by the cables should be in equilibrium with the equivalent nodal forces of external loads and can be calculated from Equation 16.3. The results are listed in Table 16.1 for finite strip analyses using different numbers of series terms. The longitudinal bending moments in MN m at a number of cross-sections are given in Table 16.2. In both

Table 16.1 *Deflections (in meters) of the girder at cable attachment points and vertical forces (in MN) of cables*

	$n_1 + n_2$	w_B	w_F	w_H	V_1	V_2	V_3	V_4
Load case 1	6+0	0.0	0.0	0.0	7.377	9.375	8.325	7.901
	15+0	0.0	0.0	0.0	7.480	9.080	8.074	7.982
	25+0	0.0	0.0	0.0	7.480	9.060	8.060	7.986
	6+9	0.0	0.0	0.0	7.480	9.041	8.043	7.990
	16+9	0.0	0.0	0.0	7.480	9.046	8.049	7.990
Theory		0.0	0.0	0.0	–	9.072	8.075	7.985
Load case 2	6+0	−0.1499	0.2761	0.5117	5.473	0.068	4.415	3.802
	15+0	−0.1498	0.2736	0.5110	5.481	0.049	4.341	3.836
	23+0	−0.1498	0.2735	0.5111	5.481	0.049	4.338	3.837
	6+9	−0.1500	0.2733	0.5112	5.482	0.045	4.332	3.838
	15+8	−0.1500	0.2733	0.5112	5.482	0.045	4.332	3.838
Theory		−0.1498	0.2735	0.5111	–	0.056	4.343	3.837

tables, n_1 is the number of eigenfunctions associated with the continuous beam over three spans and n_2 is the number of eigenfunctions related to the continuous beam with additional supports at all of the cable attachment points; hence, the total number of eigenfunctions employed in the analysis is $n_1 + n_2$.

In load case 2, the live load $q_l = 0.05$ MN m^{-1} is imposed on the entire second span only. The vertical components of the prestressing forces of the cables are $V_{01} = 7.480$ MN , $V_{02} = 9.071$ MN, $V_{03} = 8.064$ MN and $V_{04} = 7.984$ MN, consistent with the girder having zero deflection at every cable attachment point in load case 1. The results in load case 2 are also shown in Tables 16.1 and 16.2. However, in this case, every quantity listed in the tables is its increment due to live load alone.

From the results, the use of the eigenfunctions of a continuous beam with three spans only has led to a satisfactory value for the deflection of the girder at cable attachment points; their errors are less than 1.0% with only six series terms used. In this case, the results of the vertical components of the cable tensions are also acceptable, with errors of less than 3.5%. However, the accuracy of the bending moments is disappointing, the errors of maximum bending moment are still more than 22% in both load cases, despite using 45 series terms. From Table 16.2 it can be seen that the improvement brought about by the additional eigenfunctions associated with the continuous beam with extra fictitious supports at all the cable attachment points is significant. The error of maximum bending moment is less than 5% when using $n_1 + n_2 = 16 + 9 = 25$ series terms for

Table 16.2 *Bending moments (in MN m) in the girder*

	$n_1 + n_2$	M_A	M_B	M_C	M_D	M_E	M_F	M_G	M_H	M_I
Load case 1	6+0	0.018	0.021	0.011	0.001	0.005	0.016	0.021	0.020	0.019
	15+0	45.10	−34.95	20.52	−47.60	27.19	−23.68	20.93	−20.73	20.87
	25+0	51.95	−43.17	25.30	−44.20	30.76	−32.99	33.39	−32.23	31.49
	45+0	44.70	−53.05	18.82	−46.54	26.10	−41.53	24.07	−40.20	24.34
	6+9	48.27	−61.35	22.17	−44.97	29.07	−48.90	28.23	−48.24	28.69
	16+9	46.49	−70.06	21.43	−46.30	28.58	−55.55	26.71	−50.91	28.68
	36+9	46.08	−67.74	20.74	−48.45	27.36	−54.55	26.30	−53.46	26.71
Theory		46.19	−67.62	21.43	−49.51	28.08	−54.34	26.26	−53.13	26.87
Load case 2	6+0	−7.986	−11.30	−8.720	−2.618	−1.290	0.8137	4.299	8.052	9.711
	15+0	−4.858	−10.18	−12.66	−15.96	16.39	−13.42	13.49	−1.265	19.89
	23+0	−4.744	−10.01	−13.32	−14.43	19.07	−18.53	19.69	−7.063	25.37
	45+0	−4.930	−10.04	−13.12	−15.04	16.09	−23.10	15.17	−10.90	21.86
	6+9	−4.935	−9.726	−13.22	−14.97	17.76	−27.12	17.29	−14.59	23.78
	15+8	−5.055	−10.57	−13.28	−14.23	17.64	−29.20	17.02	−15.80	23.60
	36+9	−4.970	−10.10	−13.06	−15.51	16.79	−30.20	16.41	−17.24	22.83
Theory		−5.054	−10.11	−12.93	−15.74	17.07	−30.11	16.41	−17.07	22.93

load case 1, and when using $n_1 + n_2 = 15 + 8 = 23$ series terms for load case 2. As a comparison, the error of maximum bending moment is 36% with $n_1 + n_2 = 25 + 0 = 25$ terms for load case 1, and 38% with $n_1 + n_2 = 23 + 0 = 23$ terms for load case 2, when the additional eigenfunctions are not included.

In both tables, the theoretical results are obtained by means of beam theory, with the beam deflections at the corresponding cable attachment points under the external applied loads being assumed to be the same as the actual girder.

16.6.2 Double-plane cable-stayed bridge

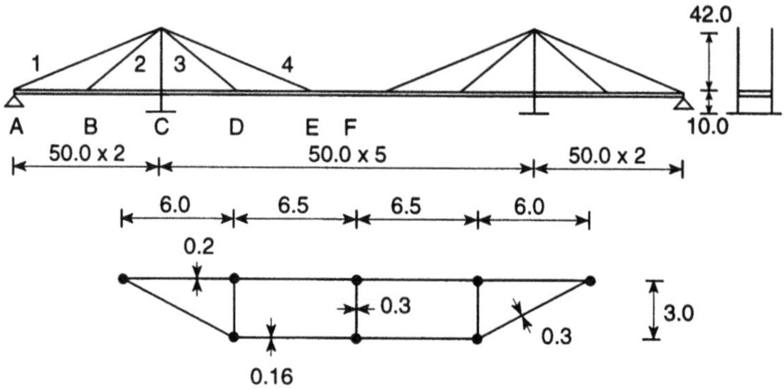

Figure 16.6 *Double-plane cable-stayed bridge (dimensions in meters).*

A double-plane cable-stayed concrete box girder bridge is shown in Fig. 16.6. The loads are the self-weight of the girder and a live load, distributed uniformly on the whole deck between towers, of intensity 4.488 KN m^{-2}. For the girder $E_g = 25 \times 10^6$ KN m^{-2}, Poisson's ratio $\nu = 0.2$ are the specific weight $\gamma_g = 24$ KN m^{-3}; for the pylons $EI_p = 2.5 \times 10^8$ m^2 KN; for the cables $E_c = 1.7 \times 10^8$ KN m^{-2}, $\gamma_c = 78$ KN m^{-3}, the areas are $A_1 = A_4 = 0.04$ m^2 and $A_2 = A_3 = 0.024$ m^2; the vertical components of cable prestressing forces under the dead load are $V_{01} = 7849.6$ KN, $V_{02} = 9066.5$ KN, $V_{03} = 8121.2$ KN and $V_{04} = 8322.7$ KN, which make the deflections of the girder at the cable attachment points under the dead load equal to zero.

Because of the symmetry, only half of the bridge, which is divided into six strips by five nodal lines, needs to be analyzed. In the finite strip analysis of the girder, the number of terms used in the eigenfunction series is 21 for n_1 and nine for n_2. Convergence is obtained within eight iterations.

The deflection profile of the girder and pylon in the plane of cables is depicted in Fig. 16.7. The vertical forces V, applied to the girder by the

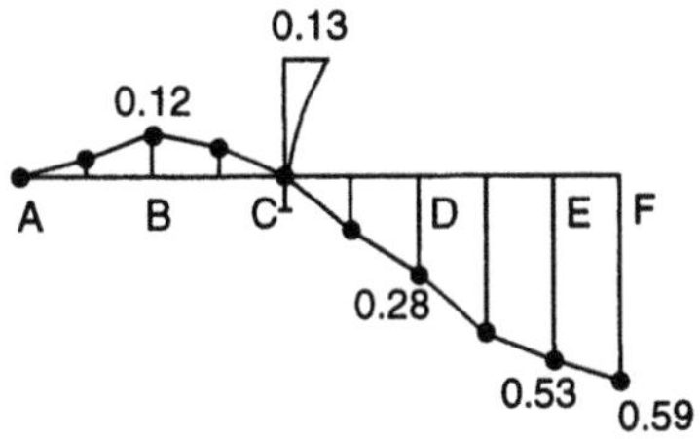

Figure 16.7 *Deflection of girder and pylon (dimensions in meters).*

Table 16.3 *Forces of cables on girder and pylon (MN)*

Cable	V	H
1	10.730	25.985
2	9.831	11.841
3	11.315	13.419
4	10.521	25.103

cables, and the horizontal components H of cable tension are given in Table 16.3. The maximum negative bending moment of the girder is at cross-section B, and the maximum positive bending moment is at the central symmetrical cross-section F. The distribution of the longitudinal bending stresses on these two cross-sections are shown in Figs 16.8 and 16.9.

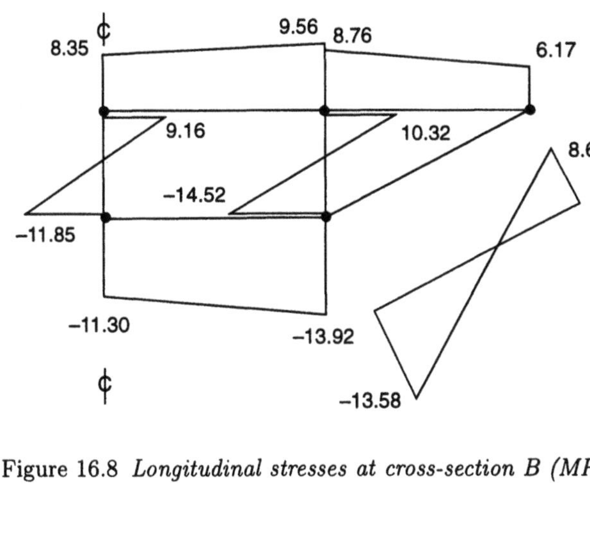

Figure 16.8 *Longitudinal stresses at cross-section B (MPa).*

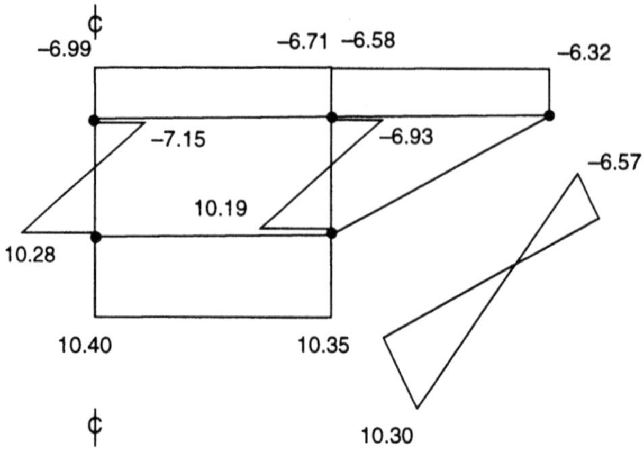

Figure 16.9 *Longitudinal stresses at cross-section F (MPa).*

CHAPTER 17

Finite strip modeling of bridges

17.1 Introduction

In the previous chapters, the fundamental theory of the finite strip method has been discussed in detail. A variety of approaches and strips has been introduced with their applications in analyzing different types of highway bridge. With the above knowledge, combined with skills in bridge design, it is not difficult for engineers to apply the available finite strip computer software to their projects. However, beginners may be troubled by some problems, such as how to select a proper approach for a given bridge, and how many finite strips and series terms should be taken. This chapters aims to guide those who are new to finite strip analysis, with advice based on the investigations and numerical examples previously described.

17.2 Selecting approach and strip

A variety of approaches and strips has been introduced in the previous chapters. For the analysis of a particular bridge, often more than one approach is applicable. All will lead to the same converged results, but the cost of the analysis can vary. Selection of the best approach and the most suitable type of strip is based mainly on the bridge geometry, namely the shape and support conditions. The load conditions should also be considered.

For a right, skew or circularly curved slab, slab-on-girder or box girder bridge with a constant cross-section and simply supported ends, the semi-analytical finite strip method introduced in Chapters 4, 5 and 13 is the most efficient method. For slab bridges, plate strips (Sections 4.3 and 5.2 etc.) are used. For other forms of bridge, flat shell strips (Sections 4.5 and 5.5 etc.) are employed. Because this method uses sine and cosine functions as longitudinal shape functions, different series terms are uncoupled. Therefore, the size and half-bandwidth of the stiffness matrix are both reduced significantly. Moreover, since all the entries of the stiffness matrix are relatively simple and can be calculated directly without using numerical integration, the time consumption in forming this matrix is also reduced to minimum. Although in some load cases, (see Section 14.5), tens or even hundreds series terms may be required to reach the desired accuracy, this method still yields good efficiencies.

If some intermediate cross-sections are supported or stiffened, as described in Sections 4.7, 12.3 and 14.3, the above method combined with the flexibility approach can be applied, irrespective of whether these supports or stiffeners are rigid or elastic, or whether an initial settlement may occur at an intermediate support. Using this approach, the cost of the analysis may still be lower than other approaches because the different series terms remain uncoupled, even though the formulation and programing are more complicated and many series terms may be required to evaluate the maximum negative bending moments over the intermediate supports.

If either end of the above-mentioned bridge is clamped or free, since the different series terms are coupled together, the semianalytical finite strip method becomes uneconomical, especially when point loads are involved in the analysis. For these types of bridge, the spline finite strip method (Chapter 6) is the best choice. If the bridge has intermediate supports and stiffeners, the B-spline compound strip method (Section 7.3) should be used.

For an arbitrarily curved slab bridge, the spline finite strip method introduced in Section 6.4 is the only choice.

For a haunched continuous bridge, the spline finite strip method (Section 15.3) is more efficient than the semianalytical finite strip method (Section 15.2).

For a cable-stayed bridge, the approach presented in Chapter 16 can be used.

If the walls of the bridge deck are very thick, the finite prism method should be called.

17.3 Generating a finite strip model

Since all the above-mentioned finite strip approaches use the same type of transverse shape function, the major considerations in generating a finite strip model are identical for all these approaches. However, depending on the objective of the analysis and the loading conditions, the suitable finite strip model may differ. For a beginner, it is advisable to start with a coarse mesh in order to find possible errors more easily. Then, one can refine the modeling and improve the accuracy step by step, until convergence occurs. In the limit, if the width of each strip (lower order) and the halfwave length of the highest order harmonic are both equal to the thickness of the strip, then finite strip analysis yields results that are as accurate as thin plate theory. From experiences in using the finite strip method for the overall analysis of a single-span slab-on-girder or box girder bridge using the lower order semi-analytical finite strip method, the following modeling will provide adequate accuracy for design use:

1. The flanges and webs are idealized as the orthotropic plates with equivalent elastic properties (see Section 14.2).

2. Each web is divided into one to three strips. The flange in each cell or overhang is divided into two to five strips (Fig. 17.1). If only the deflection and longitudinal stress are required, fewer strips are sufficient. If more accurate results are desired for the in-plane shear stress and for the transverse bending moment, more strips should be used.

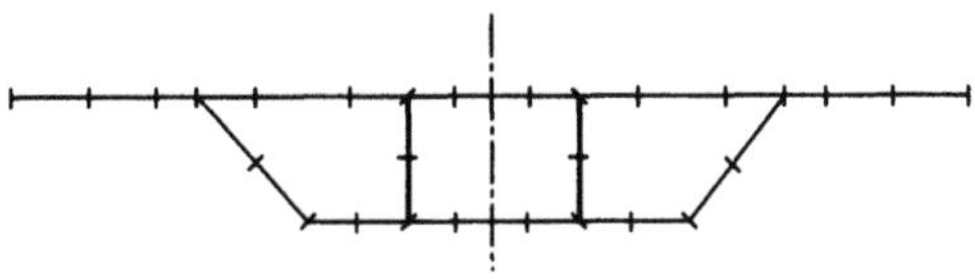

Figure 17.1 *Finite strip simulation.*

3. Each wheel load should act on a nodal line. Move a nearby nodal line to the application point of the load or add an additional nodal line underneath the load if necessary.

4. In regions with a higher stress gradient, narrower strips should be used. The width of strip should change gradually from one strip to another.

5. If the cross-section and loads are both symmetrical about the centerline, only half of the bridge needs to be analyzed (Section 4.6.3 and Figs. 4.10 and 4.11)

6. If the bridge is subjected to uniform load only, five to ten symmetrical harmonics are adequate for the analysis. If the bridge is subjected to truck loading, 20–30 series terms are necessary.

If the stresses in each longitudinal stiffener are required, the detailed finite strip simulation is then suggested (Section 14.6.2 and Fig. 14.17).

If the bridge has intermediate supports and transverse stiffeners, the models shown in Sections 12.4 and 14.6 could be a useful reference.

If some localized discontinuity or abrupt change of stress is to be simulated, such as the longitudinal stress profile over an intermediate support, the in-plane shear stress distribution at a point force, or the local bending moments due to wheel loading (Section 14.5), not less than 95 harmonics should be used in the analysis.

For other types of bridge, the reader may consult the corresponding chapters.

17.4 Numbering nodal lines and strips

In order to minimize the half-bandwidth of a stiffness matrix, in a finite strip model, the nodal lines should be numbered so as to keep the difference between the numbers of all the nodal lines within each strip at a minimum. For instance, the numbering in Fig. 17.2(a) yields better efficiency than that in Fig. 17.2(b).

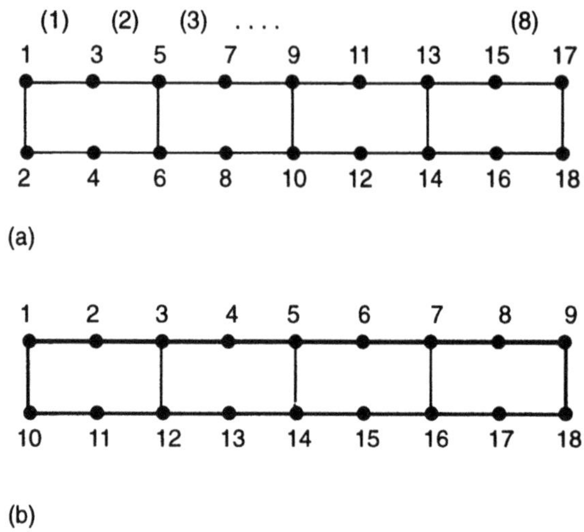

Figure 17.2 *Numbering of nodal lines and strips. (a) Efficient numbering; (b) Deficient numbering.*

The finite strips in a model can be numbered in any sequence without affecting the cost of analysis. However, some data in input file and some results in output file are arranged according to the number of each finite strip. Therefore, the strips in the same structural component, such as for the top flange, should be numbered consecutively, as shown by the numbers in parentheses in Fig. 17.2(a), in order to prepare the input data and to read the results easier.

17.5 Input data file

The following example is introduced to illustrate the basic contents in a data file.

A simply supported concrete box girder bridge with span $l = 40$ m is subjected to self-weight and eight point forces, which are symmetrical about the midspan and the centerline of cross-section and have magnitude $P = 50$ kN each, as shown in Fig. 17.3. The cross-section has a top flange of thickness 0.3 m, a bottom flange of thickness 0.5 m and two webs of thickness 0.4 m. The material properties are $E = 30000$ MPa and ν=0.15. The specific gravity is $\gamma = 24$ kN m^{-3}.

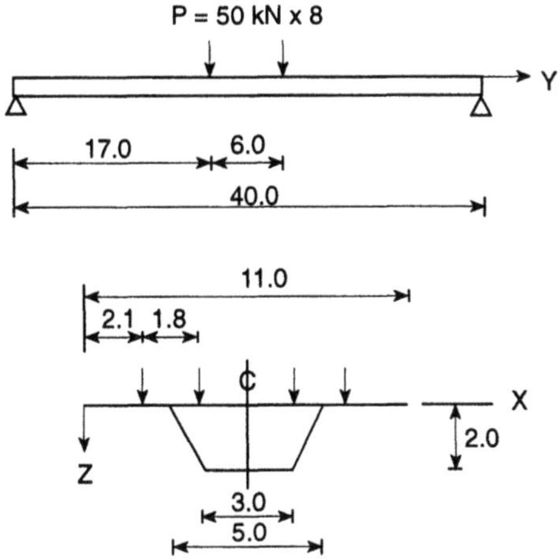

Figure 17.3 *Box girder bridge.*

Since the structure and the loads are both symmetrical about the centerline of the cross-section, only half the bridge is to be considered. The lower order flat shell strips with simply supported ends are employed. The model in Fig. 17.4 is used for an overall analysis, and 15 symmetrical harmonics are taken. In the analysis, the self-weight is applied as a uniform load to each strip. Its intensity is equal to the product of the specific gravity and the thickness of the strip.

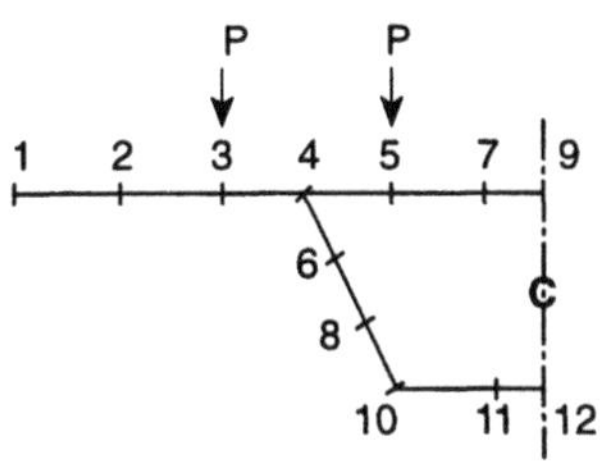

Figure 17.4 *Finite strip division.*

The input file should include the following information:

1. Analysis control parameters:
   total number of finite strips — 11
   total number of nodal lines — 12
   number of series terms — 15
   whether the terms are symmetrical to midspan — Yes

number of point forces — 4
number of restrained nodal lines — 2
number of materials — 1
number of cross-sections where the results are desired — 2
span length (m) — 40.0

2. Coordinates of nodal lines:
   nodal line 1 — x=0.0, z=0.0
   nodal line 2 — x=1.1, z=0.0
   ⋮
   nodal line 12 — x=5.5, z=2.0
3. nodal lines (I and J), material, thickness (m) and uniform load (MPa) for each strip:
   Strip 1 — I=1, J=2, material No.1, t=0.3, p_z=0.0072
   Strip 2 — I=2, J=3, material No.1, t=0.3, p_z=0.0072
   ⋮
   Strip 11 — I=11, J=12, material No.1, t=0.5, p_z=0.0120
4. Material properties (MPa for elasticity moduli):
   material No.1 — $E_x = E_y = 30000.0$, $E_{xy} = 13043.5$, $\nu_x = \nu_y = 0.15$
5. Boundary condition along nodal lines:
   Nodal line 9 — $u = 0$, $\theta = 0$
   Nodal line 12 — $u = 0$, $\theta = 0$
6. Position (m) and magnitude (MN) of point forces:
   Nodal line 3 — y=17.0, P_z=0.050
   Nodal line 3 — y=23.0, P_z=0.050
   Nodal line 5 — y=17.0, P_z=0.050
   Nodal line 5 — y=23.0, P_z=0.050
7. The y coordinates (m) of the cross-sections with output:
   $y_1 = 17.0$, $y_2 = 20.0$

It must be noted that in a data file the metric system is used and thus all the dimensions are in meters (coordinate, length and thickness) while MN is used as the unit for force. Therefore, the unit of uniform load, stress and elasticity modulus should be MPa. (If kN is used as the unit of force, then kPa must be used as the unit of uniform load etc.)

17.6 Output file

Upon completion of the analysis, the following information is given for each prescribed cross-section:

1. the displacement components of nodal lines in the global coordinate system;

2. the in-plane stresses and bending moments of each strip in its local coordinate system.

The positive directions of the displacements, stresses and moments are shown in Figs. 17.5 and 17.6, respectively.

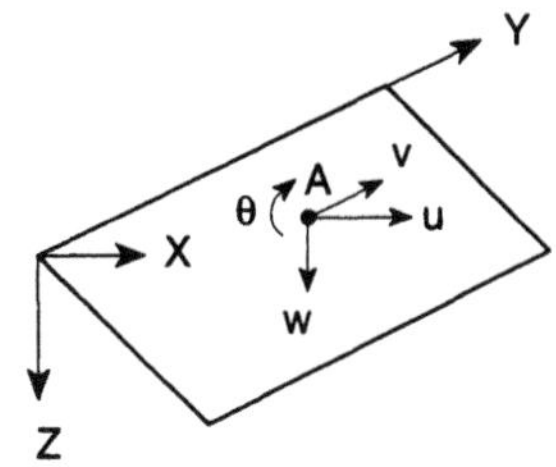

Figure 17.5 *Displacement in a global system.*

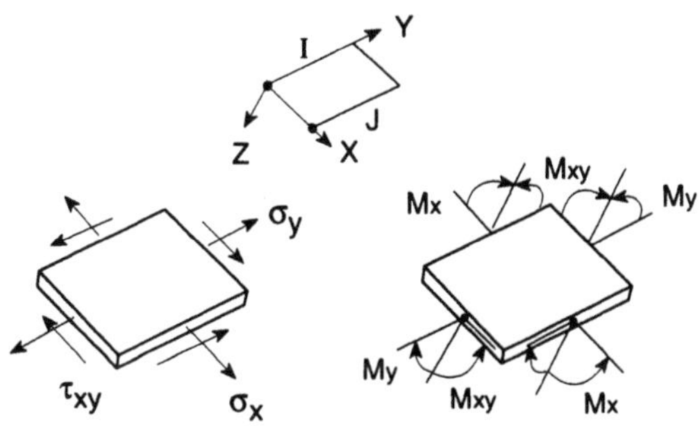

Figure 17.6 *Stresses and moments in a local system.*

For the current example, the results at midspan are depicted in Fig. 17.7. The solutions based on beam theory are also given by the numbers in parentheses. The finite strip solution for the deflection at midspan is larger than the beam theory solution because the latter does not include the effects of shear deformation. Readers may use these values to check their own results.

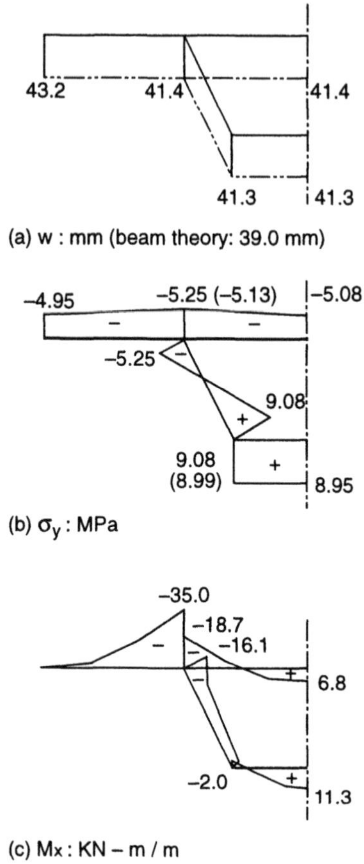

Figure 17.7 *Deflections, stresses and moments at midspan.*

References

AASHTO (1977) *Standard Specifications for Highway Bridges*, 12th edn., Association General Office, Washington, D.C.

AASHTO (1989) *Standard Specifications for Highway Bridges*, 14th edn., American Association of State Highway and Transportation Officials.

Abayakoon, S.B.S., Olson, M.D. and Anderson, D.L. (1989) Large Deflection Elastic-Plastic Analysis of Plate Structures by Finite Strip Method, *Int. J. Num. Meth. Eng.*, **28**, (2) 331–358.

Abdel–Akher, A. and Hartley, G.A., (1989) Evaluation of Boundary Integrals for Plate Bending, *Int. J. Num. Meth. Eng.*, **28**, 75–93.

Abdulla, M.A. (1977) Ph.D. Thesis, University of Dundee.

Ainsworth, M., Zhu, J.Z., Craig, A.W. and Zienkiewicz, O.C. (1989) Analysis of the Zienkiewicz-Zhu a posteriori error estimator in the finite element method, *Int. J. Num. Meth. Eng.*, **28**, 2161–2174.

Allen, D.N.de G. and Severn, R.T. (1961) Composite Action of Beams and Slabs, under Transverse Loading, *Struct. Eng.*, **39**, 235–239.

Anon (1986), *A finite Element Primer*, Dept. of Trade and Industry, National Engineering Laboratory, Glasgow, UK.

ANSYS (1982) *ANSYS Engineering Analysis System User's Manual*, Revision 4.0, Swanson Analysis, Inc., Houston, PA.

Arantes e Oliveira, E.R. (1968) Theoretical Foundations of the Finite Element Method, *Int. J. Solid Struct.*, **4**, 929.

Arbabi, F. and Li, F. (1991) Macrostrip for Variable Section Plates, *J. Struct. Eng.*, **117** (6), 1759–1779.

Arizumi, Y., Hamada, S. and Oshiro, T. (1982) Finite Strip Analysis of Curved Composite Girders with Incomplete Interaction, *Comput. Struct.*, **15** (5), 603–612.

Azizian, Z.G. and Dawe, D.J. (1985) Geometrically Non-Linear Analysis of Rectangular Mindlin Plates Using the Finite Strip Method, *Comput. Struct.*, **21**, 423–436.

Babu, P.V.T. and Reddy, D.V. (1971) Frequency Analysis of Skew Orthotropic Plates by the Finite Strip Method, *J. Sound Vib.*, **18**, 465–74.

Babuska, I. and Rheinboldt, W.C. (1978) A Posteriori Error Estimates for the Finite Element Method, *Int. J. Num. Meth. Eng.*, **12**, 1597–1615.

Babuska, I. and Rheinboldt, W.C. (1978) Error Estimates for Adaptive Finite Element Computation, *SIAM J. Num. Anal.*, 15, 736–754.

Babuska, I. and Rheinboldt, W.C. (1979) Adaptive Approaches and Reliability Estimates in Finite Element Analysis, *Comp. Meth. Appl. Mech. Eng.*, **17/18**, 519–540.

Babuska, I. and Rheinboldt, W.C. (1984) The Post-Processing Approach in the Finite Element Method — Part II: the Calculation of Stress Intensity Factors, *Int. J. Num. Meth. Eng.*, **20**, 1111–1129.

Babuska, I. and Rheinboldt, W.C. (1984) The Post-Processing Approach in the Finite Element Method — Part III: A Posteriori Error Estimates and Adaptive Mesh Selection, *Int. J. Num. Meth. Eng.*, **20**, 2311–2324.

Babuska, I. and Rheinboldt, W.C. (1984) The Post-Processing Approach in the Finite Element Method — Part I: Calculation of Displacements, Stresses and Other Higher Derivatives of the Displacements, *Int. J. Num. Meth. Eng.*, **20**, 1085–1109.

Bathe, K.J. (1982) *Finite Element Procedures in Engineering Analysis*, Prentice–Hall, Englewood Cliffs, NJ.

Bazeley, G.P. et al. (1966) *Triangular Elements in Plate Bending Conforming Solutions*, Proc. 1st Conf. Matrix Methods in Structural Mechanics, Wright-Patterson Airforce Base, Ohio, 547–576, AFFOL-TR-CC-80.

Blacker, T.D. and Stephenson, M.B. (1991) Paving: a New Approach to Automated Quadrilateral Mesh Generation, *Int. J. Num. Meth. Eng.*, **32**, 811–847.

Bonet, J. (1994) Error Estimators and Enrichment Procedures for the Finite Element Analysis of Thin Sheet Large Deformation Processes, *Int. J. Num. Meth. Eng.*, **37**, 1573–1591.

Boughan, R.B. and Baab, G.W. (1944) *Charts for Calculation of the Critical Compressive Stress for Local Instability of Idealized Web and T Stiffened panels*, National Advisory Committee on Aeronautics, A.R.R., L4H29.

Bradford, M.A. (1990) Elastic Local Buckling of Trough Girders, *J. Struct. Eng.*, 1594–1610.

Branco, F.A. and Green R. (1985) Composite Box Girder Bridge Behaviour during Construction, *J. Struct. Eng.*, **111** (3), 577–593.

Brebbia, C.A. and Connor, J.J. (1973) *Fundamentals of Finite Element Techniques*, Butterworths, London.

Brebbia, C.A., Telles, J.C.F. and Wrobel, L.C. (1984) *Boundary Element Techniques, Theory and Applications in Engineering*, Springer-Verlag, New York.

Brown, T.G. and Ghali, A. (1972) *Finite Strip Analysis of Skew Slabs*, Proc. McGill-EIC Conference in Finite Element Method in Civil Engineering, 1141–1151.

Brown, T.G. and Ghali, A. (1975) Semi-Analytic Solution of Skew Box Girder Bridges, *Proc. Inst. Civ. Eng.*, Part 2, **59**, 487–500.

Bucco, D., Mazumdar, J. and Sved, G. (1979) Application of the Finite Strip Method Combined with the Deflection Contour Method to Plate Bending Problems, *Comput. Struct.*, **10**, 827–830.

Chan, H.C. and Cheung, Y.K. (1972) Static and Dynamic Analysis of Multilayered Sandwich Plates, *Int. J. Mech. Sci.*, **14**, 399–406.

Chan, M.Y.T. and Cheung, M.S. (1978) *Static and Dynamic Analysis of Three-Dimensional Thick Plates Using Finite Strips*, Proceeding of the Fourth Symposium on Engineering Applications on Solid Mechanics, Sheridan Park Research Community (September 25 and 26, 1978).

Chen, C.J., Gutkowski, R.M. and Puckett, J.A. (1990) B-Spline Compound Strip Analysis of Stiffened Plates under Transverse Loading, *Comput. Struct.*, **34** (2), 337–347.

Chen, C.J., Gutkowski, R.M. and Puckett, J.A. (1991) B-Spline Compound Strip Formulation for Braced Thin-Walled Structures, *J. Struct. Eng.*, **117** (5) 1303–1316.

Chen, J.L. and Chong, K.P. (1984) *Vibration of Irregular Plates by Finite Strip Method with Spline Functions*, in Engineering Mechanics in Civil Engineering, Proc. of the 5th Engineering Mechanics Div., ASCE, Edited by A. P. Boresi and K. P. Chong, **1**, 256–260.

Chen, J.L. and Chong, K.P. (1985) *Buckling of Irregular Plates by Finite Strip Method with Spline Functions*, in Collection of Technical Papers-AIAA/ASME/ASCE/AHS 26th Structures, Structural Dynamics and Material Conference, Part 1, 147–151. AIAA, New York.

Chen, M.J., Tham, L.G. and Cheung, Y.K. (1984) *Spline Finite Strip for Parallelogram Plate*, Proceedings, International Conference on Accuracy Estimates and Adaptive Refinements in Finite Element Computation (ARFEC), Lisbon, **1**, 95–104.

Cheung. M. S., Akhras, G. and Li, W. (1994) Combined Boundary/Finite Strip Analysis of Bridges, *J. Struct. Eng.*, **120** (3) 716–727.

Cheung, M.S. and Chan, M.Y.T. (1978A) Live Load Field Test on the Muskwa River Bridge, *Can. J. Civil Eng.* **5**, 186–201.

Cheung, M.S. and Chan, M.Y.T. (1978B) Three-Dimensional Finite Strip Analysis of Elastic Solids, *Comput. Struct.*, **9**, 629–638.

Cheung, M.S. and Chan, M.Y.T. (1979) Finite Strip Analysis of Continuous Bridges with Intermediate Columns, *RTAC Forum*, **2** (1), 75–83.

Cheung, M.S. and Chan, M.Y.T. (1981) Static and Dynamic Analysis of Thin and Thick Sectorial Plates by the Finite Strip Method, *Comput. Struct.*, **14** (1-2), 79–88.

Cheung, M.S. and Cheung, Y.K. (1971) Flexural Vibrations of Rectangular and Other Polygonal Plates, *Proc. ASCE, Eng. Mech. Div.*, **97** (EM2), 391–411.

Cheung, M.S. and Cheung, Y.K. (1971A) Natural Vibrations of Thin, Flat-Walled Structures with Different Boundary Conditions, *J. Sound Vib.*, **18**, 325–337.

Cheung, M.S. and Cheung, Y.K. (1972) Static and Dynamic Behaviour of Rectangular Plates using Higher Order Finite Strips, *Build. Sci.*, **7**, 151–158.

Cheung, M.S., Cheung, Y.K. and Ghali, A. (1970) Analysis of Slab and Girder Bridges by the Finite Strip Method, *Build. Sci.*, **5** (2), 95–105.

Cheung, M.S., Cheung, Y.K. and Reddy, D.V. (1972) *Frequency Analysis of Certain Single and Continuous Span Bridges, Development in Bridge Design and Construction*, Crosby Lockwood, 188–199.

Cheung, M.S. and Li, W. (1988A) Finite Strip Analysis of Continuous Structures, *Can. J. Civil Eng.*, **15**, 424–429.

Cheung, M.S. and Li, W. (1988B) Analysis of Haunched, Continuous Bridges by Finite Strip Method, *Computers and Structures*, **28** (5) 621–626.

Cheung, M.S. and Li, W. (1989A) Analysis of Continuous, Haunched Box-Girder Bridges by Finite Strips, *J. Struct. Eng.*, **115** (5) 1076–1087.

Cheung, M.S. and Li, W. (1989B) A modified Finite Strip Method for Geometrically Nonlinear Analysis of Plates, *Comput. Struct.*, **33** (4), 1031–1035.

Cheung, M.S. and Li, W. (1990A) Finite Strip Method for Materially Nonlinear Analysis of RC Slabs, *Comput. Struct.*, **35** (5), 603–607.

Cheung, M.S. and Li, W., (1990B) Analysis of Haunched, Continuous Bridges by Spline Finite Strips, *Comput. Struct.*, **36** (2), 297–300.

Cheung, M.S. and Li, W. (1991) Combined Finite Strip / Finite Element Solution for Analysis of Plates, *Comput. Struct.*, **41** (5) 1119–1124.

Cheung, M.S. and Li, W. (1992) Finite Strip Method Combined with Other Numerical Methods for the Analysis of Plates, *Comput. Struct.*, **45** (1) 1–7.

Cheung, M.S., Li, W., and Jaeger, L.G. (1988) Nonlinear Analysis of Cable-Stayed Bridge by Finite Strip Method, *Comput. Struct.*, **29** (4) 687–692.

Cheung, M.S., Li, W., and Jaeger, L.G. (1990) Improved Finite Strip Method for Nonlinear Analysis of Long Span Cable-Stayed Bridges, *Can. J. Civil Eng.*, **17**, 87–93.

Cheung, M.S., Li, W. and Jaeger, L. G. (1992) Spline Finite Strip Analysis of Continuous Haunched Box-Girder Bridges, *Can. J. Civil Eng.*, **19**, 724–728.

Cheung, M.S., Ng, S.F. and Zhao, J.Q. (1993) Analysis of Curved Reinforced Concrete Slab Bridges by the Spline Finite Strip Method, *Can. J. Civil Eng.*, **20**, 855–862.

Cheung, M.S., Ng, S.F. and Zhong, B. (1989) Finite Strip Analysis of

Beams and Plates with Material Nonlinearity, *Comput. Struct.*, **33** (1), 289–294.

Cheung, P., Dade H. and Zongmu, W. (1987) Static, Vibration and Stability Analysis of Stiffened Plates Using B Spline Functions, *Comput. Struct.*, **27** (1), 73–78.

Cheung, Y.K. (1968A) The Finite Strip Method in the Analysis of Elastic Plates with Two Opposite Simply Supported Ends, *Proc. Inst. Civ. Eng.*, **40**, 1–7.

Cheung, Y.K. (1968) Finite Strip Method Analysis of Elastic Slabs, *Proc. ASCE* **94** (EM6), 1365–1378.

Cheung, Y.K. (1969A) *Analysis of Box Girder Bridges by Finite Strip Method*, Proc. Second Int. Symposium on Concrete Bridge Design, Chicago, ACI Publications SP 26, April, 357–378.

Cheung, Y.K. (1969B) *The Analysis of Cylindrical Orthotropic Curved Bridge Decks*, IABSE Publications, **29–II**, 41–52.

Cheung, Y.K. (1976) *Finite Strip Method in Structural Analysis*, Pergaman Press, Oxford.

Cheung, Y.K. and Chakrabarti, S. (1971) Analysis of Simply Supported Thick, Layered Plates, *Am. Soc. Civil Eng.*, **97** (EM3), 1039–1044.

Cheung, Y.K. and Chakrabarti, S. (1972) Free Vibration of Thick, Layered Rectangular Plates by a Finite Layer Method, *J. Sound Vib.*, **21** (3), 277–284.

Cheung, Y.K. and Cheung, M.S. (1971) *Analysis of Curved Box Girder Bridge by Finite Strip Method*, IABSE Publications, **31/I**, 1–19.

Cheung, Y.K. and Cheung, M.S. (1972) *Free Vibration of Curved and Straight Beam-Slab or Box-Girder Bridges*, International Association for Bridges and Structural Engineering, **32–II**, 41–52.

Cheung, Y.K. and Delcourt, C. (1977) Buckling and Vibration of Thin Flat-Walled Structures Continuous over Several Spans, *Proc. Inst. Civ. Eng.*, 2, **63**, 93–103.

Cheung, Y.K. and Fan, S.C. (1983) Static Analysis of Right Box Girder Bridges by Spline Finite Strip Method, *Proc. Inst. Civ. Eng.*, 2, **75**, 311–323.

Cheung, Y.K., Fan, S.C. and Wu, C.Q. (1982) *Spline Finite Strip in Structure Analysis*, Proceedings, The International Conference on Finite Element Method, Shanghai, 704–709.

Cheung, Y.K., Tham, L.G. and Chong, K.P. (1982) Buckling of Sandwich Plate by Finite Layer Method, *Comput. Struct.*, **15** (2), 131–134.

Cheung, Y.K., Tham, L.G. and Li, W.Y. (1986) Application of Spline-Finite-Strip Method in the Analysis of Curved Slab Bridge, *Proc. Inst. Civil Eng.*, , **81**, 111–124.

Cheung, Y.K., Tham, L.G. and Li, W.Y. (1988) Free Vibration and Static Analysis of General Plate by Spline Finite Strip, *Comput. Mech.*, **3**, 187–197.

Cheung, Y.K., Tham, L.G. and Li, W.Y. (1989) *Free Vibration Analysis of Arbitrary Double Curved Shell by Spline Finite Strip Method*, Research Report, University of Hong Kong.

Chong, K.P. and Chen, J.L. (1986) Buckling of Irregular Plate by Spline Finite Strips, *AIAAJ*, **24**, 534–536.

Clough, R.W. and Penzien, J. (1975) *Dynamics of Structures*, McGraw-Hill.

Cook, R.D., Malkus, D.S. and Plesha, M.E. (1989) *Concepts and Applications of Finite Element Analysis*, 3rd edn., Wiley, New York.

Crandall, S.H. (1956) *Engineering Analysis*, McGraw-Hill, New York.

CRCJ (1971) Column Research Committee of Japan.

CSA-S6 (1988) *Design of Highway Bridges - A National Standsard of Canada*, Canadian Standards Association (CSA), CAN/CSA-S6-88.

Darwin, D. and Pecknold, D. (1974) *Inelastic Model for Cyclic Biaxial Loading of Reinforced Concrete*, SRS No. 409, University of Illinois at Urbana-Champaign, Illinois, (July).

Darwin, D. and Pecknold, D. (1977) Nonlinear Biaxial Stress-Strain Law for Concrete, *J. Eng. Mech. Div.*, **103** (EM2), 229–241.

Delcourt, C. and Cheung, Y.K. (1978) *Finite Strip Analysis of Continuous Folded Plates*, Proceedings, International Association of Bridge and Structural Engineers, P–14/78, 1–16.

Desai, C.S. and Abel, J.F. (1972) *Introduction to the Finite Element Method*, Nostrand Reinhold, New York.

Dym, C.L. and Shames, I.H. (1973) *Solid Mechanics: A Variational Approach*, McGraw-Hill.

Eason, E.D. (1976) A review of Least-Square Methods for Solving Partial Differential Equations, *Int. J. Num. Meth. Eng.* **10** (5), 1021–46.

Fan, S.C. and Cheung, Y.K. (1983) Analysis of Shallow Shells by Spline Finite Strip Method, *Eng. Struct.*, **5**, 255–263.

Felippa, C.A. and Clough, R.W. (1968) *The finite element method in solid mechanics*, Symposium on Numerical Solutions of Field Problems in Continuum Mechanics, Durham, NC, Apr. 1968.

Finlayson, B.A. (1972) *The Method of Weighted Residuals and Variational Principles*, Academic Press.

Finlayson, B.A. and Scriven, L.E. (1966) The Method of Weighted Residuals - A review, *Appl. Mech. Rev.*, **19** (9), 735–748.

Fried, I. (1972) Condition of Finite Element Matrices Generated from Nonuniform Meshes, *AIAA J.*, **10** (2), 219–221.

Forray, M.J. (1968) *Variational Calculus in Science and Engineering*, McGraw-Hill.

Gierlinski, J.T. and Graves-Smith, T.R. (1984) The Geometric Non-Linear Analysis of Thin-Walled Structures by Finite Strips, *Thin-Wall. Struct.*, **2**, 27–50.

Gilbert, R.I. and Warners, R.F. (1978) Tension Stiffening in Reinforced Concrete Slabs, *J. Struct. Div.*, **104** (ST12), 1885–1900.

Graves-Smith, T.R. and Sridharan, R. (1978) A Finite Strip Method for the Post-Locally-Buckled Analysis of Plate Structures, *Int. J. Mech. Sci.*, **20**, 833–842.

Greunen, J.V. (1979) *Nonlinear Geometric, Material and Time Dependent Analysis of Reinforced and Prestressed Concrete Slabs and Panels*, Department of Civil Engineering, University of California, Berkeley, California.

Guo, C.M., Mufti, A.A., Jaeger, L.G. and Cheung, M.S. (1988) *Nonlinear Earthquake Response Analysis of Reinforced Concrete Bridges Using the Finite Strip Method*, Proceedings of Ninth World Conference on Earthquake Engineering, **VI**, 453–458, August, Tokyo-Kyoto, Japan.

Hamada, H. (1966) Statical Deflection of Parallelogram Plate with Clamped Edges Subjected to Uniform Pressure, Trans. *Jpn. Soc. Aerospace Sci.*, **9** (Nov), 84.

Hancock, G.J. (1981) Non-Linear Analysis of Thin Sections in Compression, *J. Struct. Div.*, **107**, 455–471.

Hartmann, F. (1985) *The Mathematical Foundation of Structural Mechanics*, Springer-Verlag.

Hartmann, F. (1989) *Introduction to Boundary Elements, Theory and Applications*, Springer-Verlag, Berlin, 290–293.

Higdon, H., Ohlsen, E.H., Stiles, W.B., Wcere, J.A., and Riley, W.F. (1976) *Mechanics of Materials*, Wiley, New York.

Hinton, E. and Campbell, J.S. (1974) Local and Global Smoothing of Discontinuous Finite Element Functions Using a Least Square Method, *Int. J. Num. Meth. Eng.*, **8**, 461–480.

Irons, B. and Shrive, N. (1983) *Finite Element Primer*, Ellis Horwood, Chichester, UK.

Iyengar, K.T.S. and Srinivasan, R.S. (1967) Clamped Skew Plates under Uniformly Distributed Load, *J. R. Aero. Soc.*, **71**, (Feb), 139–140.

Jategaonkar, R., Jaeger, L.G. and Cheung, M.S. (1985) *Bridge Analysis Using Finite Elements*, The Canadian Society for Civil Engineering, Montreal, Canada.

Kabir, A.F. (1976) *Nonlinear Analysis of Reinforced Concrete Panels, Slabs and Shells for Time Dependent Effects*, Ph.D. Dissertation, Division of Structural Engineering and Structural Mechanics, University of California, Berkeley, UC-SESM Report No. 76–6.

Kajita, T. and Cheung, Y.K. (1973) *Finite Element Analysis of Cable-Stayed Bridges*, Publications, International Association for Bridges, and Structural Engineering, **33–II**, 101–112.

Key, S. (1966) *A Convergence Investigation of the Direct Stiffness Method*, Ph.D. Thesis, University of Washington.

Khong, P. W. (1990) The Comparison of Lower Order and Higher Order

Finite Strip Analysis in the Stability Problems of Thin-Walled Structures, *Comput. Struct.*, **36** (1), 109–118.

Langhaar, H.L. (1962) *Energy Methods in Applied Mechanics*, Wiley, New York.

Langyel, P. and Cusens, A.R. (1983) A Finite Strip Method for the Geometrically Nonlinear Analysis of Plate Structures, *Int. J. Num. Meth. Eng.*, **19**, 331–340.

Lau, S.C.W. and Hancock, G.J. (1986) Buckling of Thin Flat-Walled Structures by a Spline Finite Strip Method, *Thin Walled Struct.*, **4** (4), 269–294.

Lau, S.C.W. and Hancock, G.J. (1989) Inelastic Buckling Analysis of Beams, Columns and Plates Using Spline Finite Strip Method, *Thin-Walled Struct.*, **7** (3–4), 213–238.

Li, W.Y., Cheung, Y.K. and Tham, L.G. (1986) Spline Finite Strip Analysis of General Plates, *J. Eng. Mech.*, **112** (1), 43–54.

Li, W.Y., Tham, L.G. and Cheung, Y.K. (1988) Curved Box Girder Bridges, *J. Struct. Eng.*, **114** (6), 1324–1338.

Lim, P.T.K. and Moffatt, K.R. (1971) *Finite Element Analysis of Curved Slab Bridges with Special Reference to Local Stress*, Developments in Bridge Design and Construction, Crosby Lock-Wood, 27–51.

Lo, S.H. (1991) Automatic Mesh Generation and Adaptation by Using Contours, *Int. J. Num. Meth. Eng.*, 31, 689–707.

Loo, Y.C. (1972) *Developments and Applications of the Finite Strip Method in the Analysis of Right Bridge Decks*, Ph.D. Thesis, University of Dundee.

Loo, Y.C. (1973) *Local Strains in a Beam and Slab Bridge Model under Concentrated Load*, Publications, International Association for Bridges and Structural Engineering, **33/I**, 167–189.

Loo, Y.C. (1975) *The Finite Strip Analysis of Multispan Box-Girder Bridges with Intermediate Stiffening Members*, British Steel Corporation Fellowship Final Report, University of Dundee.

Loo, Y.C. (1976) *Analysis of Continuous Highway Box Bridges with Intermediate Stiffening*, Eighth Australian Road Research Board Conference, 13–20.

Loo, Y.C. and Cusens, A.R. (1971) *Developments of the Finite Strip Method in the Analysis of Cellular Bridge Decks*, Developments in Bridge Design and Construction, Crosby Lockwood, London, 53–72.

Loo, Y.C. and Cusens, A.R. (1978) *The Finite-Strip Method in Bridge Engineering*, A Viewpoint Publication, Wexham Springs, Slough, UK.

Maleki, S. (1987) *Compound Strip Method for Folded Plates and Box Girders*, Ph.D. thesis, New Mexico State University, Las Cruces, NM.

Maleki, S. (1991) *Compound Strip Method for Box Girders and Folded Plates*, *Comput. Struct.*, **40** (3), 527–538.

Mebane, P.M. and Stricklin, J.A. (1971) Implicit Rigid Body Motion in Curved Finite Elements, *AIAA J.* **9**, 344–345.

Melosh, R.J. (1963) Basis of Derivation of Matrices for the Direct Stiffness Method, *AIAA J.* **1**, 1631–1637.

Meyer, C. and Scordelis, A.C. (1970) *Analysis of Curved Folded Plates Structure*, SESM Report No. 70-8, University of California, Berkeley.

Mikhlin, S.C. (1964) *Variational Methods in Mathematical Physics*, Pergamon Press, Oxford.

Mikhlin, S.C. (1965) *The Problems of the Minimum of a Quadratic Functional*, Holden-Day, San Francisco.

Mizusawa, T. (1988) Application of Spline Strip Method to Analyse Vibration of Open Cylindrical Shells, *Int. J. Num. Meth. Eng.*, **26** (3), 663–676.

Morley, L.S.D. (1964) Bending of Clamped Rectilinear Plates, *Quart. J. Mech. Appl. Math.*, **19**, 293.

Mukhopadhyay, M. (1976) *Finite Strip Method of Clamped Skew Plates in Bending, Inst. Civ. Eng.* Part II, **61**, 189–195.

Ng, S.F., Cheung, M.S. and Zhong, B. (1991) A Finite Strip Method for the Analysis of Structures with Material Nonlinearility, *J. Struct. Eng.*, **117** (2), 489–500.

Noble, B. (1969) *Applied Linear Algebra*, Prentice-Hall, Englewood Cliffs, NJ.

Oden, J.T. and Reddy, J.N. (1976) *An Introduction to the Mathematical Theory of Finite Elements*, Wiley, New York.

OHBDC (1991) *Ontario Highway Bridge Design Code*, 3rd edn, Ministry of Transportation, Ontario.

Olawale, A.O. and Plank, R.J. (1988) The Collapse Analysis of Steel Columns in Fire Using a Finite Strip Method, *Int. J. Num. Meth. Eng.*, **26** (12), 2755–2764.

Owen, D.R.J. and Hinton, E. (1980) *Finite Element in Plasticity* (Theory and Practice), Pineridge Press, Swansea.

Podolny, W. and Scalzi, J.B. (1976) *Construction and Design of Cable-stayed Bridges*, Wiley-Interscience, New York.

Powell, G.H. and Ogden, D.W. (1969) Analysis of Orthotropic Steel Plate Bridge Decks, *Proc. ASCE*, **95** (ST5), 909–922.

Prenter, P.M. (1975) *Splines and Variational Methods*, Wiley, New York.

Przemieniecki, J.S. (1973) Finite Element Structural Analysis of Local Instability, *AIAAJ*, **11** (1), 33–39.

Puckett, J.A. (1983) *Compound Strip Method for the Analysis of Continuous Elastic Plates*, Str. Research Report No. 48, Civil Eng. Dept., Colorado State Univ., Fort Collins.

Puckett, J.A. (1986) Application of the Compound Strip Method for the Analysis of Slab-Girder Bridges, *Comput. Struct.*, **22** (6), 979–986.

Puckett, J.A. and Gutkowski, R.M. (1986) Compound Strip Method for Analysis of Plate Systems, *J. Struct. Eng.*, **112** (1), 121–138.

Puckett, J.A. and Lang, G.J. (1986) Compound Strip Method for Continuous Sector Plates, *J. Eng. Mech.*, **112** (5), 498–514.

Rao, S.S. (1992) *The Finite Element Method in Engineering*, Pergaman Press, Oxford, UK.

Richardson, L.F. (1910) The Approximate Arithmetical Solution by Finite Differences of Physical Problems Involving Differential Equations with an Application to the Stresses in a Masonry Dam, *Trans. Roy. Soc. A*, **210**, 307–357.

Rosanoff, R.A., Gloudeman, J.F. and Levy, S. (1969) *Numerical Conditioning of Stiffness Matrix Formulations for Frame Structures*, Proceedings of the Second Conference on Matrix Methods in Structural Mechanics, Wright-Patterson AFB, Ohio, 1968, (AFFDI-TR-68-150, Dec. 1969; AD-703-685, N.T.I.S.), 1029–1060.

Rowe, R.E. (1976) *Concrete Bridge Design*, Applied Science Publishers Ltd., London, England.

Rushton, K.R. (1964) Electrical Analogue Solutions for the Deformation of Skew Plates, *Aeronaut. Quart.*, **15**, (May) 169–180.

Scordelis, A.C. (1971) Analytical Solution for Box Girder Bridges, Development in Bridge Design and Construction, Crosby Lockwood, London, 200–216.

Scordelis, A.C. and Davis, R.E. (1969) *Stresses in Continuous Box Girder Bridges*, Proceedings of the Second International Symposium on Concrete Bridge Design, Chicago, ACI Publication SP-26, 283–318.

Seif, S.P. and Dilger, W.H. (1990) Nonlinear Analysis and Collapse Load of P/C Cable-Stayed Bridges, *J. Struct. Eng.*, **116** (3), 829–849.

Sisodiya, R.G., Cheung Y.K. and Ghali A. (1972) New Finite Elements with Application to Box Girder Bridges, *Proc. Inst. Civ. Engrs.*, Suppl. (x), 207–225.

Smith, J.W. (1973) Finite Strip Analysis of the Dynamic Response of Beam and Slab Highway Bridges, *Earthq. Eng. Struct. Dyn.*, **1**, 357–370.

Sridharan, S. and Graves-Smith, T.R. (1981) Postbuckling Analyses with Finite Strips, *J. Eng. Mech. Div.*, **107**, 869–887.

Strang, G. (1976) *Linear Algebra and its Applications*, Academic Press, New York.

Strang, G. and Fix, G.J. (1973) *An Analysis of the Finite Element Method*, Prentice-Hall, Englewood Cliff, NJ..

Tang, M.C. (1971) Analysis of Cable-Stayed Girder Bridges, *J. Struct. Div.*, **97** (ST5), 1481–1496.

Task Committee (1977) Task Committee on Cable-Suspended Structures of the Committee on Special Structures of the Committee on Metals of the Structural Division, Tentative Recommendations for Cable-Stayed Bridge Structures, *J. Struct. Div.*, **103** (ST5), 929–959.

Taylor, R.L. (1972) On Completeness of Shape Functions for Finite Element Analysis, *Int. J. Num. Meth. Eng.*, **4**, 17–22.

Taylor, R., Maher, D.R.H. and Hayes, B. (1966) Effect of the Arrangement of Reinforcement on the Behavior of Reinforced Concrete Slabs, *Mag. Concr. Res.*, **18** (55), 85–94.

Tham, L.G., Li, W.Y., Cheung, Y.K. and Chen, M.J. (1986) Bending of Skew Plates by Spline-Finite-Strip Method, *Comput. Struct.*, **22** (1), 31–38.

Tham, L.G. and Szeto, H.Y. (1990) Buckling Analysis of Arbitrary Shaped Plates by Spline Finite Strip Method, *Comput. Struct.*, **36** (4), 729–735.

Thorpe, J. (1976) Ph.D. Thesis, University of Dundee.

Timoshenko, S.P. and Gere, J.M. (1961) *Theory of Elastic Stability*, 2nd edn., McGraw-Hill, New York.

Timoshenko, S.P. and Gere, J.M. (1972) *Mechanics of Material*, Van Nostrand Reinhold.

Timoshenko, S. and Goodier, J.N. (1970) *Theory of Elasticity*, 3rd edn., McGraw-Hill, New York.

Timoshenko, S. and Woinowsky–Krieger, S. (1959), *Theory of Plates and Shells*, 2nd edn., McGraw-Hill, New York.

Warburton, G.B. (1954) The Vibration of Rectangular Plates, *Proc. Inst. Mech. Eng.*, **168**, 371–384.

Washizu, K. (1982) *Variational Methods in Elasticity and Plasticity*, Pergamon Press, Oxford, England.

Willems, N. and Lucas Jr, W.M. (1978) *Structural Analysis for Engineers*, McGraw-Hill, New York.

Wiseman, D.L. and Puckett, J.A. (1991) Applications of Compound Strip Method for Folded Plates with Connecting Elements, *J. Struct. Eng.*, **117** (1), 255–285.

Wittrick, W. H. and Williams, F. W. (1970), *Natural Vibrations of Thin, Prismatic, Flat-Wall Structures* IUTAM Symposium on High Speed Computing of Elastic Structures, Liege.

Wittrick, W. H. and Plank, R.J. (1974) Buckling under Combined Loading of Thin, Flat-Walled Structures by a Complex Finite Strip Method, *Int. J. Num. Meth. Eng.*, **8**, 323.

Woinowsky–Krieger, S. (1955) Clamped Semicircular Plate under Uniform Bending Load, *J. Appl. Mech.*, **22** (1), 129.

Wu, C.I. and Cheung, Y.K. (1974) *Frequency Analysis of Rectangular Plates Continuous in One or Two Directions*, Earthquake Engineering and Structural Dynamics.

Zhu, D.S. and Cheung, Y.K. (1989) Postbuckling Analysis of Shells by Spline Finite Strip Method, *Comput. Struct.*, **31** (3), 357–364.

Zhu, J.Z., Hinton, E. and Zienkiewicz, O.C. (1990) Adaptive Finite Element Analysis with Quadrilaterals, *Comput. Struct.*, **40** (5), 1097–1104.

Zhu, J.Z. and Zienkiewicz, O.C. (1990) Superconvergence Recovery

Technique and a Posteriori Error Estimator, *Int. J. Num. Meth. Eng.*, **30**, 1321–1339.

Zienkiewicz, O.C. (1977) *The Finite Element Method*, 3rd edn., McGraw-Hill, UK.

Zienkiewicz, O.C. (1982) *The Finite Element Method*, McGraw-Hill, UK.

Zienkiewicz, O.C. and Taylor, R.L. (1989) *The Finite Element Method*, 4th edn., McGraw-Hill, London.

Zienkiewicz, O.C. and Too, J.J.M. (1972) The Finite Prism in Analysis of Thick Simply Supported Bridge Boxes, *Proc. Inst. Civ. Eng.*, **53**, 147–172.

Zienkiewicz, O.C. and Zhu, J.Z. (1987) A Simple Error Estimator and Adaptive Procedure for Practical Engineering Analysis, *Int. J. Num. Meth. Eng.*, **24**, 337–357.

Zienkiewicz, O.C. and Zhu, J.Z. (1989) Error Estimates and Adaptive Refinement for Plate Bending Problems, *Int. J. Num. Meth. Eng.*, **28**, 2839–2853.

Zienkiewicz, O.C. and Zhu, J.Z. (1991) Adaptivity and Mesh Generation, *Int. J. Num. Meth. Eng.*, **32**, 783–810.

Zienkiewicz, O.C. and Zhu, J.Z. (1992) The Superconvergence Path Recovery and a Posteriori Error Estimator, Part I: the Recovery Technique, *Int. J. Num. Meth. Eng.*, **33**, 1331–1364.

Zienkiewicz, O.C. and Zhu, J.Z. and Gong N.G. (1989) Effective and Practical h-p Version Adaptive Analysis Procedures for the Finite Element Method, *Int. J. Num. Meth. Eng.*, **28**, 879–891.

Zienkiewicz, O.C. and Zhu, J.Z. and Wu, J. (1993) Superconvergent Path Recovery Techniques - Some Further Tests, *Comm. Num. Meth. Eng.*, **9**, 251–258.

Index

Page numbers in **bold** refer to figures and page numbers in *italic* refer to tables.

For Product Safety Concerns and Information please contact our EU representative GPSR@taylorandfrancis.com
Taylor & Francis Verlag GmbH, Kaufingerstraße 24, 80331 München, Germany

www.ingramcontent.com/pod-product-compliance
Ingram Content Group UK Ltd.
Pitfield, Milton Keynes, MK11 3LW, UK
UKHW022008180926
475701UK00003B/259
* 9 7 8 0 3 6 7 8 6 4 5 7 6 *